S0-AHG-437

PROPIONIBACTERIA

PROPIONIBACTERIA

by

Lena I. Vorobjeva
Moscow State University,
Moscow, Russia

KLUWER ACADEMIC PUBLISHERS
DORDRECHT / BOSTON / LONDON

A C.I.P. Catalogue record for this book is available from the Library of Congress.

ISBN 0-7923-5884-8

QR
82
.P7
V668
1999

Published by Kluwer Academic Publishers,
P.O. Box 17, 3300 AH Dordrecht, The Netherlands.

Sold and distributed in North, Central and South America
by Kluwer Academic Publishers,
101 Philip Drive, Norwell, MA 02061, U.S.A.

In all other countries, sold and distributed
by Kluwer Academic Publishers,
P.O. Box 322, 3300 AH Dordrecht, The Netherlands.

Printed on acid-free paper

All Rights Reserved
© 1999 Kluwer Academic Publishers
No part of the material protected by this copyright notice may be reproduced or
utilized in any form or by any means, electronic or mechanical,
including photocopying, recording or by any information storage and
retrieval system, without written permission from the copyright owner.

Printed in the Netherlands.

To my daughter Nina

Contents

v

Preface

In a sense, propionic acid bacteria are domesticated bacteria. They might have been used for cheese making as early as 9000 years BC. In the last 40 years their practical uses have expanded to include vitamin B_{12} and propionic acid production, bread baking, starters for ensilage and some pharmaceutical preparations. New prospects for their future uses are also emerging, based on the useful properties recently discovered.

This monograph is the result of many years of investigating propionibacteria by the faculty, staff and postgraduate students in the Department of Microbiology at the Moscow State University, as well as a number of scientists in other countries. The encouragement and various contributions of my colleagues has made this book possible, which might as well be entitled "My life with propionic acid bacteria", since these bacteria were the subject of our investigations for more than 40 years.

I hope that this book will be of interest not only to scientists of biological specialties, but also to those associated with industrial firms and medical institutions.

L.I. Vorobjeva

Acknowledgements

Writing a monograph is impossible without the cooperation of many people. I am very grateful to all my postgraduate students — thirty of them — who work at present not only in Russia, but also in various other countries, including Canada, Cuba, Egypt, India, Iran and Vietnam. Many thanks are due to my colleagues who shared my scientific interests and enthusiasm in experimental research. Special thanks are to N. Baranova, E. Iordan, N. Ikonnikov, E. Khodzhaev and G. Ponomareva, my research associates, with whom I began and continued investigations at the Department of Microbiology. The assistance in preparing this book of Drs. Nina Vorobjeva and Vladimir Melnik, who were the most rigorous critics as well, is gratefully acknowledged.

Introduction
Historical Background

Three principal stages can be distinguished in studying propionic acid bacteria. At the first stage, methods of their isolation were developed, a considerable number of strains were isolated and identified by their phenotypic and physiological properties. This period started with the discovery of propionic acid bacteria by E. Freudenreich and O. Orla-Jensen in 1906. Cheese, milk and dairy products were the main sources for the isolation of propionibacteria. In the USA, J.M. Sherman (1921) recognized that the poor quality of cheese manufactured from pasteurized milk is due to the loss of some vital microflora. Sherman then isolated a number of strains, including a new one that differed from the earlier isolates by higher acid production during growth in milk or in glycerol media. In 1928, C.B. Van Niel published a remarkable work "Propionic Acid Bacteria"(his Ph.D. dissertation), in which a complete characterization of propionic acid bacteria was presented, including morphology, physiology, biochemistry of fermentation and possible practical uses.

Van Niel isolated 30 new strains, identified them with the earlier isolates and laid foundations to the systematics of propionic acid bacteria. A new genus, *Propionibacterium* (suggested by Orla-Jensen), was established with eight species and one variety named after the scientists who discovered them:

1. *P. freudenreichii*
2. *P. jensenii*
3. *P. shermanii*
4. *P. petersonii*
5. *P. pentosaceum*
6. *P. rubrum*

1

7. *P. thoenii*
8. *P. technicum*
9. *P. jensenii* var. *raffinosaceum.*

Valuable contributions to the study of propionic acid bacteria were made by C.H. Werkman and H. Kendall. They identified new species and suggested considering *P. jensenii* var. *raffinosaceum* as a new species, *P. raffinosaceum.* Two new species, named *P. zeae* and *P. arabinosum*, were described by E.R. Hitchner in 1934. New species were also isolated and described by K. Sakaguchi in 1934 and A. Janoschek in 1944.

Since then, the genus was extended primarily by transferring four species of anaerobic corynebacteria (Moore, Holdeman, 1974) from the genus *Corynebacterium.* Anaerobic corynebacteria possess unique immuno-stimulating properties and present certain perspectives for medicine. They are called cutaneous propionibacteria, since in considerable numbers they live on the surface of human skin, and also are natural components of the gut microflora of ruminants. In 1975, the first seminar *"Corynebacterium parvum*: Application in Experimental and Clinical Oncology" was held in Paris and continued on a regular basis with the founding of the International Club "Corynebacterium parvum" (presently called *P. acnes*). The propionic acid corynebacteria are given special attention in this monograph.

Another species, *Arachnia propionica*, was transferred to *Propioni-bacterium* from *Actinomyces* (Charfreitag et al., 1988). Investigations of propionibacterial taxonomy by modern methods of molecular biology (DNA hybridization, 16S rRNA sequencing, multiplex PCR) resulted in a better understanding of inter- and intrageneric relationships of the genus *Propioni-bacterium.* Modern systematics of the propionibacteria is discussed in more detail below.

The formation and studying of the collections of propionic acid bacteria proceeded simultaneously with investigations of their biochemistry, first of all, biochemistry of their unique mode of fermentation. Propionic acid fermentation was discovered by A. Fitz; later, it was studied by H.G. Wood and C.H. Werkman. It was in propionibacteria that the heterotrophic assimilation of CO_2 was discovered by H.G. Wood. Owing to the studies by Wood, Werkman and their school, then by H.A. Barker and F. Lipmann as well as E.A. Delwiche in the USA, the chemistry of this unique fermentation was elucidated. Another development at the second stage of biochemical investigations concerns the discovery of aerobic metabolism in propionic acid bacteria, previously considered anaerobic. Important contributions to this field of study were made by the school of A.H. Stouthamer in the Netherlands and in our laboratory at the Moscow State University. These investigations demonstrated a surprising lability of the metabolism of propionic acid bacteria, which were found to be well equipped for both the anaerobic and aerobic styles of life. However, this poses an intriguing question: why these bacteria show such an obvious preference for anaerobiosis?

The third stage in studying propionic acid bacteria is closely associated with the discovery of vitamin B_{12}, since these bacteria proved to be the record producers of vitamin B_{12} in nature. This stage is characterized by studying conditions and pathways of the biosynthesis of corrinoids, their metabolic functions, and by starting industrial vitamin B_{12} production based on the vital functions of propionibacteria. The results of these investigations have convinced us that the metabolism of propionic acid bacteria is tuned to the high level of corrinoids in the cell.

Any biological phenomenon can be understood in the light of evolutionary representations. Experimental data accumulated over the last 100 years of studying propionic acid bacteria by many famous scientists and laboratories allow us to consider the propionibacteria an important subject for understanding microbial evolution. In the last 10-15 years propionic acid bacteria became an interesting subject for genetic engineering. Investigating the genetics of propionibacteria can open a new stage for this important group of microorganisms.

Having reviewed the achievements of the past, we now turn our attention to modern developments in understanding the biology of this interesting bacterial group, and on various ways of their practical applications.

Chapter 1

The genus *Propionibacterium*

1.1 Distinctive Features of Propionibacteria

The propionic acid bacteria constitute the genus *Propionibacterium*, which, together with *Eubacterium*, comprises the family Propionibacteriaceae. The name *Propionibacterium* was suggested by Orla-Jensen (1909), because these bacteria produce large amounts of propionic acid during fermentation. Overall, the propionibacteria are characterized as Gram-positive, non-spore-forming, non-motile, facultatively anaerobic or aerotolerant, rodlike bacteria. They contain menaquinones, mainly MQ-9(H$_4$) (Fernandez and Collins, 1987), and C$_{15}$-saturated fatty acids in their membrane lipids. The G+C content in their DNA is in the range of 53-67 mol%.

In Bergey's Manual of Determinative Bacteriology (9th ed., Holt et al., 1994) the genus *Propionibacterium* is placed in Group 20: "Irregular, non-sporing, Gram-positive rods." Type species is *P. freudenreichii*. Recently, the genus *Propionibacterium* has been included (Stackebrandt et al., 1997) in the newly established class Actinobacteria.

Anaerobic corynebacteria are also included in the genus *Propionibacterium* after the pioneering studies of Douglas and Gunter (1946). They were transferred from the genus *Corynebacterium* since they are:

a) anaerobes (most corynebacteria are aerobes),
b) produce propionic acid as the main metabolic product,
c) peptidoglycan in their cell wall contains mainly L-diaminopimelic acid (L-DAP), whereas related forms such as aerobic corynebacteria contain *meso*-DAP, and a morphologically similar group of actinomycetes in general does not contain DAP,

d) *iso-* and *anteiso-*C_{15} saturated acids are the main fatty acids of cellular lipids,

e) unlike aerobic corynebacteria, anaerobic species do not contain mycolic acids and arabinogalactan.

In general, corynebacteria differ from propionibacteria by their high proteolytic activity and temperature optimum for growth at 37°C (which is 28-30°C for the classical propionibacteria). Placing the anaerobic coryne-bacteria in the genus *Propionibacterium* was opposed by Prévot (1976). He was convinced that the group *'acnes'*, due to its pathogenicity and reticulo-stimulating properties, should be placed in the subgenus *Coryneforms*, family Corynebacteriaceae. He summarized the differences between aerobic corynebacteria, anaerobic corynebacteria, and propionibacteria as shown in Table 1.1. Nevertheless, the argumentation listed above prevailed, and in the 1st Edition of Bergey's Manual of Systematic Bacteriology (1986) the anaerobic corynebacteria were placed in the genus *Propionibacterium*.

Table 1.1. Typical differences between corynebacteria, coryneforms and propionibacteria

Characteristic	Corynebacteria	Anaerobic coryneforms	Propionibacteria
Respiration	aerobic or facultative	strictly anaerobic	microaerophilic or facultative
Cell wall composition	arabinose galactose mannose meso-DAP alanine glutamic acid	galactose glucose glycine L-DAP	galactose glucose mannose rhamnose inositol alanine glutamic acid glycine L-DAP
DNA (G+C mol%)	51.9	58-64	65-68
Toxin	+	–	–
Reticulostimulin	–	+	–
Cobalamin and corrinoids	–	– (*)	+
Colonies	colorless	pinky	gray, creamy, yellow, brown, red
Pathogenicity	+	+	–
Biochemical properties:			
indole	+	±	–
gelatin	+	±	–
Fermentation	acetic or acetobutyric	mostly acetopropionic	propioniacetic

(*) Except for *P. acnes*.   Reproduced from Prévot (1976), with permission.

Since the main habitat of anaerobic coryneforms is human skin, they are referred to as cutaneous propionibacteria, whereas species isolated from cheese and milk are known as dairy, or classical propionibacteria. On the basis of van Niel's and other investigations, 11 species of classical propioni-bacteria were described in the 7th edition of Bergey's Manual of Determinative Bacteriology (Breed et al., 1957). After the establishment of high levels of DNA homology and similarities in the cell wall composition between the strains (Johnson and Cummins, 1972; Moore and Holdeman, 1974), the number of classical propionibacterial species was reduced to four. Following the isolation of a new species, *P. cyclohexanicum*, and the transfer of the anaerobic corynebacteria to *Propionibacterium*, the final number of species in the genus is 10 (Table 1.2).

Table 1.2. Species of the genus *Propionibacterium*

Names of species and groups accepted by 1998	Original names of species
Classical	
P. freudenreichii	*P. freudenreichii, P. shermanii*
P. jensenii	*P. jensenii, P. petersonii, P. raffinosaceum, P. technicum, P. zeae*
P. thoenii	*P. thoenii, P. rubrum*
P. acidipropionici	*P. arabinosum, P. pentosaceum*
*P. coccoides**	*P. coccoides*
P. cyclohexanicum	*P. cyclohexanicum*
Cutaneous	
P. acnes	*Corynebacterium acnes*
P. avidum	*C. avidum*
P. granulosum	*C. granulosum*
P. lymphophilum	*C. lymphophilum*
P. propionicum	*Arachnia propionica*
Propioniferax innocua	*P. innocuum*

*Proposed by Vorobjeva et al. (1983)

Cutaneous propionibacteria are also known as anaerobic coryneforms or anaerobic diphtheroids. Initially, 12 species of anaerobic corynebacteria were described (Prévot and Fredette, 1966), but later, when some 80 strains had been evaluated for their cell wall composition and DNA homology (Johnson and Cummins, 1972), the number of species was reduced to three, namely, *P. acnes*, *P. granulosum* and *P. avidum*. Cutaneous propionibacteria are not restricted to the normal skin surface. They can be isolated from facial blackheads, less frequently from stomach contents, wounds, bone marrow, blood, and tissue abscesses. Thus, classical and cutaneous propionibacteria differ principally by their typical natural habitats.

Other differences between the two groups are shown in Table 1.3. The classical propionibacteria, in contrast with the cutaneous ones, do not produce indole and cannot liquefy gelatin.

Table 1.3. Species differentiation in the genus Propionibacterium

Organism	Acid production from:					Esculin hydrolysis	Indole product.	Nitrate reduction	Gelatin liquefact.	Cell wall DAP-isomer	Hemo-lysis	G+C in DNA, mol%
	sucrose	maltose	arabinose	cellobiose	glycerol							
P. acnes[a]	-	-	-	d	d	-	d	+	+	LL, some meso	d	59±1.5
P. avidum[a]	+	+	d	+	+	+	-	-	+	LL, some meso	(+)	62±0.5
P. granulosum[a]	+	+	-	-	+	-	-	-	-	LL	(-)	62±1.0
P. lymphophilum[a]	d	-	-	-	-	-	d	-	-	no DAP, lysine		53±0.8
P. propionicum[b,c]	+	+	-	-	d	-	-	+	d	LL		63-65
P. freudenreichii[a]	-	-	+	-	+	+	-	d	-	meso	-	65±1.0
P. jensenii[a]	+	+	-	d	+	+	-	-	-	LL	+	66±0.5
P. thoenii[a]	+	+	-	d	+	+	-	-	-	LL	+	67±1.1
P. acidi-propionici[a]	+	+	+	+	+	+	-	+	-	LL	-	67±0.8
P. cyclo-hexanicum[d]	+	+	-	+	+	+	-	-	+	meso	nd	66.8
P. coccoides[e]	+	+	+	+	+	nd	-	-	-	LL	nd	63.4

Compiled from: [a]Bergey's Manual of Systematic Bacteriology; [b]Schaal (1986); [c]Charfreitag and Stackenbrandt (1989); [d]Kusano et al. (1997); [e]Vorobjeva et al. (1983). Key: +, 90% or more strains are positive; (+), 80-89% are positive; d, 21-79% are positive; (-), 11-20% are positive; -, 90% or more strains are negative.

In 1988, chiefly on the basis of comparative analysis of long stretches of 16S rRNA, *Arachnia propionica*, formerly a member of the genus *Actinomyces*, was reclassified to *Propionibacterium* under the new name *P. propionicum* (Charfreitag et al., 1988). *A. propionica* forms filamentous branched cells as distinct from rod-shaped cells of the other propionibacteria, making the genus *Propionibacterium* morphologically heterogeneous.

Vorobjeva et al. (1983) also proposed to include propionic acid cocci in the genus *Propionibacterium*, since they share some general characteristics and a high level of DNA homology with rodlike propionibacteria. The cocci can be isolated from milk and cheese at early stages of ripening. As distinct from typical propionibacteria, the cocci can grow on the surface of solid media forming orange colonies, and possess a wide range of enzymatic activities. The name suggested for the new species is *P. coccoides*.

The most important property of the genus *Propionibacterium* is the production of propionic acid as a result of the propionic acid fermentation dependent on coenzyme B_{12}. If the dependence on coenzyme B_{12} is disregarded, some clostridial strains that do not form spores may be erroneously attributed as propionibacteria. For example, *Cl. botulinum, Cl. propionicum* and some other species can produce propionic acid, but propionibacteria have the GC-type DNA (65-67 mol% G+C in classical and 53-62 mol% in cutaneous bacteria), while clostridial DNA is of the AT-type (25-30 mol% G+C).

At the same time, some anaerobic corynebacteria also produce propionic acid, have a high G+C content, but are different from propionibacteria in their lipid composition and cell walls: they do not contain significant amounts of C_{15} branched-chain fatty acids in their membrane lipids, whereas their cell walls contain arabinogalactan and mycolic acids, absent from propionibacteria.

A bacterium that shows a propionate-acetate type of fermentation was isolated from the activated sludge of sewage-processing tanks (Samain et al., 1982). It can ferment ethanol and propanol and displays a growth factor-dependent association with Gram-negative homoacetogens. The authors called it a new propionic acid bacterium. It is an obligate anaerobe, asporogenic and rodlike. However, the true propionibacteria do not ferment propanol and ethanol and, in addition, are Gram-positive. Some other microorganisms are also known to produce propionic acid: *Selenomonas ruminantum, Bacteroides ruminicola, Megasphaera elsdenii, Propionispira arboris*, but they are not related to propionibacteria.

1.2 Classical Propionic Acid Bacteria

Their main habitat is hard rennet cheese, and more than 60% of the strains isolated in Finland from Emmental cheese are classified as *P. freudenreichii* subsp. *shermanii* (Merilainen and Antila, 1976). There have been reports on the isolation of *P. petersonii* (van Niel, 1957) and *P. pentosaceum* (Prévot and Fredette, 1966) from soil, *P. zeae* from silage (van Niel, 1957; Beerens et al., 1986); classical propionibacteria were also isolated from spoiled olives (Plastorgos and Vaughn, 1957; Cancho et al., 1970, 1980) and spoiled orange juice (Kusano et al., 1997), from anaerobic sewage (Riedel and Britz, 1993; Sarada and Joseph, 1994) and from nematodes recovered from the hind gut of zebra (Krecek et al., 1992).

1.2.1 General characteristics

The classical propionibacteria are aerotolerant or microaerophilic, although some strains may prefer aerobic conditions. Rodlike cells (Fig. 1.1) of these bacteria tend to be pleiomorphic. Rudimental branching is observed in dairy propionibacteria under aerobic conditions or anaerobically at low pH values. Colonies are usually wet, round or granular, bright and oily, although we (Vorobjeva et al., 1990) described strains forming dry leathery colonies, difficult to pick from the surface of solid medium. In liquid medium these strains (apparently *P. shermanii*) form heavy viscous sediment. Colonies can be creamy, orange, red, or brown.

When examined under the microscope, propionibacteria differ from other bacteria by a peculiar 'palisade-like' arrangement of the cells, sometimes forming short curved chains and hieroglyph-like patterns (Fig. 1.1) as a result of the division with a snap. Under the scanning microscope it can be seen that the cells are uneven, with rounded ends, in some cases covered with slime or forming slimy filaments. A number of strains of *P. thoenii* and *P. jensenii* are encapsulated and representatives of other species form extracellular slime. Capsule material is represented by polysaccharides, e.g., a strain of *P. zeae* produces slime and capsular material composed of mannose (main sugar), glucose and galactose (Skogen et al., 1974). Slime capsules may defend the cell from bacteriophage invasion or other deleterious agents. Immunological properties of slime capsules have been suggested (Hettinga and Reinbold, 1972a).

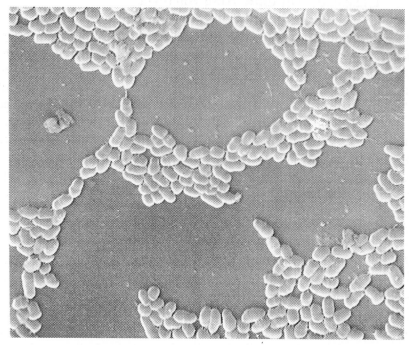

Figure 1.1. Scanning electron micrograph of *Propionibacterium shermanii*. General view. × 12,000. From Vorobjeva (1976).

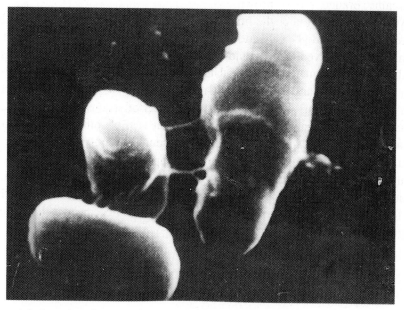

Figure 1.2. Scanning electron micrograph of *Propionibacterium freudenreichii.* × 60,000. From Vorobjeva (1976).

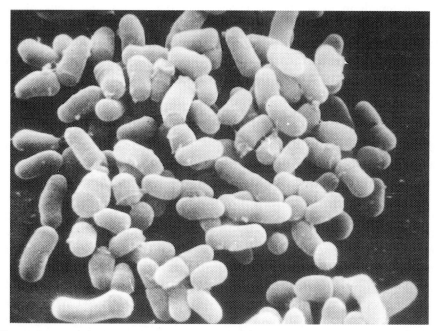

Figure 1.3. Scanning electron micrograph of *Propionibacterium pentosaceum.* × 12,660. From Vorobjeva (1976).

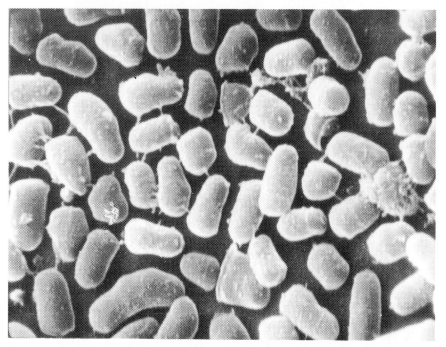

Figure 1.4. Scanning electron micrograph of *Propionibacterium technicum.* × 12,660. From Vorobjeva (1976).

Cell length in most species varies from 0.5 to 1.5 μm. Cells of *P. freudenreichii* sometimes look as cocci (Fig. 1.2), cells of *P. pentosaceum* (Fig. 1.3) and *P. technicum* (Fig. 1.4) are short rods, and those of *P. petersonii* are elongated rods of 8-10 μm, thus being 4-5 times longer than in the other species (Fig. 1.5). Cell diameter is generally in the range of 0.5-1.0 μm. Cells have a tendency for rudimental branching (Figs. 1.6, 1.7), sometimes forming local thickenings and branches.

P. petersonii differs from the others by its extraordinary pleiomorphism. Cells of irregular shape with disordered partitions persist during all growth stages, unlike corynebacteria, in which rodlike forms are followed by coccoid forms (Stevenson, 1968). *P. petersonii* produces bright-orange dry colonies under aerobic conditions and light-yellow colonies under anaerobic conditions. For this species, expansions of the outer (murein) layer and internal partitions are typical. Internal partitions divide the cell into many compartments. Expansions of the cell wall due to the oversynthesis of murein have also been described in corynebacteria (Freer et al., 1969; Barksdale, 1970). Presumably, these expansions can cause the formation of cell aggregates precipitating to the bottom of the fermentor, and may also be responsible for some highly unusual forms of *P. petersonii* appearing as a cell enclosed within another cell (Fig. 1.8) or as budding cells.

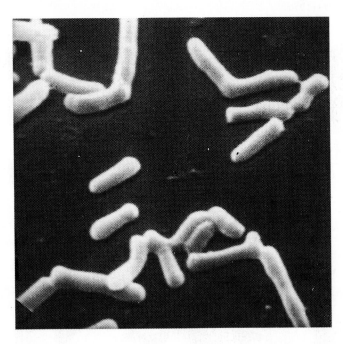

Figure 1.5. Scanning electron micrograph of *Propionibacterium petersonii*. × 12,000. From Vorobjeva (1976).

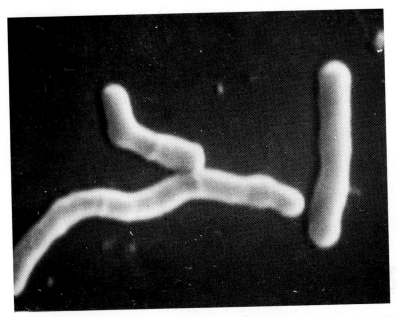

Figure 1.6. Scanning electron micrograph of *Propionibacterium petersonii*. × 24,000.
From Vorobjeva (1976).

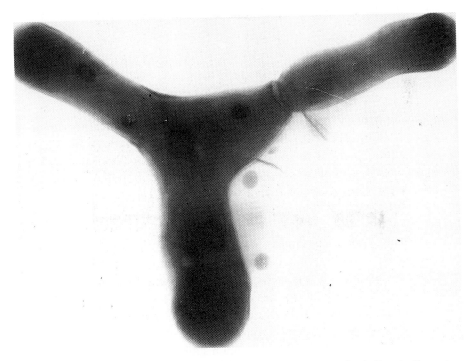

Figure 1.7. Transmission electron micrograph of negatively stained *Propionibacterium petersonii*. × 30,000. From Vorobjeva (1976).

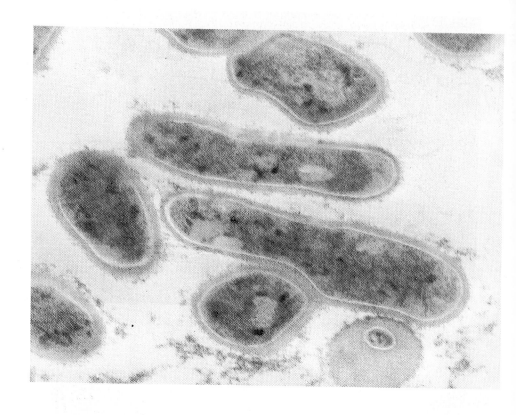

Figure 1.8. Thin section of *Propionibacterium petersonii* stained with uranyl acetate. A large murein layer is seen. × 60,000. From Vorobjeva (1976).

Cell anatomy of the classical propionic acid bacteria is typical for Gram-positive cells (Figs. 1.9, 1.10). Thickness of the cell wall varies a little with the age of culture, being in the range of 20.0-30.0 nm. In cells treated with a lead fixative two layers of the cell wall can be observed. Cell division begins with a centripetal formation of a division septum so that the new cell wall appears before the cell division is complete.

The cytoplasm (Figs. 1.9, 1.10) represents a region of intermediate electron density, where ribosomes (or polysomes) and the nuclear area of a still lower density are discernible. The nuclear area is represented by fibrils. On negative contrast numerous membranous bodies of mesosomes are found (Figs. 1.10, 1.11). In a single species all three types of mesosomes common to Gram-positive bacteria may be present: tubular, vesicular and lamellar.

The cytoplasm of propionibacteria contains a regular inclusion in the form of granules that strongly absorb osmium stains and have a high electron density. The granules are stained by methylene blue, appearing purple red under the light microscope and maintaining the same localization as that found in the electron microscope. The number of granules in the cell correlates with the content of acid-insoluble polyphosphates. So it can be assumed that these inclusions represent metachromatic granules and consist, at least in part, of polyphosphates. Similar granules have been described in mycobacteria and corynebacteria (Freer et al., 1969; Barksdale, 1970).

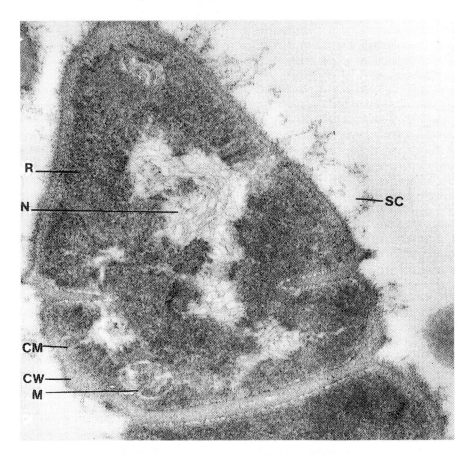

Figure 1.9. Thin section of *Propionibacterium shermanii* with negative contrast. Labeled structures: cytoplasmic membrane (cm); cell wall (cw); mesosome (m); nucleoid (n); ribosomes (r); slimy capsule (sc). × 228,500. From Vorobjeva (1976).

In propionibacteria the granules usually have a polar location in the cell. When cells are low in polyphosphates, it is possible to recognize fine grains in the structure of granules by the electron microscopy. Young cells of propionibacteria contain a lower number of granules and these are small. In aging cultures the size of granules and sometimes their numbers are significantly increased. Since polyphosphate granules occur mainly in old cultures of propionic acid bacteria, they can be considered as a storage form of pyrophosphate for its utilization during cell reactivation. Propionibacteria contain enzymes active in polyphosphate metabolism (Kulaev et al., 1973), which evidently play an important role in the metabolism of these cells.

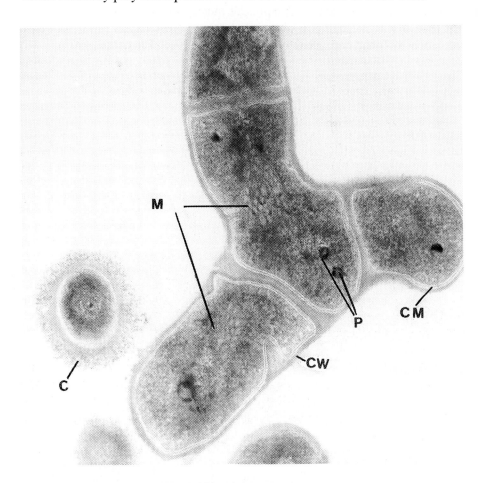

Figure 1.10. Electron micrograph of a thin section of *Propionibacterium petersonii*. Labeled structures: cytoplasmic membrane (cm); cell wall (cw); mesosome (m); polyphosphate granules (p, staining dark); capsule (c). × 80,000. From Vorobjeva (1976).

All the strains tested require vitamins for growth, i.e. pantothenic acid and biotin (Delwiche, 1949), some of them need thiamine and *p*-amino-benzoic acid as well. The growth of all strains is stimulated by Tween 80 (Johnson and Cummins, 1972). Skerman (1967) included also the ability to utilize lactate as a taxonomic character.

The main products of glucose fermentation are propionic and acetic acids and carbon dioxide. All the strains of the genus can ferment pyruvate, dioxyacetate and glycerol with the same end products as in glucose fermentation.

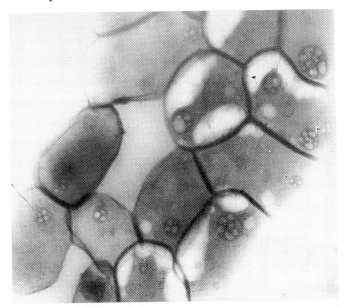

Figure 1.11. Transmission electron micrograph of negatively stained *Propionibacterium petersonii.* From Vorobjeva (1976).

All propionic acid bacteria are catalase-positive (*P. arabinosum* displays a very weak catalase activity). As a rule, members of the genus *Propionibacterium* produce significant amounts of vitamin B_{12}. We (Vorobjeva, 1976) suggested to differentiate propionibacteria by the nature of the coenzyme form of vitamin B_{12}. The presence of the isomerization reaction of succinyl-CoA $\rightleftharpoons$ methylmalonyl-CoA, which is the key reaction of propionic acid fermentation, represents another valuable taxonomic property of the genus.

1.2.1.1   Bacteriophages
Partially lyzed colonies of *P. shermanii* are sometimes observed (Borisova et al., 1973), with the type of lysis resembling phagolysis (Fig. 1.12). Under the electron microscope we found half-degraded cells with adsorbed phage coats (Fig. 1.13).

To identify the phage, the culture of *P. shermanii* was plated on corn steep-glucose agar and a drop of the culture liquid pre-filtered through a Zeitz-filter was placed in the center of a plate. The plates were incubated at 28-30°C in an anaerostat at a residual air pressure of 10-20 mm Hg for 8 days. The presence of phage on the lawns was assessed by scoring clear plaques (lysed colonies).

Figure 1.12. Colonies of *Propionibacterium shermanii* showing signs of phagolysis. Natural size. From Vorobjeva (1976).

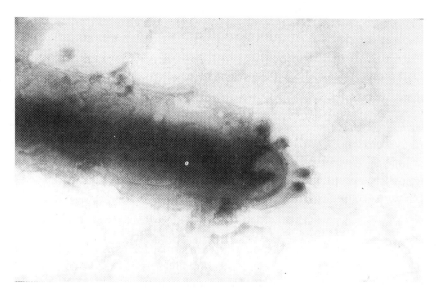

Figure 1.13. Transmission electron micrograph of negatively stained *Propionibacterium shermanii* showing several phages attached. × 80,000. From Vorobjeva (1976).

Phage concentration was estimated by differential centrifugation. The lysate was separated from unlysed cells by centrifugation at 6000 rpm for 20 min. The supernatant was centrifuged at 18000 rpm for 40 min to remove cell debris. Phage particles were pelleted by centrifuging for 1 h at 40,000 rpm. Phage particles present in the pellet were contrasted with 2% phosphotungstic acid. Under the electron microscope phage particles, homogeneous in shape and size, were found. The phage consisted of isometric heads and noncontractile tails (Fig. 1.14). The heads were approximately 500 Å in diameter, and the tails were 1700 Å in length and 100 Å in diameter. Plaques had a rounded shape and diameters of 1.5-3.0 mm, indicating a possible presence of several types of phages in *P. shermanii*. After 7 days small colonies of *P. shermanii* appeared in sterile zones. In liquid medium the phage titer reached 10^9, and a secondary growth was observed in 2 to 3 days.

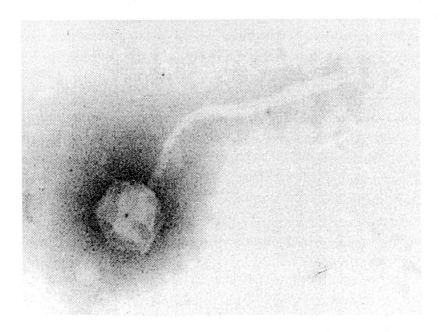

Figure 1.14. Bacteriophage of *Propionibacterium shermanii* (negatively stained). × 700,000. From Vorobjeva (1976).

It was found that the phage preferentially affects cells in the early exponential phase of growth. The phage multiplies in a wide range of pH from 2.7 to 10.5 with an optimum at pH 6.5-7.2. Under aerobic conditions the phage action is expressed weakly or is completely absent, which can be explained by its inactivation during shaking as a result of the separation of heads from the long tails. The phage is completely inactivated by heating at 100°C for 3 min or by formalin treatment for a few minutes. Hydrogen peroxide at a concentration of 1.5% inactivates the phage in 24 h, at a concentration of 3% in 1 h; a 2% formalin solution inactivates the phage in 1 h, and 90% ethanol in 3 h (Borisova et al., 1973). The phage specifically attacks *P. shermanii*. The cells of *P. shermanii* resistant to phage are shorter and acquire a more oval shape than the cells of the original strain.

In total, 19 bacteriophages have been isolated from *P. shermanii* (Gautier et al., 1995) and their morphology is very similar to that described above and also to the morphology of phages isolated from cutaneous propionibacteria (Zierdt, 1974; Webster and Cummins, 1978). The phages of *P. shermanii* had isometric heads with noncontractile tails with a terminal plate. The authors (Gautier et al., 1995) suggest that the phages belong to group B_1 in Bradley's classification (Bradley, 1967). Most of the phages had identical or similar restriction patterns and were grouped into 3 clusters according to their restriction patterns obtained by treatment with endonuclease Pst I. Some strains harbored bacteriophage DNA in their chromosomes; these bacteriophages can be used as tools for the development of engineering techniques for propionibacteria.

Sixteen out of the 32 cheeses studied were contaminated with bacterio-phages (10^7 PFU/g cheese), and the phage contamination was shown (Steffen et al., 1993) to be a chronic phenomenon. Two types of bacterio-phages that infect dairy propionibacteria were isolated from Swiss-type cheeses. One belongs to the Bradley's classification group B_1 and the other is the first infectious filamentous virus described in Gram-positive bacteria (Gautier et al., 1998). Raw milk may be the source of bacteriophages in cheese manufacturing, although these phages are sensitive to heat treatment.

Another source of phages is represented by prophages (Herve et al., 1998). Chromosomal DNA of 94 strains of dairy propionibacteria was hybridized with the genomic DNA of 6 bacteriophages isolated from Swiss-type cheese. The authors found that the DNA of 20 bacterial strains hybridized strongly with the phage DNA, confirming the presence of prophages. Phage lysogeny has been directly demonstrated in propioni-bacteria (Herve et al., 1998). Bacteriophages may have a role in the autolysis of propionibacteria in cheese and concomitant release of various peptidases and lipases that produce aromatic compounds (Gautier et al., 1995).

1.2.1.2   Cell wall composition

Two different types of peptidoglycan are found (Schleifer and Kandler, 1972). *P. shermanii* and *P. freudenreichii* contain a directly cross-linked peptidoglycan with *meso*-DAP at position 3. *meso*-DAP was also found in *P. cyclohexanicum* sp. nov., recently isolated by Kusano et al. (1997). All other species studied contain LL-DAP-glycine, i.e., glycine forms a cross link between tetrapeptides via the amino group of LL-DAP and the C-terminal alanine of a neighboring tetrapeptide (Schleifer et al., 1968). Cell wall sugars of *P. shermanii* and *P. freudenreichii* are represented by galactose, mannose, and rhamnose (the latter is not found in the cell walls of other species); glucose is absent (Table 1.4) (Cummins and Johnson, 1992). In the whole cells of *P. cyclohexanicum* the following sugars are found: galactose, mannose, glucose, ribose and rhamnose. The cell wall polysaccharides of all other propionibacteria, except *P. freudenreichii*, contain 2,3-aminohexuronic acid (Cummins and White, 1983; Cummins, 1985). The two types of murein in the genus *Propionibacterium* correspond with the two morphological cell types: two *meso*-DAP-containing species, *P. shermanii* and *P. freuden-reichii*, are more or less coccoid, while the other species form coryneform rods. Thus, for bacterial systematics the molecular basis of morphology is rather more important than the pleiomorphism of bacteria.

Table 1.4. Cell wall composition in the genus *Propionibacterium*

Organism	Sugar moieties in polysaccharide	Amino acid residues in peptidoglycan
P. freudenreichii[a]	galactose mannose rhamnose	alanine, glutamic acid, *meso*-DAP
P. jensenii[a]	galactose glucose mannose	alanine, glutamic acid, glycine, LL-DAP
P. thoenii[a]	galactose glucose	alanine, glutamic acid, glycine, LL-DAP
P. acidipropionici[a]	galactose glucose mannose (in some strains)	alanine, glutamic acid, glycine, LL-DAP
P. cyclohexanicum[b]	galactose, glucose mannose, rhamnose ribose	alanine, glutamic acid, *meso*-DAP
P. acnes[c,*]	galactose glucose mannose	alanine, glutamic acid, (glycine), LL-DAP, meso-DAP in some strains
P. avidum[c,*]	galactose glucose mannose	alanine, glutamic acid, (glycine), LL-DAP, *meso*-DAP in some strains
P. granulosum[c]	galactose mannose	alanine, glutamic acid, glycine, LL-DAP
P. lymphophilum[c]	galactose glucose mannose	alanine, glutamic acid, lysine (no DAP)
P. propionicum[c,d]	galactose glucose mannose	alanine, glutamic acid, glycine, LL-DAP

*In both *P. acnes* and *P. avidum* two serological types are present, differing in cell wall sugars. Compiled from: [a]Cummins and Johnson (1986); [b]Kusano et al. (1997); [c]Schaal (1986); [d]Charfreitag et al. (1988); [d]Charfreitag and Stackebrandt (1989).

It is known that teichoic acids (and capsules), forming part of the cell wall, may determine the serological properties of Gram-positive bacteria. Moore and Cato (1963), on the basis of serological similarity, showed that *P. acnes* is close to *P. arabinosum, P. jensenii, P. pentosaceum, P. rubrum, P.*

thoenii, P. zeae, but not to *P. shermanii* and *P. freudenreichii*. The latter showed a weak cross-reaction with the other species of propionibacteria, and this is associated with the sugar composition of the cell walls of *P. shermanii* and *P. freudenreichii* (Johnson and Cummins, 1972). Serological tests showed that *P. shermanii* and *P. freudenreichii* are closely related and comprise a separate group in the genus *Propionibacterium*. Therefore, the choice of *P. freudenreichii* by van Niel (1928) as the typical species of the genus now seems less certain.. Actually, this species is atypical since: (i) it contains *meso*-DAP instead of L-isomer in the other species, (ii) its cell wall contains rhamnose which is absent from the other species, (iii) it is distant from the other species with respect to DNA homology (Johnson and Cummins, 1972), (iv) its lipids contain *anteiso*-C_{15} fatty acid (12-methyl-tetradecanoic), while the other species (*P. arabinosum, P. jensenii, P. pentosaceum, P. thoenii*, and *P. zeae*) basically contain *iso*-C_{15} acid (13-methyltetradecanoic).

In most bacterial species C_{15} fatty acids are present in trace amounts, so their elevated levels (2-3 times higher than any other fatty acid) in propionibacteria can serve as a diagnostic marker. However, this marker should be taken with caution, since the levels of free fatty acids in bacteria depend on media composition, age of culture and the level of vitamin B_{12} in the cells. Addition of isoleucine to the medium increases the synthesis of *anteiso*-C_{15} acid by propionibacteria. In the presence of L-leucine they produce more *iso*-C_{15} acid by decreasing *anteiso*-C_{15} acid (Moss et al., 1969). In the cells of young active cultures usually the level of straight-chain mono-unsaturated acids ($C_{16:1}$, $C_{18:1}$) is higher. The content of mono-unsaturated fatty acids is higher than the branched-chain fatty acids in cultures deficient in vitamin B_{12}. With the cell free extract of *C. simplex* it was shown that vitamin B_{12} deficit leads to a decrease in the activity of transmethylase system and in the rate of the transformation of mono-unsaturated acids to CH_3-branched fatty acids (Fujii and Fukui, 1969). A distinct fatty acid composition was found (Kusano et al., 1997) in *P. cyclohexanicum*. The major fatty acid was ω-cyclohexyl undecanoic acid, while *iso*- and *anteiso*-C_{15}, C_{16}, and C_{17} fatty acids were also present, but in a small amount.

The phospholipid composition also is an important taxonomic indicator of the genus. Biochemical activity of the cell depends considerably on the association of some specific proteins with phospholipids. In bacterial membranes one major phospholipid, accounting for 50% or more of the total, is usually found (Salton and Owen, 1976). The major phospholipid class in propionibacteria is represented by glycolipids, which amount to about 40% of the total lipids in *P. shermanii*, with smaller amounts in *P. freudenreichii* and *P. arabinosum* (Shaw and Baddiley, 1968). The exact composition of the propionibacterial glycolipids has been, however, a matter of controversy. First reported as being composed primarily of mannose-containing glycolipids (Prottey and Ballou, 1968), similar to those found in mycobacteria, the phospholipid composition of *P. freudenreichii* was re-investigated using more advanced methods, with very different results (Sutcliffe and Shaw, 1993).

In contrast with the previous reports, no phosphatidylinositol mannosides could be detected (Sutcliffe and Shaw, 1993). The major phospholipids identified in *P. freudenreichii* were bis(phosphatidyl)glycerol, phosphatidyl-glycerol and phosphatidylinositol, with small contributions from phosphatid-ylethanolamine and lysophosphatidylinositol. The relative abundance of the two phosphatidylglycerol species is perfectly consistent with the taxonomic position of the propionibacteria among the Gram-positive eubacteria, especially their close relations with actinomycetes.

Sensitivity to antimicrobial agents. All the strains tested are highly resistant to sulfamides; they are more resistant to semisynthetic penicillins, such as oxacillin, than to penicillin (Reddy et al., 1973). Some strains grow in the presence of sulfadiazine (1000 µg/ml). Nisin displays a strong inhibitory action (Galesloot, 1957). Only *P. freudenreichii* is sensitive to lysozyme, the other strains are relatively sensitive only to certain muralytic enzymes such as muranolysin and achromopeptidase.

Vitamin requirements. All the strains require biotin and pantothenic acid (Delwiche, 1949). Thiamine and nicotinamide show a growth-stimulatory effect. Some strains require *p*-aminobenzoic acid. Tween 80 stimulates growth of all strains. (Holland et al., 1979).

1.2.1.3   Growth media
A medium that supports good growth of all propionibacteria has the following composition (%): trypticase (BBL), 1; yeast extract (Difco), 0.5; glucose, 1; $CaCl_2$, 0.002; $MnSO_4$, 0.002; NaCl, 0.002; K-phosphate buffer, 0.05 M; Tween 80, 0.05; sodium dithionite, 0.05; $NaHCO_3$ (added as sterile solution along with inoculate), 0.1%; final pH 7.0 (Cummins, Johnson, 1981).

In our studies (Vorobjeva, 1976) good growth was observed in corn steep-glucose medium containing (%): corn-steep liquor, 2; glucose, 1; K_2HPO_4, 0.2; $(NH_4)_2SO_4$, 0.5; $CoCl_2·6H_2O$, 10 mg/l; pH 6.9-7.0 (with a saturated solution of $NaHCO_3$). This medium ensures high levels of vitamin B_{12} synthesis and high biomass yields. Minimal medium used in physio-logical experiments has the following composition (%): sodium lactate or glucose, 1.5; $(NH_4)_2SO_4$, 0.3; $MgSO_4$, 0.02; KH_2PO_4, 0.1; NaCl, 0.002;

MnSO$_4$, 0.0005; FeCl$_3$·6H$_2$O, 0.001; ZnSO$_4$·7H$_2$O, 0.01; Ca-pantothenate, 1.0 mg/l; thiamine, 0.2 mg/l; biotin, 1.0 µg/l; pH 7.0. Distilled water to volume.

The minimal pH at which growth occurs and the lethal pH values depend on the acid used for adjusting the pH. The strains of *P. technicum*, *P. jensenii* and *P. acidipropionici* can grow down to pH 5.5, 5.6 and 5.3 but die at pH 4.3, 4.4 and 3.9 in lactic, propionic and hydrochloric acid, respectively (Rehberger and Glatz, 1995).

For long-term storage it is not recommended to include glucose in the medium; it should be substituted by any other carbon source.

Holdeman et al. (1977) suggested to use meat broth for storage; it is poured into test tubes and inoculated under N$_2$ by the method of Hungate (Holdeman et al., 1977). Viability is retained for many months at room temperature. However, at low temperatures (in a refrigerator) cultures can be lost in a short time. In Difco medium 'Brewer anaerobic agar 279' propioni-bacteria remain viable for four months (Niethamer and Hitzler, 1960). Generally, propionibacteria are preserved by periodic subculturing (once every 5-6 months) in liquid medium containing lactate, or by lyophilization. Lyophilized cultures remain viable for 15-20 years (Arkadjeva et al., 1988).

1.2.2 Individual classical species

Differentiation between the species is shown above, in Table 1.3. Electrophoretic separation of cellular proteins for the identification and differentiation of propionic acid bacteria was suggested by Baer (1987). Every strain was characterized by a specific spectrum of proteins that was compared with the spectrum of 'standard strains'. This method is more rapid and accurate than the classical microbiological analysis of propionibacteria, although it could not differentiate between the strains of *P. freudenreichii* subsp. *freudenreichii* and *P. freudenreichii* subsp. *shermanii*. Modern methods of molecular biology are being used at present for the identification of propionibacteria (see below).

1.2.2.1 *Propionibacterium freudenreichii*

The bacterium is usually isolated from Swiss-type cheese and raw milk. It produces large amounts of propionic acid that confers a specific aroma to cheese. Cells look like short rods, often as cocci. They differ from the other species by a higher thermal resistance. The major fatty acid is 12-methyltetradecanoic acid (approx. 43%). The major sugar in peptidoglycans is galactose; mannose and rhamnose are also present, but in smaller amounts; glucose is completely absent. *P. freudenreichii* can ferment a limited number of carbohydrates: fructose, galactose, glucose and mannose;

sucrose and maltose are not fermented (cf. Table 1.3). Peptidoglycans contains *meso*-DAP instead of L-DAP present in the other strains. The G+C-content is 64-67 mol%. On the basis of the capacity for lactose fermentation and nitrate reduction three subspecies can be distinguished:

	Nitrate reduction	Lactose fermentation
P. freudenreichii subsp. *freudenreichii*	+	–
P. freudenreichii subsp. *shermanii*	+	+
P. freudenreichii subsp. *globosum*	–	–

All the strains are not hemolytic.

1.2.2.2 *Propionibacterium jensenii*

This species incorporates the strains originally described as five separate species, hence there are interstrain variations in morphology, physiology and biochemistry. Some strains grow aerobically as well as anaerobically. L-DAP and glucose are the major carbohydrates in peptidoglycans. 13-Methyltetradecanoic acid is the major fatty acid in lipids (as in the other species described below). The G+C content is 65-68 mol%. Isolated from milk products, silage and sometimes from infected lesions. Some strains may be hemolytic. Typical strain is ATCC 4868.

1.2.2.3 *Propionibacterium thoenii*

Usually forms orange or brownish red colonies on solid agar. Causes hemolysis of human, cow, pig, sheep and rabbit blood. Cell wall contains L-DAP and the following carbohydrates: glucose, galactose and mannose. The G+C content is 66-67 mol%. This species was formed by combining *P. rubrum* and *P. thoenii* on the basis of their high DNA homology. They can be differentiated by the nature of sugar fermentation: *P. rubrum* ferments raffinose and mannitol, but not sorbitol, whereas *P. thoenii* ferments sorbitol, but not raffinose and mannitol.

At the same time, Riedel and Britz (1992) found a high degree of similarity between the strains of *P. jensenii* and *P. rubrum* by comparing electrophoretic protein profiles. A similar finding was also reported by studying the 16S rRNA sequences (De Carvalho et al., 1995), and the authors proposed to reclassify *P. rubrum* as a β-hemolytic biovar of *P. jensenii*.

1.2.2.4 *Propionibacterium acidipropionici*

This species also is composite; in contrast with the other propionibacteria, it has a weak or negative catalase activity. Strains grow well under aerobic conditions. Contains L-DAP. In peptidoglycan glucose, galactose and/or mannose are present. The G+C content is 66-68 mol%. It is isolated from milk products. Typical strain is ATCC 25562. The species incorporates the strains originally described as *P. pentosaceum* and *P. arabinosum*. Whereas *P. arabinosum* does not ferment xylose and rhamnose, *P. pentosaceum* can; but this difference is insufficient for species formation. The strains have a high degree of DNA homology.

1.2.2.5 Propionic acid cocci

We (Vorobjeva et al., 1990) suggested that propionic acid cocci might be related to the dairy propionibacteria. The cocci were isolated from Soviet cheese at an early stage of ripening (manufactured in the Altai region) and show a similarity with propionic acid bacteria in fatty acid composition, i.e. they contain *anteiso*-C_{15} saturated fatty acid as the main type of fatty acids. The highest similarity in fatty acid composition was found with *P. jensenii*. The similarity between propionicocci and propionibacteria was reflected in the structure of their genome. The DNA-DNA hybridization indicated a relatedness with *P. jensenii* (*P. raffinosaceum*). The G+C content was 63.4 mol%, similar to the propionibacterial 65-67 mol%. Propionic acid cocci can ferment lactate, producing propionic and acetic acids and CO_2 as the end products. It was shown that propionic acid arises from methylmalonyl-CoA, formed in the isomerization reaction dependent on coenzyme B_{12}. In small amounts the cocci synthesize cobalamins (Alekseeva et al., 1973b), catalase, superoxide dismutase and peroxidase (Vorobjeva and Kraeva, 1982; Vorobjeva et al., 1983). The cocci ferment mono-, tri- and polysaccharides: glucose, fructose, mannose, galactose, raffinose, maltose, sucrose, lactose, trehalose, dextrins, starch, sorbitol, mannitol, salicin and glycerol. Xylose, rhamnose and dulcitol are not metabolized. Arabinose and cellobiose are fermented to a variable degree. Indole, hydrogen sulfide and ammonia are not produced; gelatin is not liquefied, nitrate is not reduced. Catabolism is anaerobic.

Table 1.5. DNA-DNA homology between propionic acid cocci, propionic and lactic acid bacteria. From Vorobjeva et al. (1983).

Organism	Homology to reference DNA from: (%)			
	P. coccoides 15		P. jensenii 1861	
	Incubation temperature, °C			
	70°	80°	70°	80°
P. coccoides 15	100 ± 15	100 ± 7	49 ± 3	48 ± 2
P. jensenii (P. raffinosaceum)	50 ± 3	48 ± 5	100 ± 15	100 ± 6
P. thoenii	4 ± 1	2	2	1
P. acidipropionici	4 ± 1	1	5 ± 1	1
P. freudenreichii subsp. *shermanii*	15 ± 3	5 ± 1	19 ± 1	6 ± 1
Lb. leichmannii	0	0	0	0
S. lactis	0	0	0	0

It is known that classical, or dairy propionibacteria exhibit a very high DNA homology between strains of a given species, up to 87-96 mol%, while the homology between different species varies from 8 to 53 mol% (Cummins and Johnson, 1986). Propionic acid cocci, however, could not be attributed to any of the described groups of propionibacteria on the basis of DNA homology, although the closest similarity was observed with *P. jensenii*. With the latter the cocci had the highest phenotypic similarity as well.

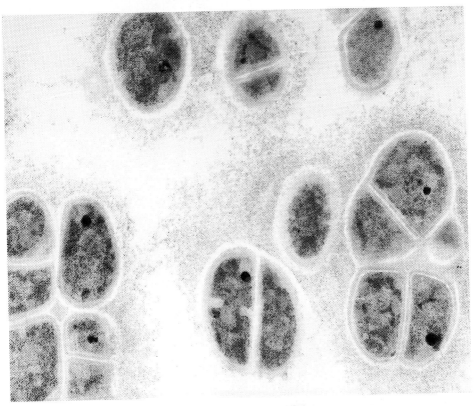

Figure 1.15. Electron micrograph of a thin section across dividing cells of *Propionibacterium coccoides.* × 16,000. From Vorobjeva (1976).

Unlike propionic acid bacteria, the cocci retain spherical shape at all stages of growth, both under aerobic and anaerobic conditions. The cells have a diameter of 0.6-1.2 μm. Colonies are smooth, creamy to yellow. Facultative anaerobes. Temperature optimum for growth is 17-22°C, but they can grow at 8-10°C. Halotolerant, can grow in the presence of 6.5% NaCl, i.e., under conditions restrictive for the growth of other propionibacteria. Cell division is irregular and in different directions (Fig. 1.15). Cells have a very large capsule, polyphosphate granules, and mesosomes (Fig. 1.16). On the basis of the above mentioned criteria we proposed to include propionic acid cocci into the genus *Propionibacterium* (Vorobjeva et al., 1983). Having a peculiar morphology and biochemistry, the cocci should be treated as a separate group within the genus, with the same taxonomic status as the species described earlier. We suggested the name *P. coccoides* for the new species (Vorobjeva et al., 1983).

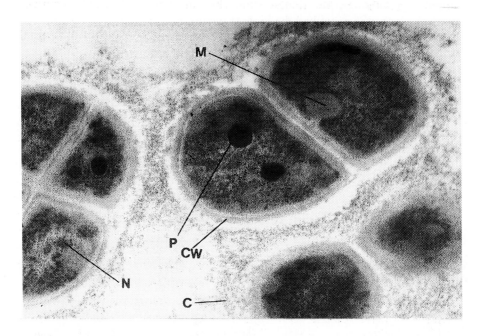

Figure 1.16. Transmission electron micrograph of an ultrathin section of *Propionibacterium coccoides*. Labeled structures: cytoplasmic membrane (cm); cell wall (cw); nucleoid (n); polyphosphate granules (p); mesosome (m); capsule (c). × 100,000. From Vorobjeva (1976).

Still, *P. coccoides* is not included in Bergey's Manual of Determinative Bacteriology (1994), despite its recognition by Cummins and Johnson (1992) as a probable member of the genus *Propionibacterium*.

To further characterize the new species, we sent a sample to Dr. Riedel (Dept. of Microbiology, University of the Orange Free State, Bloemfontein, South Africa), who had used numerical analysis as well as PCR/RFLP methods for a taxonomical study of propionibacteria (Britz and Riedel, 1995). Dr. Riedel confirmed that the hybridization rates between propionibacteria and *P. coccoides* are relatively low (personal communication). By sequencing the 16S rRNA of *P. coccoides* he came to the conclusion that it is closely related to the new species of the genus *Luteococcus*, having the phenotypic characteristics closely resembling those of propionibacteria, but being phylogenetically different. Currently, there is only one species (*L. japonicus*) in the genus *Luteococcus* described so far (Tamura et al., 1994).

The authors describe *Luteococcus japonicus* gen. nov. sp. nov. as a Gram-positive coccus with the G+C content in DNA of 67 mol%, containing menaquinone MQ-9(H$_4$) and peptidoglycan composed of LL-DAP, alanine, glycine and glutamic acid, like in typical propionibacteria. Propionic acid is formed as the major product of glucose fermentation. Facultatively anaerobic. Catalase and oxidase positive. Colonies are circular and smooth, cream-colored to yellow. But a partial 16S rRNA sequence showed that the typical strain IFO 12422T represents a distinct line among Gram-positive bacteria with high G+C content (Tamura et al., 1994).

Two strains of *P. coccoides* tested could be differentiated from the type strain of *L. japonicus* by acid production from ribose, adonitol, L-sorbose, inositol, α-methyl-D-glucoside, N-acetyl-glucosamine, salicin, xylitol and some other characteristics. At the same time, a very high degree (more than 90%) of DNA homology was observed between *L. japonicus* and *P. coccoides*. In addition, the RFLP (restriction fragment length polymorphism) profiles were identical (Riedel et al., 1998), indicating that the organisms are closely related at the molecular level. The two strains of *P. coccoides* were clustered together (84%) and with the *L. japonicus* strain at an overall similarity level of 83%. This cluster was linked with the four major clusters of classical propionibacteria at a level of 72% (K.-H. Riedel, personal communication).

1.2.2.6 *Propionibacterium cyclohexanicum* sp. nov.

A non-spore-forming, coryneform bacterium, strain TA-12^T, was isolated from pasteurized but spoiled, with an off-flavor, orange juice (Kusano et al., 1997). It is Gram-positive, aerotolerant, nonmotile, catalase-negative, pleiomorphic and rodlike, 1.5 to 3.0μm long and 1.1 to 1.6 μm wide. The colonies on the surface of PYG agar plates after 3 days of anaerobic growth are round, white to creamy, translucent, and 0.2 to 0.5 mm in diameter. The cells grow at pH 3.2 to 7.5 with an optimum at pH 5.5-6.5. Optimal temperature is 35°C. This organism produces lactic, propionic and acetic acids as a result of glucose fermentation in a molar ratio of 5:4:2, respectively. The production of lactic acid distinguishes the strain TA-12^T from all other species except *P. propionicum*.

It is an unusual propionibacterium in that it can grow at such a low pH and can withstand heat treatment at 90°C for 10 min. *P. freudenreichii* (which has the highest DNA homology with the new strain and is the most heat resistant of the propionibacteria) can survive for 10 min at 80°C, but can not survive at 90°C. The catalase, oxidase, and Voges-Proskauer tests are negative, and the methyl red test is positive. Urea is hydrolyzed, nitrate is not reduced to nitrite, and indole is not produced.

The strain TA-12^T can metabolize glycerol, D-xylose, galactose, D-glucose, D-fructose, D-mannose, amigdalin, arbutin, esculin, salicin, cellobiose, maltose, lactose, sucrose, trehalose, melecitose, D-furanose, acetic acid, lactic acid, malic acid, citric acid, and succinic acid. The G+C content in DNA is 66.8 mol%. The major menaquinone of this strain is MQ-9(H$_4$). The cell wall sugars are galactose, mannose, glucose, ribose and rhamnose. The cell wall contains *meso*-diaminopimelic acid, glutamic acid and alanine at a molar ratio of 1:1:2. ω-Cyclohexyl undecanoic acid is the major cellular fatty acid that accounts for 52.7%. Straight-chain and *anteiso*-branched-chain fatty acids include 16.8% *n*-C$_{15}$, 6.4% *anteiso*-C$_{15}$, 2.8% *n*-C$_{16}$ and 5.3% *n*-C$_{17}$. This fatty acid composition is unlike any other previously reported in a Gram-positive bacterium with the high G+C-content, except for *Curtobacterium pusillum*.

The results of a phylogenetic analysis of the 16S rRNA gene of this strain indicated that it has the highest homology with *P. freudenreichii* DSM 20271 (97%), with other propionibacteria being around 95 mol%. (The level of homology with other bacteria was lower.) But in contrast with *P. freudenreichii* it produces large amounts of lactic acid and has a distinct fatty acid composition, acid tolerance and heat resistance. On the basis of these findings the authors (Kusano et al., 1997) suggested the name *Propionibacterium cyclohexanicum* sp. nov. for this organism. Type strain is TA-12^T (= IAM 14535 = NRIC 0247). The position of *P. cyclohexanicum* on an unrooted phylogenetic tree is shown in Fig. 1.17.

1.2.3   Numerical evaluation of intraspecies groupings among the classical propionibacteria

A total of 147 strains of classical propionibacteria were used by Britz and Riedel (1995) for numerical analyses. The authors determined 74 phenotypic traits using computer analysis and PCR/RFLP restriction profiles. The results showed a good separation between the four classical propionibacterial species, confirming the major species suggested by Cummins and Johnson (1986). But the results also revealed that each major grouping consisted of several subgroupings. It was found that the major group '*jensenii*' contained at least 10 separate clusters (Fig. 1.18). The major group '*acidipropionici*' contained 14 closely related clusters. The major group '*thoenii*' consisted of two large and four smaller clusters. The major group '*freudenreichii*' was also found to consist of six clusters and a single-strain cluster (Fig. 1.18).

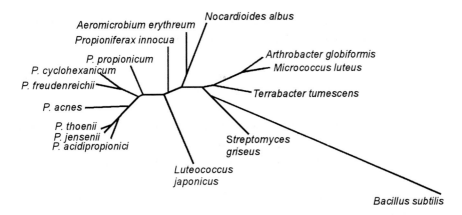

Figure 1.17. Unrooted phylogenetic tree showing the position of *P. cyclohexanicum* TA-12^T and related species. Based on the 16S rRNA gene sequences of strain TA-12^T, various propionibacteria and other related groups. Reproduced from Kusano et al. (1997), with permission.

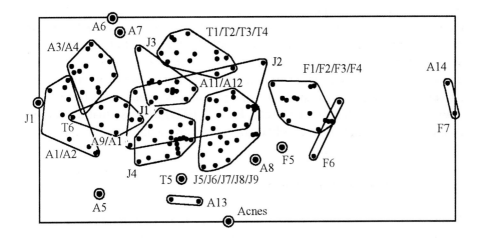

Figure 1.18. Two-dimensional plot, based on the dendrogram distances, showing the relationship between the *Propionibacterium* clusters. A = strains identified as *P. acidipropionici*; T = *P. thoenii*; J = *P. jensenii* and F = *P. freudenreichii*. Reproduced with permission from T. Britz and K-H. J. Riedel, *Lait* 75: 309-314, 1995 © Elsevier Science.

Ribotyping has also been used to evaluate the relationships of the classical propionibacteria (Riedel and Britz, 1995). Only one ribotype profile was found in all strains phenotypically identified as *P. freudenreichii*. Nine ribotype profiles were observed in all 58 strains phenotypically identified as *P. jensenii*, and the strains of either *P. rubrum* or *P. sanguineum*. *P. thoenii* was also heterogeneous, consisting of four ribotype profiles. A large degree of heterogeneity, observed in the ribotypes of *P. jensenii* and *P. thoenii*, underscores the complex structure of the existing species. The authors suggested the existence of additional groupings and proposed to revise the description of the latter two species.

1.3   Cutaneous Propionibacteria

The anaerobic corynebacteria, which now are called cutaneous propioni-bacteria, have been known since 1897, when E. Roux described *Corynebacterium pyogenes* causing purulent abscesses in cattle and for this reason named it *C. pyogenes bovis*. In 1908-1909 Ungano at the Pasteur Institute studied fecal microflora and described three species of corynebacteria: *C. liquefaciens, C. diphtheroides* and *C. granulosum*. The first indication of their pathogenicity was made in 1913 by Massini, who isolated *C. anaerobium* from patients with septicemia, adenitis and mastitis. *C. anaerobium* is the most pathogenic among anaerobic corynebacteria. *C. parvum* was first described by Mayer in 1926; this bacterium is known by its ability to stimulate the reticuloendothelial system (RES). In 1938, A.R. Prévot at the Pasteur Institute started investigating enzymatic properties, pathogenicity and RES stimulation of this bacterial group; the latter two properties are interconnected (Prévot, 1960). Having studied about 600 strains of anaerobic corynebacteria, Prévot recognized their great variability and regarded it as their most surprising feature. For example, almost all the strains, even those isolated as pure cultures from infections, lost pathogenicity after the first passages. Prévot (1975) believed that due to this peculiarity of anaerobic corynebacteria some investigators were unable to confirm their pathogenicity.

1.3.1   General characteristics

Habitat. Bacteria of three species: *P. acnes, P. granulosum* and *P. avidum* are found on different areas of normal human skin, where cutaneous propionic acid bacteria are the only anaerobic microorganisms (McGinley et al., 1978). *P. acnes* (formerly *C. parvum*) prefers moist, oily areas of the skin; on the forehead, for example, there is about 10^4 cells/cm^2, but the number of bacteria in different persons may vary from 10^2 to 10^6 cells/cm^2. The disease acne vulgaris is accompanied by a large increase in the number of *P. acnes* on human skin, although the question of the pathogenic role of *P. acnes* has not been positively resolved. It is known that the development of acne vulgaris requires the combination of the occlusion of sebaceous glands, an enhanced secretion of fat, and the presence of *P. acnes* (Holland et al., 1981). Although the disease may be caused by endocrine disorders, there are indications that *P. acnes* could be the primary pathogen (Horner et al., 1992; Ramos et al., 1995).

Even if *P. acnes* is not the causative agent of the disease, it may still contribute to its course by producing fatty acids from triglycerides by the action of various extracellular lipases. Metabolic products of *P. acnes* can cause skin irritation and inflammation.

Anaerobic corynebacteria (*C. liquefaciens, C. parvum*) were isolated from human bone marrow (Saino et al., 1976). Investigating 59 strains of *C. parvum* Cummins and Johnson (1974) found that most of them (51) actually were *P. acnes*. Since *C. parvum* shows a considerable stimulation of the immune system and can inhibit the growth of tumors in mammals and humans (Halpern, 1975), it was suggested that propionic acid bacteria can play an immunostimulating role for humans. It was also suggested (Shuster, 1976) that acne may serve as a biological defense against cancer development. When sebaceous glands are occluded, *P. acnes* colonizes them and promotes some general immunologic responses. Antibodies against *P. acnes* were found in sera of normal human subjects as well as acne vulgaris (Wolberg et al., 1977).

In general, *P. acnes* is accompanied by *P. granulosum*, for which the secretions of sebaceous glands serve as a good substrate. But these bacteria can also be found in moist areas of the skin, for example, in rectum. *P. avidum* is found only in skin areas with a high moisture content, the main source for its isolation is the intestinal tract.

Isolation. Cutaneous propionibacteria are generally isolated from skin or other epithelial surfaces, using special methods (Kabongo et al., 1981). From the surface of skin the bacteria are isolated using a brush moistened in a detergent solution. From sebaceous glands the bacteria are isolated by scraping. Agar media based on blood or peptone-yeast extract and containing glucose and Tween 80 are used for their isolation. Plates are incubated anaerobically in the atmosphere of H_2+CO_2. *P. acnes* can be isolated using the media recommended for the classical propionibacteria. A simple medium (Kabongo et al., 1982) was suggested for growing *P. acnes*, containing lecithin that inhibits growth of *Staphylococcus aureus*, thereby providing for selectivity. Lecithin (phosphatidylcholine) can be used as the sole source of carbon, energy and fatty acids by *P. acnes*, which secretes exoenzymes that hydrolyze phosphoglycerides.

Nutritional requirements. *P. avidum* is relatively insensitive to oxygen, but oxygen inhibits the growth of *P. acnes* and *P. granulosum*. The latter two species have more complex nutritional requirements than *P. avidum*, which may be explained by the different places of their habitation. *P. acnes* and *P.*

granulosum inhabit areas rich in secretions of sebaceous glands, thus having many metabolites in a readily available state, e.g. amino acids and fatty acids. After a few passages, *P. avidum* can grow in simple media with glucose, salts and vitamins (Fergusson and Cummins, 1978). All the strains require pantothenic acid, thiamine and biotin; nicotinamide stimulates growth. *P. acnes* and *P. granulosum* require a full complement of amino acids. Their growth is stimulated by three organic acids: lactic, pyruvic and α-ketoglutaric acids as well as by Tween 80 (Holland et al., 1979).

Morphology. In young cultures *P. acnes* forms long irregular rods similar to the cells of *C. diphtheria*. The cells of *P. acnes* (0.5-1.5 μm long) are slightly curved (Fig. 1.19). In old cultures all the strains form spherical cells. Colonies are typical for propionic acid bacteria, and they are formed only under anaerobic conditions. The cell anatomy of *P. acnes* is similar to that of *P. shermanii* (Fig. 1.20).

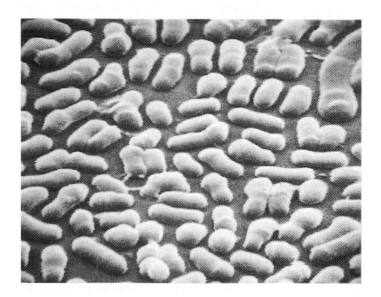

Figure 1.19. Scanning electron micrograph of *Propionibacterium acnes*. General view. × 12,660.

Extracellular enzymes. Cutaneous propionibacteria secrete nucleases, neuraminidases and hyaluronidases (Ingham et al., 1979; Höffler, 1979; Holland et al., 1979; von Nicolai et al., 1980), acid phosphatases (Ingham et al., 1980), lecithinases (Werner, 1967) and other lipases (Ingham et al., 1981) (Table 1.6). The presence of *P. acnes* in blackheads is associated (Holland et al., 1981) with its capacity to produce the above mentioned enzymes *in vivo*, as well as with the coincidence of the secreted enzymes' acidic pH optima and the acidic pH of the skin.

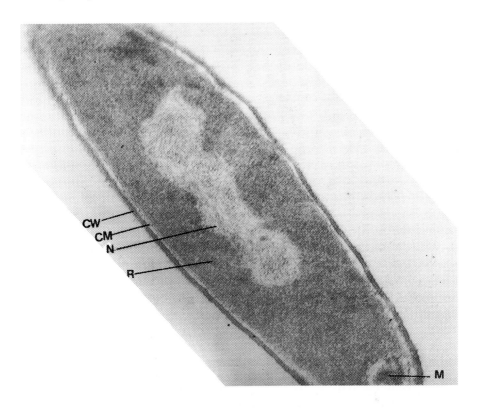

Figure 1.20. Transmission electron micrograph of a thin section of *Propionibacterium acnes*. Labeled structures: cytoplasmic membrane (cm); cell wall (cw); nucleoid (n); ribosomes (r); mesosome (m). × 100,000. From Vorobjeva (1976).

It was suggested (Puhvel and Reisner, 1972) that hyaluronidase can split extracellular substances of the cell wall of sebaceous ducts and thus increase the permeability of epithelial follicles. Neuraminidase can damage cell and tissue membranes, affecting the sialic acid residues on the surface of the cells. Under the action of proteases of *P. acnes*, which also possesses keratinolytic activity, small chemotactic peptides are produced that may have a role in the onset of inflammation. Proteolytic activity may be significant (Ingham, 1983) in complement activation. It was shown (Ingham, 1983) that a preparation of extracellular proteases from *P. acnes* P-37 contained at least three types of proteases with different molecular masses. The secretion of proteinases is used as a criterion in the classification of these bacteria.

Table 1.6. Extracellular enzymes produced by cutaneous propionibacteria

Enzyme activity	Organism		
	P. acnes	*P. granulosum*	*P. avidum*
Hyaluronidase	+	nd*	nd
Lipase	+	nd	nd
Chondroitin sulfatase	+	+	+
Neuraminidase	+	low	+
Phosphatase	+	+	+
Proteinase	+	no	+
Gelatinase	+	low	+
Lecithinase	–	+	+
Hemolysin	+	low	+
DNAse	+	+	+
RNAse	nd	nd	nd

*nd, not determined.   Reproduced from Holland et al. (1981), with permission.

Pathogenicity. Anaerobic corynebacteria can be pathogenic to laboratory animals (Prévot et al., 1968), an effect most likely based on the virulence of their cell wall. A substance isolated from the cell wall was named reticulostimulin since it caused stimulation of the RES when administered in small doses. In large doses it led to historeticulosis and lethal outcomes. In general, cutaneous propionibacteria may be characterized as opportunistic pathogens. So, although *P. avidum* is found in infectious chronic fistulae and abscesses, it is always accompanied by other microorganisms.

Reticulostimulin (RS). It was shown (Prévot, 1975) that aerobic corynebacteria do not synthesize RS, which is produced only by anaerobic forms, if they are not lysogenic. Stimulation of the RES by RS is accompanied by the stimulation of synthesis of several serum proteins, i.e. proceeds on the basis of natural RES defenses (nonspecific immunity) by both cellular and humoral mechanisms.

Histamine production. Growing cultures of *P. acnes* produce histamine (Allaker et al., 1986) and its synthesis increases with increasing growth rate. The growth of *P. acnes* is optimal at pH 6.0, while histamine synthesis has two pH-optima of 4.5 and 7.5. Histamine is formed as a result of histidine decarboxylation.

Holland et al. (1978) suggested that a second enzyme may be required for histamine synthesis and/or stability (histidine decarboxylase was shown to be unstable). Histamine production can have a significant effect on the course of acne vulgaris. If the pH inside the follicle drops to 4.5, approaching the first pH-optimum, histamine production *in vivo* will increase, serving as a metabolic response to the bacterial environment; since amines reduce the acidity, histamine production creates more favorable conditions for bacterial growth (Holland et al., 1978). In follicles of the sebaceous glands the pH is 5.0-6.2, but in the follicles colonized by *P. acnes* the pH can be about 7.5, being close to the tissue pH (Greenman et al., 1981). Under such conditions the synthesis of histamine will increase again, thereby aggravating an ongoing inflammation.

Bacteriophages. In *P. acnes* several phages have been described (Prévot and Thouvenot, 1961; Webster and Cummins, 1978), which do not lyse *P. avidum* and *P. granulosum,* and this difference may have a diagnostic significance. Bacteriophages of the two latter species have not been studied. Spontaneous as well as UV- and mitomycin C-induced production of phages was observed in strains of *P. acnes* (Jong et al., 1975). The authors assumed that at least 38% of the strains of *P. acnes* tested were lysogenic: 46 bacterial strains were lysed by the total of 28 isolated phages, whereas 7 strains were phage-resistant.

Sensitivity to antimicrobial agents. *P. acnes* is sensitive to penicillin, erythromycin and novobiocin, and resistant to streptomycin and sulfamides (Pochi and Strauss, 1961). All strains are especially resistant to sulfamides, being able to grow in the presence of more than 500 µg/ml of these agents. Insensitive to lysozyme.

Cell wall composition. Johnson and Cummins (1972) have found two types of the cell wall in cutaneous bacteria depending on the presence of three sugars: galactose, glucose and mannose. Diaminoacid is mainly represented by L-DAP (cf. Table 1.4). Two serological types, differing in the composition of polysaccharide antigens and the structure of cell walls, are distinguished in *P. acnes*, two in *P. avidum* and one in *P. granulosum. P. acnes* type 2 shows a strong cross-reaction with *P. avidum* type 2. The cell wall of *P. granulosum* contains pyruvate (Cummins and Hall, 1985), which is absent from *P. acnes* and *P. avidum*.

Fatty acids. Most strains contain 31-40% of C_{15} branched-chain fatty acids, while the content of the *iso*-type fatty acids was reported as ranging from 40 to 50% of the total (Moss et al., 1967). However, by using alternative methods of extraction and analysis, C_{15} and C_{17} fatty acids of the *iso*-type were found to account for 70-90% of the total in all the strains of cutaneous bacteria studied (Saino et al., 1976). Thus, the presence of fatty acids of the *iso*-type in the cells of propionibacteria was confirmed as the general characteristics of this genus.

1.3.2   Individual cutaneous species

Diagnostic features of the species are summarized in Table 1.7

Table 1.7. Characters used for identification of cutaneous propionibacteria

Character	P. acnes type 1	P. acnes type 2	P. avidum type 1	P. avidum type 2	P. granu-losum	P. lympho-philum
β-Hemolysis of rabbit blood (68%, 5 days, 37°C)	+	–	+	+	(–)	–
Colonies at 4 days, pigmentation of colonies	small, <1 mm, semi-opaque, white to gray	large, 1–2 mm, opaque, creamy	intermediate, ca. 1 mm, opaque, white to creamy		ca. 1 mm, opaque, white to gray	ca. 1 mm, white
Acid production from:						
sucrose	–	–	+	+	+	d
L-arabinose	–	–	d	d	–	(–)
maltose	–	–	+	+	+	+
glycerol	d	d	+	+	+	–
Esculin hydrolysis	–	–	+	+	–	–
Gelatin liquefaction	+	+	+	+	d	d
Homology within group, %:						
P. acnes type 1	97					
P. acnes type 2				51		nd
P. avidum type 1	50		90		17	
P. granulosum	12		15		95	

Symbols are same as in Table 1.3.
Compiled from: Cummins and Hall (1985), Bergey's Manual of Determinative Bacteriology (1986), Cummins and Johnson (1992), Bergey's Manual of Determinative Bacteriology (1994)

1.3.2.1   *P. acnes*

The bacterium is catalase-positive, liquefies gelatin, produces indole, reduces nitrate, ferments glucose, sucrose, maltose, galactose, fructose, mannose, lactose, starch, dextrins, dulcitol, glycogen, xylose, rhamnose, salicin and does not ferment sorbose. The major long-chain fatty acid is methyltetradecanoic. Some strains produce bacteriocin-like substances that inhibit other species (Fujimura and Nakamura, 1978).

P. acnes does not grow under aerobic conditions, similar to *P. granulosum*, although the cells are not sensitive to oxygen. On the surface of solid media it grows slowly, with colonies appearing after 4-5 days. Many strains are hemolytic for human, rabbit and horse blood (Höffler, 1977). The main products of fermentation are acetic and propionic acids in a ratio of 1.5:1 (Sizova and Arkadjeva, 1968), succinic acid and traces of lactic and formic acid are also produced (Moore and Cato, 1963). *P. acnes* isolated from the cow rumen produces 4-5 µg/l of vitamin B_{12} (Vorobjeva, 1976).

On the basis of fermentation of inositol, maltose, mannitol and sorbitol 72 strains were divided into 8 biotypes (Pulverer et al., 1973). These 8 biotypes comprise 11 serotypes; the biotypes and serotypes do not coincide. A combination of bio- and serotypes was suggested to be used for a further subdivision of the species. In this context, the lysotypes established on the basis of investigating the lytic action of phages on 69 strains of *P. acnes*, may have a special significance (Pulverer et al., 1973). The DNA G+C content ranges from 59 to 60 mol%.

1.3.2.2   *P. granulosum*
The species does not grow under aerobic conditions, is mostly nonhemolytic, and contains more active lipases than *P. acnes*. It was isolated from the same skin areas as *P. acnes*, but in smaller numbers. Indole is not produced, gelatin is not liquefied, nitrate is not reduced; unlike *P. acnes*, it can ferment glucose, sucrose and maltose (cf. Table 1.3). *P. granulosum* has a low (about 12-15%) DNA homology with *P. acnes* and with classic propionibacteria. The DNA G+C content has been reported at 61-63 mol% (Johnson and Cummins, 1972).

1.3.2.3   *P. avidum*
It grows well under aerobic conditions, is hemolytic, produces gelatinase and deoxyribonuclease, but no lecithinase, hyaluronidase and chondroitin sulfatase, in contrast with *P. acnes* (Höffler, 1977). A strong serological cross reaction is observed between *P. avidum* II and *P. acnes* II. Both serological types contain glucose and mannose, but not galactose in the cell wall. Contains *meso*-DAP instead of LL-DAP, as distinct from other cutaneous strains. The DNA G+C content is 62-63 mol%.

1.3.2.4   *P. lymphophilum*
An anaerobic coryneform, produces propionic and acetic acids as the main products of glucose fermentation, although the mechanism of propionate formation is unclear. Differs from the other species by its cell wall composition: it contains lysine instead of DAP. The DNA has a low G+C content of 53-54 mol%. Only a few strains have been studied. One was isolated as an infectious agent of human urinary tract, another one from monkey mesentery gland; several similar strains were isolated from human Hodgkin lymphoma cells (Torray, 1916).

1.3.2.5   *P. innocuum* **sp. nov.**
A new species isolated from human skin was named *P. innocuum* by Pitcher and Collins (1991). The type strain is NCTC 11082. Six strains of *P. innocuum* tested produced luxuriant growth on nutrient agar at 37°C when

grown aerobically. Anaerobic growth on this medium was less effective. It grows at 10°C or 40°C, but not at 45°C. *P. innocuum* possesses many of the characteristics of the classical propionibacteria, including coryneform morphology, i.e., the cells are pleiomorphic rods. The bacterium produces propionic acid with small amounts of succinic and lactic acids as the end products of glucose fermentation. The DNA has a high G+C level (approx. 59-63 mol%), close to that of *P. acnes* (59 mol%). *P. innocuum* differs from the other propionibacteria inhabiting human skin by its cell wall composition, having polysaccharides in which only arabinose and mannose could be detected, but not galactose. Peptidoglycan is composed of L-DAP. Although cutaneous propionibacteria are anaerobes, *P. innocuum* possesses a primary respiration. Corynebacteria contain *meso*-DAP in combination with arabinose as the main sugar of the cell wall as well as mycolic acids that are not found in *P. innocuum*. The main metabolic product in *P. innocuum* is propionic acid rather than lactic acid in the genus *Corynebacterium*.

The authors (Pitcher and Collins, 1991) described the bacteria of the new species as Gram-positive, non-motile and non-spore-forming, containing metachromatic granules. Most strains can reduce nitrate and some of them also nitrite. Some strains can hydrolyze urea and starch. Mycolic acids are not found. The major respiratory quinone is MQ-9(H_4).

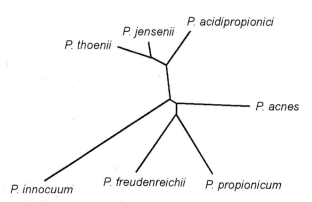

Figure 1.21. Phylogenetic network showing the position of *Propionibacterium innocuum* within a radiation of propionibacteria. The tree is based on a comparison of 1230 bases. Reproduced with permission from D.G. Pitcher and M.D. Collins, *FEMS Microbiol. Lett.* 84: 295-300, 1991 © Elsevier Science.

DNA hybridization of *P. innocuum* with other propionibacteria did not reveal significant genomic homologies, but the 16S rRNA sequence of strain NCTC 11082 had the highest homology with *Propionibacterium* species. An unrooted phylogenetic network based on the K_{nuc} values is shown in Fig. 1.21, indicating that *P. innocuum* forms a distinct line within the genus.

On the other hand, however, Yokota et al. (1994) emphasized such properties of *P. innocuum* as the aerobic growth potential and the presence of arabinose in the cell wall. Convinced that *P. innocuum* should not be classified with the authentic *Propionibacterium* species, they proposed to transfer it to a new genus, *Propioniferax*, under the name *Propioniferax innocua* gen. nov., comb. nov. The results of 16S rDNA analysis indicate *Luteococcus japonicus* as the phylogenetic neighbor of the new species.

1.3.3   *Propionibacterium propionicum (Arachnia propionica)*

The genus *Arachnia* was created for the microorganism formerly called *Actinomyces propionicus* (Pine and Georg, 1969). *Arachnia propionica* resembles actinomycetes, especially *A. israelii*, in being pathogenic and having a filamentous morphology (Slack and Gerencser, 1975), but differs in that propionic and acetic acids are the major metabolic products; no gas is produced (Schaal, 1986a, b). Moreover, propionic acid is formed in the pathway involving methylmalonyl-CoA, like in the other propionibacteria (Allen and Linehan, 1977). The cell wall contains L-DAP, whereas in actinomycetes diaminoacids are represented by lysine or ornithine. Like the propionibacteria, *A. propionica* (now called *P. propionicum*) contains *iso-* and *anteiso-* branched-chain fatty acids, i.e., 12-and 13-methyltetradecanoic acids (C_{15}) as the major fatty acid components, with small amounts of $C_{16:1}$ and $C_{18:1}$ acids (Cummins and Moss, 1990). Three of the four strains tested also contained an unusual mono-unsaturated straight-chain C_{18} acid, accounting for 2-22% of the total. In contrast with *P. propionicum*, members of the genus *Actinomyces* contain mainly $C_{16:0}$ and $C_{18:1}$ acids, with branched C_{15} acids either not present or found in trace amounts, thus further confirming a close relationship of *A. propionica* with propionic acid bacteria.

Bacteria of the genera *Arachnia* and *Propionibacterium* are also similar in relation to menaquinones. Methylmalonyl-CoA transcarboxylase was found in *A. propionica*, as well as four other enzymes involved in propionic acid fermentation. However, these genera were attributed to absolutely different families according to numerical systematics (Schofield and Schaal, 1981). Nevertheless, sequence comparison of 744 bases of the 16S rRNA (Charfreitag et al., 1988) in *A. propionica, Act. bovis, Act. viscosus, P. freudenreichii* and *P. acnes* confirmed that *A. propionica* is more closely related to *Propionibacterium* than to any other taxonomic group. On an unrooted phylogenetic tree *A. propionica* is placed within the genus *Propionibacterium*, being almost equidistant to *P. freudenreichii* and *P. acnes* (Fig. 1.22).

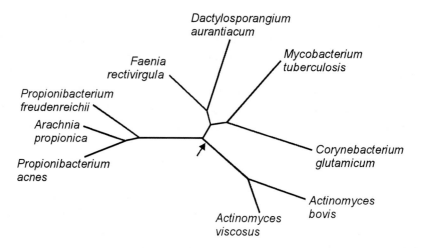

Figure 1.22. Unrooted phylogenetic tree showing the positions of the test and reference strains. The tree is based on the comparison of 16S rRNA sequences, as a result of which *Arachnia propionica* was reclassified as *Propionibacterium propionicus*. The arrow indicates the estimated rooting point when *Bacillus subtilis* was used as a remotely related reference. Reproduced from Charfreitag et al. (1988), with permission.

Although these results place *A. propionica* in the genus, the relatedness with *P. freudenreichii* is only 1-5%. But similarly low relatedness values were reported between *P. freudenreichii* and *P. acnes*. The relationship with the members of *Actinomyces* studied was remote, although they share filamentous morphology. The inclusion of filamentous strains in the genus *Propionibacterium* makes it very heterogeneous with respect to morphology, but similar examples are known for such actinomycetous groups as *Actinomyces*, *Mycobacterium* and *Rhodococcus*. Taking into account the additional evidence, a revised diagnosis of the genus was presented (Charfreitag et al., 1988).

Gram-positive, nonmotile, non-acid-fast, branched diphtheroid rods (0.2-0.8 μm by 1-5 μm), may be branched, filamentous or not. Filamentous forms are more typical for young cultures and for clinical material. Branched filaments can be up to 5-20 μm long. Several strains form spherical cells resembling spheroplasts with diameters of 5-20 μm. Cells grow well in standard complex media. In liquid media it usually grows in floccular or granular masses of variable size. Filamentous microcolonies may be found. Colonies on solid media may vary from smooth, convex types to rough types. Major diaminoacids in peptidoglycan are L-DAP or lysine.

Tetrahydrogenated menaquinones with nine isoprene units, i.e. MQ-9(H_4) are the major respiratory quinones. Long-chain fatty acids are of the straight-chain saturated and *iso-* and *anteiso-*methyl branched-chain types (12- or 13-

methyltetradecanoic acids, or $C_{15:0}$). Monounsaturated acids may also be present in small amounts. G+C content of the DNA is 63-65 mol%.

P. propionicum is a normal inhabitant of human mouth and can be isolated from dental plaque as well as from cervical smears. It may be found in various organs affected by actinomycoses (Brock et al., 1973), and can be occasionally involved in infections. It grows both aerobically and anaerobically, but a better growth and more biomass are obtained under anaerobic conditions. Temperature optimum for growth is 35-37°C. All strains are catalase-, indole- and Voges-Proskauer negative. Nitrate is reduced to nitrite and starch is hydrolyzed. Most strains can ferment adonitol, sorbitol, mannitol, fructose, sucrose, lactose, trehalose, glucose, galactose, mannose and raffinose (Slack and Gerencser, 1975; Schofield and Schaal, 1981). The bacterium is sensitive to beta-lactam antibiotics, tetracyclines, chloramphenicol, macrolides such as erythromycin, and to vancomycin. It is highly resistant to aminoglycosides such as gentamycin, to nitroimidazoles such as metronidazole, and to peptide antibiotics such as colistin (Schaal and Pape, 1980; Niederau et al., 1982).

1.3.3.1   Subtypes of *P. propionicum*

By using fluorescent antibodies and gel diffusion tests two serotypes were identified (Holmberg and Forsum, 1973; Slack and Gerencser, 1975; Schaal, 1986b). Two serovars of *P. propionicum* are represented by the type strain ATCC 14157 (serovar 1) and ATCC 29326 (serovar 2, WVU 346, F. Lentze strain Fleischman). No cross reaction between these strains was detectable in agglutination tests (Johnson and Cummins, 1972), and the serovar 2 strain (ATCC 29326 (VP1 5067) showed a very low DNA homology (1%) to ATCC 14157. Since the two serovars form distinct subclusters in numerical phenetic analysis (Schofield and Schaal, 1981) it cannot be excluded that the two serovars should in fact be distinguished as different species.

A. propionica is called *P. propionicum* now, although Cummins and Johnson (1992) consider the name *P. arachniforme* more appropriate. Nevertheless, in the 9th edition of Bergey's Manual of Determinative Bacteriology it is retained in the genus *Arachnia* (for determinative purposes).

1.4   Phylogenetic Relationships of Propionibacteria

The results of DNA hybridization of various strains of all seven species of the genus revealed that it is not a homogeneous group (Johnson and Cummins, 1972). One group includes a cluster composed of the strains of *P. acnes* and *P. avidum* and a more remote *P. granulosum*, and the second group composed of *P. jensenii, P. thoenii, P. acidipropionici* and a distantly related *P. freudenreichii*. The relationship of *Arachnia propionica* to the propionibacteria is at the same level that separates the former two groups.

Although the genus *Propionibacterium* is heterogeneous, it forms a distinct subline within the actinomycetous branch (Stackebrandt and Woese, 1981), that subline being the principal ancestral line of Gram-positive bacteria with the high G+C type of the DNA. It was confirmed (Charfreitag and Stackebrandt, 1989) that *P. jensenii* and *P. thoenii* form a phylogenetically tight cluster (the level of homology is 98.3%). Together with *P. acidipropionici*, they form a cluster, which embraces those classical propionibacteria that contain *meso*-DAP and ferment sucrose and maltose. This cluster is well separated from the fourth species, *P. freudenreichii*, and differs from the latter by the characteristics mentioned above. The cutaneous species *P. acnes* and *P. propionicum* display the same level of evolutionary relationship as with the classical propionibacteria (Fig. 1.23A). It is essential that the data obtained by two different methods, i.e. determination of 16S rRNA sequence and DNA hybridization, are in complete agreement.

To establish intergeneric relationships, a 16S rRNA gene fragment, consisting of 490 bases, was sequenced in four strains of *Propionibacterium* and fifteen members of various actinomycetes (Charfreitag and Stackebrandt, 1989). (The method of DNA hybridization could not be used because of low DNA homology.) It was shown (Fig. 1.23B) that the genus *Propionibacterium* is distantly related with *Nocardioides* (Charfreitag and Stackebrandt, 1989) and *Terrabacter tumescens* (Collins et al., 1989). At the same time, it was found that the propionibacteria are not even remotely related with the genus *Corynebacterium* as suggested by van Niel, and therefore, the formation of the subgenus *Coryneforms* containing anaerobic corynebacteria, proposed by Prévot (1976), would not be appropriate.

Recently, by 16S rDNA sequencing, *P. thoenii*, *P. jensenii* and *P. acidipropionici* were clustered (Dasen et al., 1998) in a group distinct from the two closely related clusters of *P. freudenreichii* and *P. cyclohexanicum*, respectively. Within the cutaneous propionibacteria three groups are clustered: (i) *P. granulosum*, (ii) *P. avidum*, *P. acnes* and *P. propionicum*, and (iii) *P. lymphophilum*. The latter is the most divergent of the whole genus. 16S rRNA sequence of *P. lymphophilum* shows more similarity with *Luteococcus japonicus* than with the other propionibacteria (Figure 1.24).

Alternatively, a method based on multiplex PCR (MPCR) was used (Dasen et al., 1998) to differentiate the genus *Propionibacterium* from other genera. The method is based on the use of a gene probe specific for *Propionibacterium*. Using this method, the authors detected an additional 900 bp-sized fragment (primers gd1 and bak4) specific to propionibacteria, whereas in other microorganisms only the universal 1500 bp PCR fragment could be amplified (primers bak11w and bak4). A specific nucleotide probe, gd1 (20 nucleotides), was constructed and used in the MPCR assay (Dasen et al., 1998). 150 strains of propionibacteria were tested with this probe, and the specific PCR fragment was always amplified, whereas the analysis of several species, superficially related to propionibacteria or involved in dairy fermentation, revealed no positive signals.

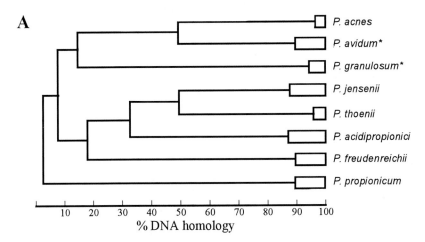

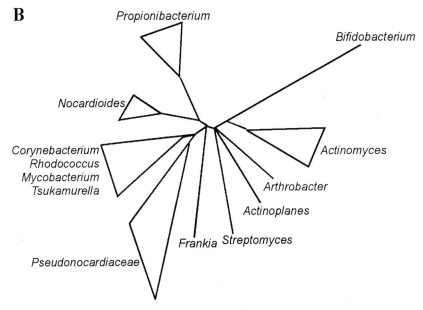

Figure 1.23. (A) Dendrogram of interspecies DNA similarities, derived from average linkage clustering (Johnson & Cummins, 1972). Boxes indicate the range of intraspecies DNA similarities. Species marked with an asterisk were not included in rRNA sequence analysis. (B) Unrooted phylogenetic tree showing the intergeneric relationships of *Propionibacterium* and various actinomycete genera, based on comparison of 490 selected nucleotides of the 16S rRNA. Reproduced with permission from O. Charfreitag and E. Stackebrandt, *J. Gen. Microbiol.* 135: 2065-2070, 1989 © Society for General Microbiology.

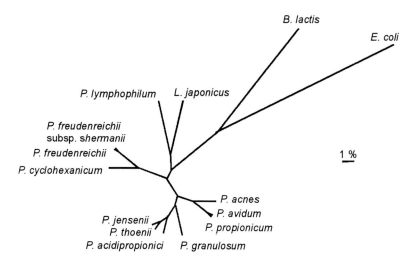

Figure 1.24. Phylogenetic tree of the genus *Propionibacterium* based on 16S rDNA sequences. Reproduced from Dasen et al. (1998), with permission.

Chapter 2

Genetic Studies in Propionibacteria

Owing to the absence of toxicity and valuable biotechnological properties, propionic acid bacteria are widely used in industry (food industry, vitamin B_{12} production). Genetic approaches may help increase strain productivity, change nutrient requirements, confer phage resistance.

Strains isolated through mutations or developed by the methods of genetic engineering are being used in industry. Although mutagenesis was frequently used in relation to propionibacteria, genetic engineering is still at an early stage. Nevertheless, some basic information, necessary for genetic manipulations, is now available.

Naturally, theoretical interest in the genetics of propionic acid bacteria has always existed, but the first report of a genetic investigation appeared only in 1986.

2.1   Plasmids

Plasmid DNA was isolated (Rehberger and Glatz, 1987a, b, c) from 17 strains of propionic acid bacteria, representing all the species essential for food industry. By restriction analysis seven distinct plasmids were identified, numbered consecutively pRG01 through pRG07, with molecular weights of 4.4, 6.3, 25, 30, 35, 5.6 and 6.3 MDa, respectively (Rehberger and Glatz, 1990).

Seventeen strains contained identical (pRG01) plasmids, three strains had pRG02, plasmids pRG04 and pRG06 also were found in more than one strain. Some strains contained two different plasmids. In *P. freudenreichii* either pRG01 or pRG03 and pRG07 were found; in five strains of *P.*

acidipropionici there was only pRG01; in *P. jensenii* three strains contained pRG02, one contained pRG01 and pRG05 and another one only pRG01.

Detailed restriction maps of four plasmids (pRG01–pRG04) were constructed (Rehberger and Glatz, 1990), showing that their structural heterogeneity is accompanied by some compositional similarities. Like the chromosomal DNA, the plasmid DNA is characterized by a high content of GC pairs. Plasmids of propionic acid bacteria are better cut by restrictases specific for sites carrying unmodified guanine and cytosine (G and C). Comparison of the restriction maps allowed to identify conserved regions in pRG01 and pRG02 plasmids.

Using Southern hybridization, a homology of these plasmids was demonstrated: pRG02 probes hybridized with the two Sal I fragments of pRG01. However, many restriction sites present in pRG02 are absent in pRG01. The two plasmids contain an identical fragment of 1800 bp. Two plasmids (pRG01 and pRG02) are homologous with pRG07 and with a similar sequence of pRG05, suggesting a common ancestor. Plasmid pRG03 has a partial homology with pRG07 only.

Curing of plasmids pRG01, pRG02 and pRG05 from the respective strains had no effect on their capacity to synthesize bacteriocin, or to ferment 21 carbohydrates, as well as on their resistance to 21 antibiotics, including ampicillin, bacitracin, cephalotin, chloramphenicol, cloxacillin, erythromycin, fusidic acid, gentamycin, kanamycin, lyncomycin, metycillin, nalidixic acid, neomycin, novobiocin, oxacycline, penicillin, rifampicin, streptomycin, tetracycline, trimethoprim, and vancomycin.

However, at present it may be premature to make definite conclusions on the apparent absence of linkage of plasmids with the above mentioned properties, since aplasmidic derivatives were not tested following the reintroduction of the respective plasmids that might have integrated with the chromosome during acriflavin treatment.

There is evidence that the plasmid pRG05 in *P. jensenii* encodes the formation of cell aggregates. In *P. freudenreichii* subsp. *globosum* P93 two plasmids, pRG02 and pRG03, were found (Rehberger and Glatz, 1987a), with molecular masses of 6.3 and 22 MDa, respectively. This strain differs from the closely related *P. freudenreichii* subsp. *shermanii* by its capacity to ferment lactose.

Following the acriflavin treatment, 15% of the population lost this capacity, being converted into Lac⁻ strains that did not contain the plasmid pRG03. Cells that did not loose the capacity to ferment lactose after the acriflavin treatment also retained the pRG03 plasmid. It was found that pRG02 did not hybridize with pRG03, and the latter did not hybridize with a plasmid of 4.7 MDa present in 11 strains, or with two plasmids of 6.3 MDa present in two other strains of propionibacteria. Thus the genes responsible for lactose fermentation in *P. globosum* were localized to plasmid pRG03, which is unrelated to the other plasmids of propionic acid bacteria.

If plasmids are absent in the strain, as in the case of *P. technicum* (Yongsmith and Cole, 1986), new genetic material can be transferred by protoplast transformation. A method of protoplast preparation for these bacteria has been developed (see below). Chromosomal genes of propionic acid bacteria have been successfully transferred to an unrelated species. Specifically, a fragment of chromosomal DNA of *P. technicum* was ligated *in vitro* into the pBR322 plasmid of *E. coli* (Youngsmith and Cole, 1986) and *E. coli* cells were transformed with the recombinant plasmid. The transformants acquired morphological and physiological characteristics different from those inherent in the original strain of *E. coli*.

Of the 109 propionibacterial strains tested, 11 were found to be inhibitory to certain indicator organisms and 16 possessed autolytic properties of more than 90% (Miescher et al., 1998a). Among the inhibitory and autolytic strains, 14 out of 26 were found to contain plasmids sized between 2 and 50 kb. Three plasmids, pLME101, pLME106 and pLME108, were further characterized by DNA-DNA hybridization, restriction endonuclease analysis and curing experiments. The plasmids did not hybridize with chromosomal DNA of the host strains. Only the plasmids pRG01 (used as a reference) and pLME106 showed strong cross-hybridization signals with each other. The amino acid sequence deduced from pLME108 showed a 34.1% similarity with plasmid pIJ101 encoding a Rep protein from *Streptomyces lividans* (PIR A31844).

2.1.1 Preparation and transformation of protoplasts

To obtain protoplasts with an yield of 99.9%, it is recommended to use cells in the log-phase, treated with lysozyme (20 mg/ml) in an osmotically stabilized medium (Baehman and Glatz, 1986; Pai and Glatz, 1987). Transformation of rodlike cells into protoplasts (round forms) can be observed under the light microscope. Protoplast regeneration occurs in 21 days and only in semi-liquid agar medium. The suspension of protoplasts is spread on the surface of regenerative agar medium, then a semi-liquid regenerative agar containing NLA (Na-lactate agar), sucrose, gelatin and starch is overlaid. Regeneration of protoplasts is stimulated by gelatin and starch, and inhibited by $MgCl_2$ and $CaCl_2$. Regeneration rate is determined as the ratio of the number of regenerates to the total number of initial cells. Regeneration rates can reach 30 to 40%, depending on the number of initial cells, lysozyme concentration and anaerobic conditions (Pai and Glatz, 1987).

Protoplasts of *P. technicum* were successfully transformed with plasmids pE194, pT181 and pUB110 from *Staphylococcus aureus* and plasmid pAMb1 from *Streptococcus faecalis*. Erythromycin-resistant bacterial colonies were obtained following the transformation with plasmid pE194 (from *Staphylococcus aureus*), although the intact plasmid was not detected in the recipient cells. These results show that genes of taxonomically distant bacterial species can be made functional in the cells of propionic acid bacteria.

2.1.2   Molecular cloning of propionibacterial genes

At the Institute of Genetics and Selection of Industrial Microorganisms in Moscow a genomic library of *P. shermanii* was generated by cutting the chromosomal DNA with Sau A restrictase and selecting fragments of 4-8 Kbp for cloning (Pankova and Abilev, 1993). The fragments thus obtained were ligated to the vector pVZ361 (derivative of plasmid RSF1010) and competent cells of *E. coli* C600 were transformed with the ligation mix. The vector enabled a direct selection of streptomycin-resistant clones containing an insert at the BamH1 site. The Rec A and threonine genes were present in the fragments that were cloned in *E. coli*. A full complementation of thr B function was observed. The thrB gene of *P. shermanii* was expressed in *E. coli* cells from 5 Kbp and 1.8 Kbp fragments on plasmids pSPt4 and pUC19, respectively.

The recA gene of *P. shermanii* was used to complement the function of recA in different mutants of *E. coli*: in the survival tests as well as in the tests of the SOS system induction. It was shown (Pankova et al., 1993b) that the genome of *P. shermanii* contains a functional recA gene capable to complement the deficient constitutive and inducible functions in recA mutants of *E. coli*. The molecular mass of the protein product of the cloned gene was about 39 kDa, which corresponds with the other RecA proteins. The authors (Pankova et al., 1993a) made plasmid constructs carrying both the recA and lexA genes of *E. coli* and also the recA gene of *P. shermanii* and the lexA gene of *E. coli*. Treatment with cell-damaging mutagens led to the induction of *P. shermanii* recA activity, which was necessary for recombination, reparation and a further induction of the SOS system.

With the genomic library of *P. freudenreichii* now available (Murakami et al., 1993), the search is in progress for the genes encoding the remaining vitamin B_{12} and other enzymes and proteins essential for practical use, and further elucidation of some metabolic pathways of these bacteria. For example, *cob*A gene (encoding uroporphyrinogen III methyltransferase, the key enzyme in the biosynthetic pathway of vitamin B_{12} and siroheme) of *P. freudenreichii* subsp. *shermanii* was cloned, sequenced and overexpressed in *P. freudenreichii* (Sattler et al., 1995). Also, Murooka et al. (1995) have cloned *P. freudenreichii* genes involved in the synthesis of 5-aminolevulinic acid (ALA) and expressed them in *E. coli*. ALA is the common precursor in the synthesis of vitamin B_{12} and various hemes. The cloning was undertaken to improve economic efficiency of the vitamin B_{12} production.

2.2   Natural Mutagenesis

Unfortunately, the well-known saying that "the new is the once forgotten old" relates also to scientific matters. In 1988, French authors (Naud et al., 1988) reported that 'natural' mutants arose in populations of *P. freudenreichii* placed under harsh conditions permitting no growth or subjected to UV shock. The mutants of a spherical shape differed from the original rodlike cells in many properties, which will be discussed in detail below.

The facts, reported by these authors, made us to return to our almost forgotten results from the late 1960s (Vorobjeva and Baranova, 1969) that also concerned the natural genetic variability of propionic acid bacteria. In those years some microbiologists in Russia were unreceptive to the facts given below, since they did not conform to the established view of propionic acid bacteria. But now these data may have a considerable scientific and practical interest.

Working many years with *P. shermanii* strain VKM-103, we observed that after a series of passages in liquid medium this anaerobic culture acquired the ability to grow on the surface of solid medium, forming yellowish colonies (colonies of the original culture were creamy in color). The ability to grow on the surface of solid medium was preserved after a prolonged storage of the bacteria in liquid medium at 4°C.

The strain of *P. shermanii* forming yellowish colonies (we called it 'yellow') under aerobic conditions or conditions unfavorable for growth (e.g., on dried solid medium or at high levels of metabolic end-products) was converted into a new apigmented form that synthesized corrinoids in very small quantities (less than 100 µg/g biomass). The original strain produced about 800 µg/g biomass. No reversion to the apigmented form was observed when the medium was enriched with phosphate (Table 2.1). Phosphate is known to display a regulatory function in switching from respiration to fermentation. At the same time, adding citrate to the medium (the key metabolite of the Krebs cycle) increased the conversion of the original form to the apigmented form.

Cell morphology of the original, 'yellow' and apigmented forms differed little; cells were rodlike (or micrococcal), 0.5-0.6 × 0.4 µm in the first two cases and 0.6-0.8 × 0.4 in the apigmented form. However, the original and apigmented forms grew differently in liquid medium: the original form grew in the whole column of the liquid, while the apigmented formed sediment on the bottom of the test tube.

Table 2.1. Yields of the apigmented form under different growth conditions of *P. shermanii*. From Vorobjeva and Baranova (1969).

Medium composition	Apigmented colonies, %
Peptone agar	5
Peptone agar (dried plates)	80
Lactate-peptone agar	5
Lactate-peptone agar (dried plates)	50
Glucose-peptone agar	50
Glucose-peptone agar + phosphate	0
Glucose-peptone agar + citrate	30
Glucose-peptone agar + Tween 80	20
Glucose-peptone agar + Tween 80 (dried plates)	100

They also differed with respect to the fermentation of some carbon sources (Table 2.2). Unlike the original form, the apigmented form grew well on sorbitol, mannitol and glycerol, utilized Ca-gluconate and did not reduce nitrate. The apigmented form utilized lactose and raffinose better compared with the original form, producing more biomass and lowering the pH more, apparently on account of a high accumulation of acetic acid and CO_2 (propionic acid was present in low quantities). The original form produces propionic, acetic acid and CO_2 in a ratio 2:1:1. Thus, a shift to the oxidative metabolism was observed in the apigmented form. Similar changes were observed when the synthesis of corrinoids was limited by removing Co-ions from the medium (see below).

Table 2.2. Profiles of carbohydrate fermentation by different forms of *P. shermanii*

Carbohydrate	Form	
	Original	Apigmented
D-glucose	+	+
Sucrose	±	−
D-fructose	+	+
Galactose	+	+
Lactose	+	+
Maltose	+	+
Raffinose	−	±
Xylose	−	±
Arabinose	−	−
Sorbitol	−	+
Mannitol	±	++
Glycerol	−	++
Ca-gluconate	−	+
Starch	−	−
Nitrate reduction	+	−

'−' indicates no growth; ± , variable; + , normal; ++ , vigorous growth.
From Vorobjeva and Baranova (1969).

Sometimes on top of colonies of the apigmented form pink colonies appeared, possessing all the properties of the original form, i.e., a reversion from the apigmented to the original form occurred. It cannot be excluded that the described conversions were associated with the insertion or excision of plasmid DNA from the chromosome, but no genetic analysis was performed at that time (Vorobjeva and Baranova, 1969). Almost 20 years later, similar observations were made by French researchers (Naud et al., 1988) investigating "natural" mutants of *P. freudenreichii*, a close relative of *P. shermanii*. They found that *P. freudenreichii* ATCC 13673 (original strain) can exist in two stable forms: coryneform rods (P) and large (about 1 μm) thick-walled cocci (strain E1). Conversion of the original cells into spherical ones (E1) was induced by placing the rodlike cells in a medium unfavourable for growth or by subjecting the cells to an UV shock. The changes occurring in these conditions were expressed in a selective medium containing chloramphenicol or erythromycin, since in this medium only the colonies formed by the cocci could grow, but not the original form.

However, if cells pretreated as above were inoculated into a medium without antibiotics, colonies of the rodlike cells reappeared, reversing into the original, "non-induced" state. Therefore, *P. freudenreichii* strain E1 behaved as a natural mutant. No mutagenesis was used to induce this strain, and the high yield of spherical forms excluded completely the involvement of mutational mechanisms.

Experimental evidence shows (Hall, 1989) that the frequency of spontaneous mutations is not an inherited property of an organism but is subject to variations due to environmental factors. Also, expression of certain mutations depends on their selective advantages. It should be noted that spontaneously occurring Lac⁻ (unable to ferment lactose) and chloramphenicol-resistant (LM-C) mutants were also observed in the population of rodlike cells. But even after many generations without chloramphenicol the reversion of spherical cells into rods was induced by the absence of glucose. At present there is no explanation for this phenomenon. Spherical cells also converted into rods when glucose was replaced by sodium lactate. The spherical form lost the ability to ferment some polyalcoholes (erythritol, adonitol, inositol) and acquired the ability to ferment disaccharides, such as maltose, sucrose, trehalose, i.e., acquired α-glycosidase activity (Table 2.3).

Interestingly, the generation time (g) of the spherical form growing under aerobic conditions was 2 h, and that of the original form was 4.5 h. Under anaerobic conditions both strains grew at the same rate (g = 4.5 h), which shows that the spherical form is better adapted to the aerobic style of life (similar to the apigmented strain in our investigations). In contrast with the original strain possessing D(–)-lactate dehydrogenase activity (LDH), only L(+)-LDH-activity was found in strain E1. The resistance to chloramphenicol of the spherical form was accompanied by the loss of resistance to 15 out of 26 drugs tested, to which the original form and the revertant R₃ were resistant. The spherical form was more sensitive than the rod-shaped form to a number of aminoglycosides, indicating an increased permeability of the spherical cells to these positively charged molecules.

Table 2.3. Carbohydrate fermentation profiles by different forms and strains of *P. freudenreichii*

Carbohydrate	Strain		
	ATCC 13673 (original, P)	E₁ (rounded cells)	R₃ (revertant, E₁ → P)
Galactose	+	+	+++
D-glucose	++++	++++	++++
D-fructose	–	++++	–
D-mannose	++++	+	+++
Maltose	–	+++	–
Sucrose	–	++++	–
Trehalose	–	++++	–
Glycerol	+	+	+
Erythritol	++	–	++
Adonitol	+++	–	+++
Inositol	++++	–	++++
Gluconate	++	–	+

The no. of + indicates the estimated intensity of reaction.
Adapted from Naud et al. (1988), with permission.

Alterations in fermentative properties and resistance to antibiotics in spherical cells were linked by the authors (Naud et al., 1988) with the structure of the cell wall. So the loss of the ability to ferment some polyoles may be caused by the inactivation of membrane-linked transport systems, namely, PEP-phosphotransferase system (Dills et al., 1980). Acquisition of a new enzymatic activity may be linked with an increased permeability to an inducer of protein synthesis, and conversely, an inducible alpha-glycosidase in the original cells may be blocked by insertion of a plasmid or phage at the enzyme locus in homological DNA sites. Therefore, one can expect that the enzyme activity will be restored following the removal of foreign DNA. In this respect, it is important that two plasmids were found in spherical cells, but no plasmid was found in the original strain (Naud et al., 1988). Moreover, the reversion of the spherical form into the rodlike form was accompanied by the loss of plasmid DNA, which could be either lost from the cell or integrated with the chromosome. When rodlike cells were transformed with a fragment of DNA extracted from an LMC strain (resistant to chloramphenicol), spherical cells were obtained.

2.3   Chemically-Induced Mutagenesis

Mutagenic studies in propionic acid bacteria have been carried out mainly with the purpose of increasing the productivity in vitamin B_{12} (Vorobjeva et al., 1973; Grusina et al., 1973, 1974; Glatz and Anderson, 1988), since the production of this vitamin in many countries is based on the vital functions of propionic acid bacteria.

When we began these investigations in 1970, we had no predecessors, and first thing that came to our attention was a wide natural variation between strains with respect to vitamin B_{12} productivity. Differences in the vitamin content reached 80% (Fig. 2.1), while according to Grusina et al. (1974) it varied from 0 to 160%. So by selecting for active forms already present one could increase considerably the yield of the desirable product, in our case vitamin B_{12}. But this approach is limited by the bacterial genotype; therefore, it was necessary to employ induced mutagenesis.

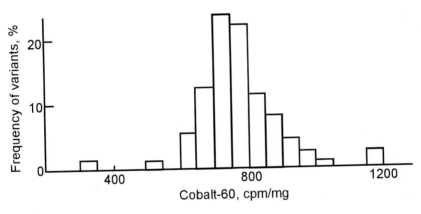

Figure 2.1. Spontaneous variability of corrinoid production by *P. shermanii*.
From Vorobjeva (1976).

In microbiological practice the most widely used mutagens are N-nitroso-substituted compounds: N-nitrosomethylurea (NMU), N-nitrozoethylurea (NEU), N-methyl-N'-nitro-N'-nitrosoguanidine (NTG). These are called supermutagens, since they produce high mutagenic effects while showing a low lethality. Supermutagens induce mutations not found in nature, they induce repeated mutations and strongly increase their frequency. We obtained an especially interesting mutational spectrum by using NTG.

The mechanism of action of NTG is complex. NTG specifically interacts with DNA. The fact that lesions induced by NTG are repaired in the same way as those induced by UV light, permits to classify NTG as an UV-mimetic agent. Mutants of *Haemophilus influenzae* having a defective reparation system and highly sensitive to UV light are also highly sensitive to NTG (Kimball et al., 1971). *E. coli* mutant strains resistant to radiation are also resistant to the action of NTG. It is known that the dark reparation of DNA lesions induced by UV light is inhibited by acriflavin (AF). AF inhibited the reparation of lethal DNA lesions in *B. subtilis* 202 induced by NTG (Yoshido and Yuki, 1968). A specific effect of NTG on DNA is indicated by the fact that the transformational activity of the DNA isolated from the cells of *B. subtilis* 202 treated with NTG is 50% lower than the transformational activity of the DNA of untreated cells. In the DNA treated with N-(C^{14})-methyl-N'-nitro-N'-nitrosoguanidine the label was found in 7-methylguanine.

In addition, NTG can react with organic anions, forming reaction products that contain nitrosoguanidine groups (McCalla, 1968). It was shown (Cerdá-Olmedo and Hanawalt, 1967) that NTG inhibits the synthesis of functional proteins, in particular the synthesis of beta-galactosidase in *E. coli* TAU bar or its intrinsic activity (if the enzyme was present in cells before the NTG treatment). On the basis of experimental evidence it was suggested that NTG could affect protein synthesis by causing mistakes during translation, thus impairing the synthesis of enzymes, catalyzing DNA-replication. It is possible that NTG acts on DNA polymerase, thereby causing mistakes in DNA replication. This alternative explanation is based on the observed effect of NTG on protein synthesis.

An important role in the induction of different mutations is played by the genotype of organism. The type of DNA (GC or AT) and irregular distribution of DNA pairs determines the presence of 'hot spots' and specific mutational spectra. The GC-rich DNA is highly sensitive to X-rays and alkylating agents, and the DNA of propionic acid bacteria is of the GC-type. Mutational spectra also depend on the physical state of DNA. Denatured regions in the DNA are highly susceptible to the action of mutagens.

A number of interesting mutations induced by dimethylsulfate (DMS), diethylsulfate (DES), ethyl methanesulfonate (EMS) and NTG were described in *P. shermanii* by Pędziwilk et al. (1984), including apigmented mutants very low in vitamin B_{12} and differing from the parental strain metabolically: the mutants accumulated formic, lactic, acetic and propionic acids. Of considerable interest is the study by Seiler (1973) of the mutagenic action of benzimidazole and its derivatives, since 5,6-dimethylbenzimidazole (DMB) is added to the growth media of vitamin B_{12} producers as vitamin B_{12} precursor. It was concluded (Seiler,1973) that the mutagenic action of benzimidazoles is due to their incorporation in DNA, replacing purines, the conclusion based on the identification of DNA hydrolysis products in *E. coli* grown in the presence of 2-^{14}C-benzimidazole. Once incorporated in DNA instead of purine bases, benzimidazoles produce mostly transitions, although transversions are also possible since these molecules do not contain side groups needed for base pairing.

In mutagenesis the main condition of success is the development of a sensitive and rapid method of mutant selection. Vitamin B_{12} production by propionic acid bacteria correlates with acid production. But the method of selection based on picking colonies with the maximal diameter of dissolution zones of calcium carbonate (added to solid medium) did not give positive results, since the size of colonies is not identical while acid production is dependent on the quantity of biomass. Selection based on the intensity of pink color that correlates with vitamin B_{12} production has a disadvantage in that selection of small mutations is made step by step, when differences in color can hardly be noticed by eye. Another method used was based on direct determination of vitamin B_{12} in single colonies. This method is labor-intensive and severely limits the testing of large numbers of colonies.

We (Vorobjeva, 1976) have developed a method based on using a medium containing radioactive cobalt ($^{60}Co^{2+}$), since cobalt is incorporated in vitamin B_{12} molecule. Mutagenized cells and control (untreated) cells were plated on solid medium with radioactive cobalt. Plates were incubated under anaerobic conditions for nine days at 25°C. The colonies were weighed and their radioactivity was measured. Radioactivity of colonies was calculated per mg wet weight according to the formula:

$$x = \frac{\dfrac{2000}{t_1} - \dfrac{400}{t_2}}{m}, \qquad (2.1)$$

where x is ^{60}Co activity in cpm per mg, t_1 is counting time (2000 counts), t_2 is background counting time (400 counts), m is wet colony weight in mg.

The weight of colonies grown from untreated cells varied from 1.5 to 5.0 mg. On average, the activity of colonies grown from untreated cells was 600-700 cpm/mg; σ was less than 120 cpm/mg, and CV was 16%. Addition of cold cobalt to the medium decreased the incorporation of label to 200-300 cpm/mg. Counting error was less than 3%. Colonies with high ^{60}Co activities were picked out, subcultured and re-plated on solid medium containing ^{60}Co. For each variant at least 100 colonies were counted, the most active ones being further tested for vitamin B_{12} production in liquid medium. Finally, the active variants grown in 9 l of an industrial liquid medium were re-tested for vitamin B_{12} production at a pilot plant, using a 250-l fermenter.

The proposed selection method for microorganisms forming cobalt-organic substances has the capacity of testing at least 400-500 colonies per day with high precision.

2.3.1   Mutagenic treatments

We used NMU, NTG and DMS as chemical mutagens. *P. shermanii* was grown in corn steep-glucose medium for 14-18 h. The cells were separated by centrifugation, washed with phosphate buffer (pH 7.0) and treated with different doses of mutagens. The following morphological and cultural properties were evaluated: (i) size and pigmentation of colonies, (ii) character of growth in liquid medium, (iii) vitamin B_{12} production.

N-nitrosomethylurea (NMU). Cells were treated with 0.1 or 0.5% solutions of NMU in phosphate-citrate buffer for 30 min, 2, 3, 4 and 5 h.

Dimethylsulfate. We used a 1:6000 dilution (0.017% solution). Treatment time was 12 to 16 h. Alternatively, cells were treated with the mutagen vapor according to the method of Kupenov (1974). In this case, a small vessel containing the mutagen liquid was placed on the bottom of a tube that contained a thin layer of cells deposited on the walls by vacuum drying. The action of the DMS vapor was stopped by removing the vessel from the tube.

N-methyl-N'-nitro-N'-nitrosoguanidine (NTG). NTG is unstable in solution. At pH 7.0 nitrosoguanidine is converted to diazomethane, at pH 5.0 to nitric acid. Diazomethane and nitric acid also are mutagenic but less effective than NTG. NTG is relatively stable at pH 5.5-6.5. Therefore, cells were treated with 0.05% solution of NTG in phosphate buffer at pH 6.0. Treatment time was 20, 40 and 60 min.

2.3.2   Characterization of mutagen-treated cells

Survival of cells treated with chemical mutagens depends not only on the genotype of an organism, conditions of treatment and mutagen concentration, but also on the time of recovery after the treatment. Recovery is most successful under conditions of growth delay, when treated cells are suspended and incubated in a buffered solution for a certain time. This segregation period is necessary to repair cell damage and to express new metabolic characteristics. Mutagenized cells were routinely incubated for a recovery period of 2 h before being plated.

Cell survival, presence of morphological mutants and variability of ^{60}Co incorporation were determined. For each variant the results were statistically analyzed by calculating the mean value ($\bar{x}$), standard deviation (σ) and covariation (*CV*) (Vorobjeva et al., 1973).

Nitrosomethylurea (NMU). Both positive and negative variants in vitamin B_{12} synthesis induced by NMU were unstable, returning to the initial level after several passages. Morphological mutants were not found. Roslyakova (1974) reported positive results by using nitrosoethylurea (NEU) as a mutagen. Namely, adding NEU to the medium at 0.25% resulted in a three-fold increase in biomass and vitamin B_{12} yield as compared with untreated controls at the end of a 24-h incubation. Thus nitroso compounds are active also when added at low concentrations to cells growing in nutrient broth.

Dimethylsulfate (DMS). Cell survival after the DMS treatment was high enough, declining progressively with increasing mutagen concentrations and time of treatment (Table 2.4). DMS induced no morphological mutants, but increased the range of variability in ^{60}Co accumulation (Fig. 2.2).

Table 2.4. Effect of DMS treatment on survival of *P. shermanii*

DMS concentration, %	Treatment time, h	Survival, %
0.017	2	46.0
0.017	4	21.0
0.017	6	4.0
0.034	2	0.0
0.050	2	0.0
Vapor	3	39.0

From Vorobjeva (1976).

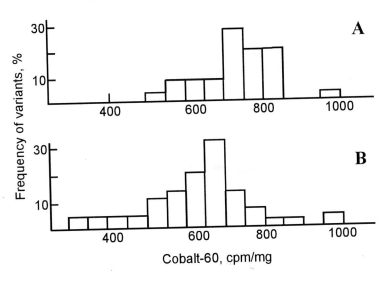

Figure 2.2. Variability of ^{60}Co-incorporation by *P. shermanii* induced by dimethylsulfate. A: spontaneous variability; B: mutagen-induced variability. From Chan (1973).

The yield of positive mutants (activity $\geq \bar{x} + 3\sigma$) was 1.2%. As a result of the DMS treatment, variants 2^{5e} and 1^{5z} were isolated, superior to the parental strain in vitamin B_{12} production by 30% (Table 2.5).

Table 2.5. Incorporation of ^{60}Co and production of vitamin B_{12} by DMS-induced mutants of *P. shermanii*

Strain	^{60}Co activity, cpm/mg		Vitamin B_{12}, µg/ml
	glucose medium	lactate medium	
19^{5a}	1960	630	nd*
25^{5a}	2480	nd	7.5
14^{5b}	2250	600	nd
29^{5b}	2520	nd	8.0
11^{5d}	1570	645	nd
2^{5e}	3000	730	10.1
1^{5z}	2150	720	9.1
21^{5a}	3500	640	nd
24^{5a}	2080	nd	nd
30^{5a}	2550	580	9.4
4^{5l}	1750	620	nd
Parental strain	2010	570	7.0

*nd, not determined.
Cells were grown in submerged culture. From Vorobjeva et al. (1973).

The selection procedure for active variants of *P. shermanii* induced by the DMS treatment is schematically represented as follows:

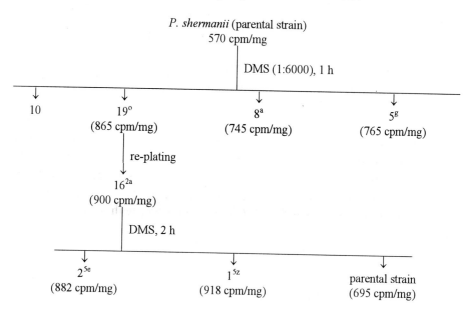

No stable negative mutants were obtained by the DMS treatment. The positive mutants remained highly active in vitamin B_{12} production when cultivated in larger volumes of liquid medium.

Nitrosoguanidine (NTG). NTG is a strong mutagen that causes a broad spectrum of mutations in different microorganisms. As shown for auxotrophic and cold-resistant mutants by Glatz and Anderson (1988), different strains required different doses of NTG. For some strains the highest yield of auxotrophic mutants among surviving cells was observed at a survival rate of about 10%, a condition that was met by treating the cells with 1.0 mg/ml of NTG at pH9.0 for 90 min. For the other strains it was sufficient to expose the cells to 100 μg/ml of NTG at pH 6.0 just for 30 min. In most experiments the yield of mutants among surviving cells was only about 1%. In our experiments (Chan, 1973) treatment of cells with a 0.05% NTG solution for 20 min dramatically reduced the incorporation of ^{60}Co (Table 2.6), although the survival rate was still high enough, i.e., 3.5%. After 60 min of treatment only 0.2% of the cells remained viable, while essentially no accumulation of ^{60}Co was observed (Table 2.6) (Vorobjeva et al., 1973). Thus the effect of reducing cell viability by NTG was proportional to the time of treatment, as was the ^{60}Co incorporation, which was low in virtually all the colonies counted, except for a few positive mutants (Fig. 2.3).

Table 2.6. Nitrosoguanidine treatment of *P. shermanii* strain 19°

Treatment time, min	Survival, % of control	Colonies counted	^{60}Co activity, cpm/mg	$N_{3\sigma}$*	σ	CV
20	3.5	105	474	0	0	0
40	1.8	176	365	1	0	0
60	0.2	70	574	0	69 ± 5	12.0 ± 1.0

* $N_{3\sigma}$, number of colonies with ^{60}Co activity $> \bar{x} + 3\sigma$. From Vorobjeva et al. (1973).

Positive mutant strains were isolated by treating suspensions of the strain 19° with 0.05% NTG for 40 and 60 min. Colonies grown from the suspensions treated with NTG, differed in appearance from the colonies of the parental strain. Colonies of the parental strain are round, convex, creamy and brilliant with diameters of 2 to 5 mm. As a result of the NTG-treatment new types of colonies were found: (a) tiny, colorless, with very low counts of ^{60}Co; (b) flat, colorless, yellow or creamy; (c) rough; (d) bright-yellow; and (e) typical for propionibacteria (Vorobjeva et al., 1973). The tiny white colonies induced by NTG were numerous, but after a number of passages reverted to the normal synthesis of corrinoids.

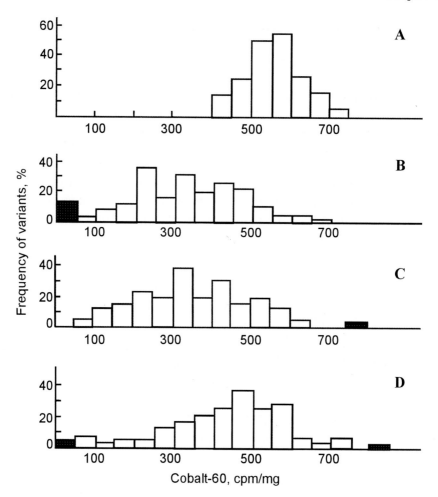

Figure 2.3. Variability of ^{60}Co-incorporation by *P. shermanii* strain 19° induced by nitrosoguanidine. A: spontaneous variability; B, C and D: nitrosoguanidine-induced variability. Treatment time was: 20 min (B), 40 min (C), and 60 min (D). From Chan (1973).

In general, five groups of NTG-induced mutants can be recognized:

1) similar in morphological, cultural and physiological characteristics to the parental strain, but having a higher activity in vitamin B_{12} production;
2) negative in biosynthesis of vitamin B_{12}, differing from the parental strain by a slower growth;
3) auxotrophic, negative in vitamin B_{12} biosynthesis;
4) morphological mutants, being mostly negative in vitamin B_{12};
5) morphological mutants with high ^{60}Co accumulation.

Group 1 mutants are active producers of vitamin B_{12} and retain this property in the next generations.

Group 2 is represented by strain 30^{3d}. It is similar to the parental strain in morphological and cultural properties, grows well in synthetic media with three vitamins (biotin, thiamine, pantothenic acid), but synthesizes only traces of corrinoids (Table 2.7). Fermentation of various substrates is not affected. Liquid cultures produce the same amounts of propionic and acetic acids as the parental strain.

Table 2.7. Biosynthesis of corrinoids by mutant strains of P. shermanii

Strain	Dry biomass, g/l*	Corrinoids, μg/ml
9^{ld}	2.37	2.38
16^{3d}	1.08	0.24
Black	2.66	0.14
Rough	2.27	0.17
30^{3d}	2.20	0.10
Parental	3.48	4.84

*Cells were grown in rich medium with glucose.
From Vorobjeva et al. (1973).

Typical of group 3 mutants is strain 16^{3d}. In morphologic properties it is similar to the parental strain, but grows very slowly even in rich medium. In synthetic media it does not grow. Requirements for growth factors were tested according to Holliday's scheme (1956), and strain 16^{3d} was found to require uracil (Table 2.8), to form only 0.24 μg/ml of corrinoids and to accumulate metabolic products typical of propionic acid fermentation (propionic and acetic acids).

Table 2.8. Effects of nucleotide bases on the growth of P. shermanii mutants

Added at 0.05% to synthetic medium	Dry biomass, g/l			
	16^{3d}	9^{ld}	rough	black
Adenine	nd*	0.64	0.40	0.48
Guanine	nd	0.79	0.52	0.51
Adenine+guanine	nd	1.6	0.50	1.5
Uracil	0.44	nd	nd	nd

*nd, not determined.   From Vorobjeva et al. (1973).

Group 4 comprises morphological mutants, among them of special interest are rough and black. Cells of the rough mutant are typical for P. shermanii (Fig. 2.4), but colonies have a rough surface with a trough in the centre. The rough is deficient in the purine bases adenine and guanine, produces only 0.17 μg/ml of corrinoids with normal quantities of propionic and acetic acids in corn steep-glucose medium with yeast extract. The black mutant is similar to the parental strain in cell morphology and in the shape and dimensions of colonies, but it produces a dark pigment so that the colonies after several days of culture are almost black in color. The *black* mutant is deficient in guanine (Table 2.8), grows well in corn steep-glucose medium, produces only 0.14 µg/ml of corrinoids and ferments glucose forming propionic and acetic acids. Biosynthesis of vitamin B_{12} by these auxotrophic mutants can be restored by adding purine bases to the medium, i.e. purine deficit has a marked influence on the productivity for vitamin B_{12}.

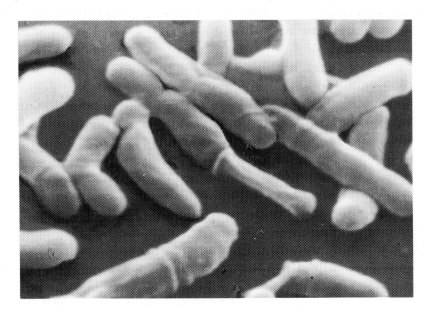

Figure 2.4. Scanning electron micrograph of the *rough* mutant cells deficient in adenine and guanine. × 24,000. From Vorobjeva et al. (1973).

At the same time mutant strain 9^{1d}, although deficient in adenine and guanine, in rich medium produced the same quantities of vitamin B_{12} as the parental strain and was indistinguishable from it in morphological and cultural properties. Thus, auxotrophy for purine bases is not always associated with an impaired synthesis of vitamin B_{12}.

In the mutants studied mainly the complete vitamin B_{12} is formed. A notable exception is *black* that produces insignificant quantities of vitamin B_{12}, but instead forms a corrinoid having the absorption maxima in the regions of 300-304, 355, 430-455, 488, 520-530 nms (Chan, 1973). In media deficient in purine bases the *rough* and *black* mutants formed more porphyrins than the parental strain. Apparently, when the synthesis of corrinoids is blocked, ALA is utilized for porphyrin synthesis. Formation of another tetrapyrrole compound, catalase, by the *rough* and *black* mutants was strongly reduced, in contrast with the prototrophic negative mutant 30^{3d} showing the enzyme activity as high as the parental strain.

By adding a cell extract of the parental strain *P. shermanii* to the culture of 30^{3d} the synthesis of vitamin B_{12} was restored, although incompletely. The activity of ALA synthetase and ALA dehydratase in the strain 30^{3d} was not reduced, the rate of PBG synthesis was normal, therefore, the block in corrinoid synthesis occurred at a step distal to PBG formation, where the pathways of porphyrin and corrinoid synthesis diverge. It should be noted that the ratio of propionic to acetic acid in negative mutants remained in the range of 2:1; formic acid was produced only by two cultures, 30^{3d} and *rough*. Mutant 9^{1d} produced a low level of acetic acid, and volatile acids were represented basically by propionic acid alone.

A special group is represented by bright-yellow mutants, isolated by mutagenic treatment of positive variants, which accumulated greater quantities of ^{60}Co as compared with the parental strain (Table 2.9). Strain 21^{3d}, showing the ^{60}Co activity of 1135 cpm/mg, was subcultured and plated out on solid medium containing ^{60}Co, and characteristics of the variation line were determined. The results were as follows: $n = 118$, $\bar{x} = 1237$ cpm/mg, $\sigma = 336$ cpm/mg and $CV = 27.1\%$. The high coefficient of variation indicates a large spontaneous variability with respect to ^{60}Co incorporation.

Table 2.9. Incorporation of ^{60}Co by bright-yellow mutants of propionic acid bacteria

Strain	^{60}Co activity, cpm/mg	Strain	^{60}Co activity, cpm/mg
29^n	969	27^4	2284
1^h	1404	28^{4a}	1269
15^h	4485	30^4	3051
21^{3d}	1135	9^{4a}	1999
Parental	773	Parental	574

From Vorobjeva et al. (1973).

The most active colonies, isolated by subculturing the strain 21^{3d}, were picked out and assayed for corrinoid production (Table 2.10). Bright-yellow mutants grew slower in liquid medium, than the parental strain. Their biomass quickly darkened on exposure to the air. In this case the method of corrinoids isolation from the cells by extraction with a 3:1 mixture of 0.01% KCN (ethanol solution) and water was ineffective. After such an extraction the cells remained dark in color, i.e., extraction was incomplete. Corrinoids can be in a free state or bound to cellular proteins. To isolate bound corrinoids, more stringent conditions are needed, namely, extraction with 80% ethanol containing 0.01 M KCN at 70°C for 30 min. This method was used for the isolation of corrinoids from bright-yellow mutants. Corrinoids were determined microbiologically, using *E. coli* 113-3 as the test organism. The mutants examined were not uniform with respect to the biosynthesis (Table 2.10).

Table 2.10. Production of corrinoids by bright-yellow mutants of *P. shermanii*

Strain	0.01% KCN in 30% ethanol, for 3 h			0.01M KCN in 80% ethanol at 70°C, for 30 min		
	Vitamin B_{12}					
	µg/ml	µg/g	ΔA_{367nm}	µg/ml	µg/g	ΔA_{367nm}
28^4	6.3	220	0.535	14.6	570	0.710
	8.2	285	0.590	13.6	580	0.650
27^4	7.6	226	0.690	13.6	692	0.770
	4.4	202	0.530	13.6	692	0.770
30^4	8.2	264	0.600	15.8	728	0.600
	8.0	291	0.680	14.0	512	0.680
16^{4b}	16.4	585	0.840	18.4	736	0.940
	11.8	539	0.530	12.0	500	0.650
Parental	12.6	458	0.700			
	11.0	492	0.686			

From Chan (1973).

Cell extracts of the mutant 21^{3d} were reddish-brown in color. The extracts were treated with acidified ethanol to remove protein and salts, then the residue was isolated by centrifugation. The supernatant was red and the pellet was black. The pellet was washed several times with acidified ethanol to remove the last traces of corrinoids and finally was dissolved in water at an alkaline pH. The resulting solution was reddish-brown in color and showed biological activity in *E. coli* 113-3 (vitamin B_{12} assay). The activity dropped considerably after 10 min of hydrolysis under alkaline conditions. It was suggested that the bright-yellow mutant 21^{3d} represented a type of mutants, in which corrinoids, produced in excessive amounts, are tightly bound in a complex. Undoubtedly, studying such mutants is of great theoretical and practical interest.

The mutants 9^{4a} and 27^4 had the wild-type phenotype, suggesting that they were the result of a single mutation. Revertants 30^4 and 15^h still formed small white colonies, so they probably contained more than one point mutation. Overall, the bright-yellow mutants exhibited low viability, which precluded their complete characterization.

Miscellaneous mutagens. Roslyakova (1974) added nitrosoethylurea (NEU) and nitrosomethylurea (NMU) directly to the growth medium at final concentrations of 0.01-0.05%, then inoculated the medium with bacteria. In this case the yield of biomass and vitamin B_{12} was two and three times higher, respectively, than in the controls. However, if cells were treated with the same dose of mutagen (NMU) in non-growing conditions, the subsequent yield of vitamin B_{12} and biomass was not different from the control.

Mutagenic effects of N-nitrosomethylbiuret (NMB), a combination of NMB and UV-light, and 3-(3-chloropropoxyphenyl)-5,6-dihydroimidazo-

(2,1)-thiazole hydrochloride (CTH) on the vitamin B_{12}-producing capacity of *P. shermanii* M-82 were studied (Ganicheva and Vorobjeva, 1991). Morphological variants were selected according to the intensity of pink color. Variants were tested for vitamin B_{12} production and grouped into classes at 20% intervals. Table 2.11 shows the frequency distribution of variants with increased productivity. Analysis of the results shows that the treatment of cell suspensions with 0.1% NMB or CTH at pH 6.0 for 30 min resulted in 89 and 32% survival, respectively, while the number of plus-variants was virtually the same. Mutagenesis led to a two-fold increase in the number of plus-variants compared to the control at almost the same level of survival. The variants isolated by mutagenesis fall principally into two classes: 81-100% and 101-120%, whereas the controls fall into three classes: 61-80, 81-100 and 101-120%. The distribution of mutations was shifted towards high-activity variants, as shown by the absence of variants from the 61-80% class. The arithmetic mean increased slightly from 94 to 100%. Conducting the NMB treatment in slightly acidic (pH 6.0) or alkaline (pH 9.0) conditions showed little difference in mutagenic effects with respect to the induction of plus-variants.

Table 2.11. Effects of different mutagenic treatments on *P. shermanii*

Mutagen	Treatment	Survival, %	Number of colonies assayed	$\bar{x}$	σ	CV	Number of plus-variants*
Control	None	100	393	94	9.5	10.1	24.4
NMB	0.1%, 30 min pH 6.0	89	107	97.8	11.3	11.6	45.5
CTH	0.1%, 30 min pH 6.0	32	97	100	11	11	42.2
NMB + UV-light	0.1%, 30 min pH 6.0	0.05	119	87.5	14.4	16.5	25.3

From Ganicheva and Vorobjeva (1991).

At higher NMB concentrations there was greater variability, especially at pH 9.0 (up to 33.6%). The mutations were shifted towards minus-variants. The combined treatment with NMB and UV-light was lethal, with survival rates reduced to 0.05%. This combination was three orders of magnitude more lethal than NMB alone. The variability of mutations (CV) was also enhanced up to 16.5% due to the increased number of minus-variants.

In spite of the large number of high-activity mutant strains isolated, none of them preserved its original activity after further selection. Since nitrosoalkylamides (NAA) were used to derive the parental strain M82 (Grusina et al., 1974), it is possible that the genome of this strain is already saturated with NAA-inducible point mutations, so that the treatment with NMB, another NAA mutagen, could not elicit stable mutants in this strain.

The number of plus-mutants resulting from the combined action of the two mutagens was reduced to 25.3% compared with 45.5% after the treatment with NMB alone. High-productivity mutant strains selected after NMB or CTH treatment generally lost their initial productivity in further experiments. Only one mutant strain isolated after CTH treatment retained the capacity for cobalamin synthesis at the level of 17% above that of the parental strain. Populations of propionic acid bacteria are highly heterogeneous (at least with regard to the biosynthesis of corrinoids). To obtain highly productive strains it may be necessary to stabilize the population with the help of antimutagens (see below).

Of the various mutagens, namely UV rays, ethylenimine + UV, DES, NMU, NEU, NTG, NMB and CTH, tested on *P. shermanii* cells in doses resulting in different lethal effects, nitroso compounds are recognized as having the highest mutagenic potential. One of these, NTG, induces the largest number of morphological and biochemical mutations and the highest variability with respect to vitamin B_{12} production.

2.4   Antimutagenesis

Propionic acid bacteria, like all living organisms, are constantly and for a long time subject to the action of exogenous and endogenous mutagens. Natural sources of exogenous mutagens include radiation, radon gas, cosmic rays, ^{40}K and other radionuclides (nuclei of unstable isotopes that disintegrate spontaneously). At present, only 1% of the chemical compounds found in the biosphere has been studied, and among them thousands of mutagens were found. Living organisms themselves are a source of endogenous mutagens, producing mutagenic (carcinogenic) compounds during their normal metabolism. For example, such mutagenic compounds as H_2O_2, nitrosamine, H_2S, formaldehyde and some antibiotics are normally produced by bacteria as metabolic products.

The most frequent mutations are point mutations, induced by chemical changes in nucleotides (transitions and transversions) during the DNA replication and reparation. Point mutations tend to reverse spontaneously. In the processes of genetic recombination and DNA reparation some deletions and insertions of nucleotide bases occur, giving rise to frameshift mutations. Since in natural environments microorganisms (as well as plants and animals) are continuously subject to the action of mutagens, they have formed endogenous and exogenous defense mechanisms: all living organisms synthesize molecules capable of antimutagenesis. The term 'antimutagenesis' means (Kada et al., 1986) the process of reducing the frequency of spontaneous and induced mutations. Antimutagens regulate the rate of spontaneous mutations, stabilize the mutational process but do not completely inhibit it, and are inactive when the background level of mutations is very low, because otherwise organisms could not evolve.

Three classical types of DNA repair include photoreactivation, excisional reparation and postreplicative recombinative reparation. In the case of considerable damage to the DNA, when replication becomes impossible, the SOS reparation system is switched on. The SOS system is the system of emergency reparation, but it is prone to errors during DNA replication.

Antimutagenesis was discovered relatively recently, in 1952, when it was found (Novic and Scillard, 1952) that the addition of purine nucleosides to the medium resulted in the reduction of spontaneous mutations in *E. coli* by 60-70%. The action of antimutagens can be displayed at any level, from detoxification of mutagens to reparation of damaged DNA. According to their mode of action, antimutagens are separated (Kada et al., 1986) into desmutagens and bioantimutagens. **Desmutagens** act by inactivating mutagens and carcinogens by various means. **Bioantimutagens** act by stimulating the nonmutagenic DNA reparation, inactivating the mutagenic reparation and SOS reparation (prone to errors) or interfering with the expression of mutations.

Antimutagens stimulate enzymatic systems active in detoxification of chemicals entering the cell, and can affect the redox potential. Many mutagens generate free radicals. In so doing they are opposed by antioxidative enzymes, e.g., Se-containing compounds and such antioxidants as tocopherol, ascorbic acid, phylloquinones. Some antimutagens ($CoCl_2$, for instance) ensure the accuracy of one type of reparation (Kuroda, Inoue, 1988) and activate the DNA reparation in general. All these processes can significantly reduce the rate of mutations.

About 300 individual compounds with antimutagenic properties are known. The frequency of mutations can be reduced by some amino acids (arginine, histidine, methionine, cysteamine etc.), vitamins and provitamins (α-tocopherol, ascorbic acid, retinol, β-carotene, phylloquinone, folic acid), enzymes (peroxidase, NADPH oxidase, glutathione peroxidase, catalase), complex compounds of plant and animal origin, chemical substances with antioxidant properties (derivatives of gallic acid, ionol, oxypyridines, selenium salts and others).

Many substances display physiological specificity, i.e., at low concentrations they act as antimutagens, but at high concentrations as mutagens. This is typical for arginine, glutamic acid, sodium selenide, streptomycin, and derivatives of gallic acid. At the same time other antimutagens do not change their mode of action even if their concentration is varied by several orders of magnitude (α-tocopherol, β-carotene, phylloquinone and some others). The narrower the limits of concentrations defining the reversal of the effect, the smaller degree of physiological specificity is displayed by the compound.

For the evaluation of mutagenic and antimutagenic properties of chemical substances about 100 different test systems have been proposed, although none of them is universal enough as to be capable of detecting all types of genetic lesions and corresponding reparations. In laboratory practice the Ames' test is used most frequently, in which the test system *Salmonella* /microsomes is employed to quantitate gene mutations. Other bacteria such as *E. coli* and *B. subtilis*, lower eukaryotes such as yeasts *Saccharomyces cerevisia* and *S. pombe*, fungi *Aspergillus nidulans* and *Neurospora crassa*, and higher eukaryotes, namely, the insect *Drosophila melanogaster*, some mammalian cell cultures (mouse tumour cells and Chinese hamster ovary cell lines), including primary human cell cultures (peripheral blood lymphocytes) are used to quantitate chromosomal abberations and sister chromatide exchanges as well. The whole organisms rather than isolated cells can also be used.

In 1989, using the Ames' test, we (Vorobjeva et al., 1991) have for the first time revealed the antimutagenicity of propionic acid bacteria against sodium azide (NaN_3) and NTG. Although this property is not directly related to the genetics of bacteria, it concerns regulation of mutations and, importantly, a possibility of reducing genetic alterations in other living organisms. This work opened a new field in biological studies, namely, bacterial antimutagenesis. Until recently, only higher plants have been studied as sources of antimutagens/anticarcinogens, i.e., citrus fruit juices (Bala and Grover, 1989), fruits and vegetables (Shinohara et al., 1988; Stolz et al., 1984), some edible seaweeds (Yamamoto et al., 1982, 1986), mushrooms (Grüter et al., 1990) and certain marine animals (Sasaki et al., 1985). Reports on antimutagenic properties of various eukaryotes have been reviewed by G. Hocman (1988, 1989).

Prokaryotes have seldom been considered as potential sources of antimutagens, although the similarity of basic metabolic reactions in pro- and eukaryotes, on the one hand, and the ability of prokaryotes to carry out some unique reactions, on the other, do not exclude bacteria as possible sources of new and valuable antimutagens (anticarcinogens). This suggestion is supported by the results of investigations in *Streptomyces* (Osawa et al., 1986), in which two strains were found that produce peptides highly effective as antimutagens.

There is another important reason to investigate antimutagenic and mutagenic properties of those bacteria that are used for food, animal feed or as food additives—bacteria as a source of bioantimutagens or desmutagens may be used for the pretreatment of food and animal feed in order to neutralize mutagenic/carcinogenic substances, and also as probiotics. In this respect, propionic acid bacteria are widely used in practice, i.e., in cheese-making, in vitamin B_{12} production, in the veterinary preparation 'Propiovit', in ensilage (Vorobjeva et al., 1984).

2.4.1 Antimutagenicity of cell extracts

Cell extracts of *P. shermanii* VKM-103 (wild type), *P. shermanii* KM-82 mutant (superproducer of vitamin B_{12}), *P. acnes* CCM 322 and propionic acid cocci (*P. coccoides*) were tested as possible sources of antimutagens. The first two strains represented classical propionic acid bacteria, and *P. acnes* represented cutaneous propionibacteria. In the assay for antimutagenic activity *S. typhimurium* TA 1535 (*his⁻* mutant) was used as the test organism (base pair substitution mutations tester).

The procedure used for the determination of antimutagenic activity is the same as that used for the determination of mutagenic potential of a substance as suggested by Maron and Ames (1984). A brief outline of the protocol is as follows: to 2 ml of top agar containing 0.5 mM histidine/biotin were added 0.1 ml of a fresh *Salmonella* culture, 0.1 ml of NaN_3 or NTG and 0.1 ml of a cell extract and mixed thoroughly before pouring on top of a minimal glucose agar plate. The plates were incubated for 2 days at 37°C. A positive control (with mutagen but no extract) and a negative control (with extract but no mutagen) were run separately, adjusting the total volume to 0.4 ml with sterile distilled water. A number of modifications of the basic procedure were run as follows: (i) extract and mutagen were preincubated together for 20 min at 37°C; (ii) extract was heated at 100°C for 10 min or at 70°C for 10 min; (iii) extract was diluted with 50 mM Na-phosphate buffer (pH 7.4) to different concentrations; (iv) adenosylcobalamin (vitamin B_{12}-coenzyme) were tested versus NaN_3.

From the linear portions of the dose-response curves the concentrations of NaN_3 and NTG were chosen at 3.0 µg/0.1 ml and 10 µg/0.1 ml per plate, respectively. In each case revertant colonies (*his⁺*) were scored. Antimutagenic effect was expressed as percent decrease in the frequency of reverse mutations as follows:

$$\text{antimutagenic effect (\%)} = \frac{(a-b) \times 100}{a-c}, \qquad (2.2)$$

where *a* is the number of *his⁺* revertants induced by the mutagen; *b* is the number of *his⁺* revertants induced by mutagen in the presence of extract; and *c* is the number of *his⁺* revertants induced in the presence of cell extract alone (negative control).

For viable cell counts 0.1 ml portions of 10^{-3} dilution were mixed with 2 ml of top agar containing 0.1 mg L-histidine and 1.5 μg D-biotin and were poured on Vogel-Bonner agar plates.

Regardless of the different mechanisms of mutagenic action of NaN_3 (metabolic formation of a mutagen, azidoalanine) and NTG (methylation of purine bases in DNA), in both cases the antimutagenic factor(s) at low concentrations displayed a strong inhibitory activity, but at high concentrations enhanced the mutagenic potential. Preincubation of the dialysate with NTG significantly enhanced the inhibitory effect. Effects of the dialysates of different strains of propionibacteria on the mutagenicity of sodium azide were studied. The dialysates of *P. shermanii* VKM-103, *P. shermanii* KM-82 and *P. acnes*, which represent both classical and cutaneous propionibacteria, showed high inhibitory effects of 64.9, 99.0 and 99.3%, respectively. It was also shown that the antimutagenic effect of the dialysate of *P. shermanii* KM-82 is reduced from 36.9 to 7.4% by heating at 70°C for 10 min. Upon heating the dialysate at 100°C for 10 min the antimutagenic activity was completely lost.

On the other hand, the dialysate of propionic acid cocci enhanced the mutagenicity of NaN_3, although it had no mutagenic effect of its own on *Salmonella* TA 1535. On the basis of phenotypic characteristics, fatty acid composition and the high degree of homology between the DNA of the cocci and *P. jensenii* we proposed earlier (Vorobjeva et al., 1983) that these cocci should be placed into the genus *Propionibacterium* as a new species. However, several differences exist between the cocci and other species of propionibacteria. One of the main differences of the cocci from propionic acid bacteria is in the low level of cobalamins in their cells (Vorobjeva et al., 1983). Therefore, the antimutagenicity of coenzyme B_{12} and vitamin B_{12} was determined and it was found that coenzyme B_{12} at concentrations of 0.001-1.0 μg/0.1 ml per plate (corresponding to the coenzyme B_{12} concentrations in the dialysates, including those inducing strong inhibitory effects) showed no antimutagenicity and had no mutagenic effects per se.

The fact that the antimutagenic activity of the dialysate is abolished by heating at 100°C together with having a molecular weight exceeding 12 kDa (exclusion limit of the dialysis bag) suggested that the antimutagenicity of dialysates is associated with proteins. After the precipitation of proteins from cell dialysates of *P. shermanii* KM-82 by ammonium sulfate, antimutagenic activity was found only in the protein fractions. Proteins were separated by gel filtration on Toyopearl. Fig. 2.5 shows the elution profiles of the proteins of *P. shermanii* KM-82 and antimutagenic effects of the fractions. The proteins were separated into three major fractions and the antimutagenic activity was found in fractions 2 and 3. The antimutagenic effect of each fraction did not always match the protein content. Fig. 2.5 shows that the dialysate of *P. shermanii* KM-82 contains a set of antimutagenic proteins with different molecular weights. We suggest that at least some of the proteins possess enzymatic activity and produce strong inhibitory effects at low concentrations.

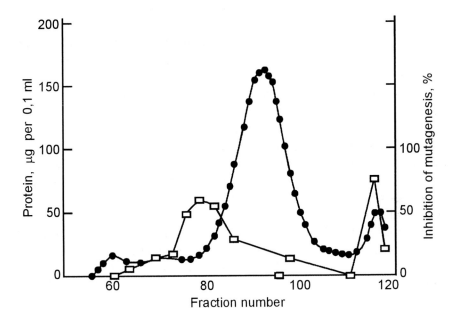

Figure 2.5. Antimutagenic effects of protein fractions from the dialysate of *Propioni-bacterium shermanii* KM-82. Proteins were separated (●) by gel filtration on Toyopearl. 0.2 ml aliquots of the resulting fractions were used in the Ames test (□) with NaN_3 at 3.0 μg/0.1 ml per plate as the mutagen and *Salmonella typhimurium* TA1535 as the test organism. Experimental points represent mean values from 3 replicate plates. Reproduced with permission from L.I. Vorobjeva et al. *Mutat. Res.* 251: 233-239, 1991 © Elsevier Science.

As shown by Owais et al. (1988), the mutagenicity of NaN_3 is mediated by L-azidoalanine produced metabolically by the action of *o*-acetylserine (thiol)-lyase. Mutagenesis appears to proceed by a mechanism similar to that of NTG by a direct mispairing pathway (Owais et al., 1988). Most of the mutations induced by NTG are produced by alkylated bases (especially O_6-methylguanine) (Walker, 1984). Although our data are insufficient to speculate on the mechanism of antimutagenic action of the dialysates of *P. shermanii* VKM-103 against the mutagenicity of NTG, we suggest, on the basis of the enhancing effects of preincubation, the possibility of mutagen binding and/or prevention of its transport into the cell by a substance in the dialysate.

In separate experiments we showed (Vorobjeva et al., 1993a) that superoxide dismutase (SOD) reduces the mutagenic activity of sodium azide and nitrosoguanidine (NTG) in *S. typhimurium* TA 1535. A correlation was observed between the antimutagenic and SOD activity. However, the antimutagenic effects of the cell extract and an SOD preparation (Sigma) with similar SOD activities (0.180 and 0.160 U/ml) were 65 and 11%, respectively. Therefore, apart from SOD, other proteins possessing antimutagenic activity are present in cell extracts of propionic acid bacteria.

It is known that H_2O_2 is formed by the interaction of O_2^- with protons. H_2O_2 also has mutagenic properties, although much weaker than O_2^- (Dahl et al., 1988). But catalase did not show any antimutagenic effects against NaN_3, which indirectly indicated a weak mutagenicity of H_2O_2 or/and insignificant accumulation of the latter.

Antimutagenic effects of propionibacteria can find its application in biotechnology for stabilization of microbial fermentations. It was shown (Vorobjeva et al., 1992) that preincubation of dialyzed cell extracts of *P. shermanii* with cells of the same species, as well as with germinating spores of actinomycetes and conidias of *Penicillium chrysogenum*, results in a considerable increase in their viability and a decrease in the induced and spontaneous variability of vitamin B_{12} production (*P. shermanii*), the antibiotics kanamycin (*Actinomyces kanamyceticus*) and penicillin V (*Penicillium chrysogenum*). The dialysate had no effect on the viability of untreated (control) cells.

2.4.2   Desmutagenic effects of culture liquid

The culture liquid obtained in propionic acid fermentation displayed antimutagenic activity with respect to NQO-induced mutagenesis in *S. typhimurium* TA100 (Table 2.12).

Table 2.12. Antimutagenic effects of the culture liquid of *P. shermanii* as a function of cultivation and preincubation time

Cultivation time, h	Incubation time, h	Number of revertants per plate*		Inhibition, %
		mutagen alone	mutagen + antimutagen*	
24	5	973 ± 10	970 ± 10	0
24	24	742 ± 11	547 ± 11	22.6
24	48	493 ± 15	378 ± 10	23.3
24	72	664 ± 6	390 ± 55	41.3
48	24	742 ± 6	593 ± 31	20.0
48	48	668 ± 27	504 ± 40	24.6
72	24	745 ± 17	689 ± 27	1.0

*Spontaneous background was 40 ± 10 revertants per plate.
Plates of *Salmonella typhimurium* TA-100, containing the mutagen NQO at 0.125 µg/ 0.1 ml per plate, were incubated with or without 0.1 ml per plate of the culture liquid from *P. shermanii* grown for the indicated periods of time (cultivation time).   From Vorobjeva et al. (1993b).

The activity was highest in the logarithmic and early stationary phases (24-48 h of growth). In aged cultures the activity dropped. The duration of mutagen and antimutagen preincuba-tion considerably influenced the antimutagenic effect, which was not yet manifested after 5 h but intensified with increasing the preincubation time from 24 to 72 h.

The culture liquid of *P. shermanii* taken after 48 h of cultivation was concentrated 6-fold by lyophilization. Increasing the culture liquid concentration led to a reduction in the number of revertants (Fig. 2.6).

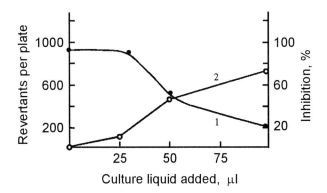

Figure 2.6. Dose-response dependence of concentrated culture liquid of *P. shermanii* against mutagenicity of 4-nitroquinoline-1-oxide. The Ames test was carried out with 4-nitroquinoline-1-oxide as the mutagen and *Salmonella typhimurium* TA100 as the test organism. Sixfold concentrated culture liquid was added at doses indicated. From Vorobjeva et al. (1993b).

In an approach to study the nature of antimutagenic substances, *P. shermanii* culture liquid was dialyzed, heated, and treated with trypsin. Table 2.13 shows that dialysis of the culture liquid leads to a complete loss of antimutagenic properties, indicating that substances possessing the antimutagenic activity have molecular weights of less than 12 kDa (deduced from the exclusion limit of the dialysis membrane). Both heating a non-dialysed culture liquid at 92°C for 10 min or trypsin treatment had no effect on the antimutagenicity, indicating the presence of thermostable substances of nonpeptide nature. High antimutagenicity of the culture liquid of the young culture suggests that the antimutagenic substances bear a relation to the lowering of the redox potential of the medium, the condition necessary for propionic acid bacteria to start growing. In subsequent phases of growth these substances may be consumed by the culture or may undergo various modifications, e.g., oxidation. Many sulfur-containing compounds, among them glutathione, cysteine, and cysteamine, possess reducing properties.

Chemical mutagens NQO, 2-(2-furyl)-3-(5-nitro-2-furyl)acrylamide (AF-2) and others are inactivated when heated with cysteine or cysteamine.

Table 2.13. Effects of dialysis, heat and trypsin treatment on antimutagenic activity of the culture liquid from *P. shermanii*. From Vorobjeva et al. (1993b).

Mutagen	Pretreatment of culture liquid*	Colonies per plate	Inhibition, %
None	not added	34 ± 8	–
+	not added	982 ± 15	–
+	sterile medium	880 ± 10	10.4
+	undialyzed	68 ± 10	93.1
+	dialyzed	1027 ± 9	0
+	undialyzed, heated at 92°C for 10 min	75 ± 24	92.4
+	undialyzed, treated with trypsin	45 ± 15	95.4

*Culture liquid was concentrated 6-fold prior to the use or treatment. Antimutagenic activity was assayed as shown in Table 2.12, using the incubation time of 96 h.

Intensification of the antimutagenic effect upon increasing the incubation time of NQO with the culture liquid indicates that the reduction in mutagenicity of NQO is probably due to its interaction (binding, modification) with a desmutagenic factor produced by the propionic acid bacteria. The formation of a desmutagenic factor apparently is not confined to a particular strain, since it was found also in cutaneous propionic acid bacteria. Production of metabolites displaying antimutagenic activity may be important in maintaining the genetic stability of the species. Industrial production of propionibacterial biomass for vitamin B_{12} production and for other purposes yields large volumes of culture liquid as by-product, which may prove a valuable source of antimutagenic substances.

The antimutagenic activity of various species of propionic acid bacteria is shown in Table 2.14. Approximately similar effects of the strains tested were observed with both culture liquids and bacterial cells, except for propionic acid cocci and *P. thoenii* (Vorobjeva et al., 1995a). The culture liquid of *P. coccoides* showed an inhibitory action of 46%, whereas the effect of the other strains was almost 100%. *P. thoenii* showed a poor antimutagenic effect. The highest inhibitory effect was found in the logarithmically growing cultures, which suggests that it is the young cultures that synthesize and secrete compounds protecting the cell from the action of mutagens. The question of which components of the culture liquid are desmutagens arises. Several amino acids (arginine, histidine, methionine, cysteine and glutamic acid) (Alekperov, 1984) that are released into the medium by bacteria, including propionic acid bacteria (Vorobjeva et al., 1979), exhibit antimutagenic activity. These amino acids are utilized in the late phases of growth.

Table 2.14. Antimutagenic effects of culture liquid and isolated cells of different propionic acid bacteria

Source of activity	Number of revertants per plate	Inhibition, %	Medium thiols, μM
P. shermanii			
Culture liquid	66 ± 5	100	12.0
Cells	80 ± 5	99	
P. pentosaceum			
Culture liquid	65 ± 4	100	13
Cells	111 ± 9	96	
P. thoenii			
Culture liquid	944 ± 35	18	6.2
P. acnes			
Culture liquid	75 ± 6	100	14
Cells	121 ± 6	95	
P. coccoides			
Culture liquid	613 ± 21	49	13
Cells	106 ± 13	97	
Growth medium	733 ± 25	38	

*Spontaneous background was 71 ± 2 revertants per plate. Mutagen alone produced 1142 ± 47 revertants. Antimutagenic activity was assayed as shown in Table 2.12, using the cultivation and preincubation times of 24 h. From Vorobjeva et al. (1995a).

It is known that actively growing bacteria, both aerobic and anaerobic, decrease the redox potential of the medium and cells as a result of the synthesis and secretion of reducing compounds. These compounds may vary in different strains, but often they are represented by sulfur-containing compounds. Almost all the strains of propionic acid bacteria were shown to release small amounts of thiols into the medium. Fig. 2.7 shows that the greatest desmutagenic effects and the maximal accumulation of thiols in the medium are found in the 24-h culture of *P. shermanii*. After 24 h of cultivation the thiol content declined in parallel with the desmutagenic effect of the liquid. These results favor the suggestion that the antimutagenic effects of propionibacteria are due to the secretion of thiols interacting with NQO (the interaction between electrophilic groups of NQO and sulfur-containing compounds). This suggestion is supported by the apparent stoichiometry of antimutagen-mutagen interactions, i.e., varying the mutagen doses from 0.125 to 0.25 μg/0.1 ml per plate at a fixed low antimutagen dose (culture liquid) led to a corresponding reduction of the antimutagenic effect.

The whole cells of propionibacteria, as shown in Table 2.14, also possess antimutagenic activity. When the cells were used as modulators, they were separated by centrifugation after incubating them with a mutagen and removed. Then the suspension of the test culture (*S. typhimurium* TA100) was added and plated. 48 h later the number of revertants was counted and the antimutagenic effect was expressed as shown above. It was found (Vorobjeva et al., 1995a) that the antimutagenicity can be induced by two successive passages of *P. shermanii* in a medium containing 0.2-1.0 µg/ml of NQO (Table 2.15).

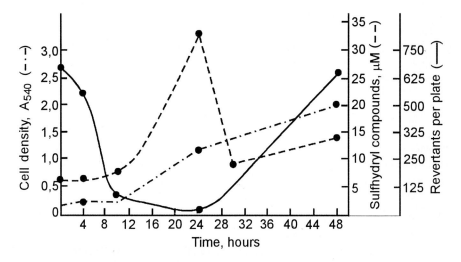

Figure 2.7. Antimutagenicity of and content of sulfhydryl compounds in culture liquid during growth of *P. shermanii*. Antimutagenic effects were assayed as in Fig. 2.5. Curve 1 - biomass; 2 - thiol group content; 3 - number of revertants per plate. From Vorobjeva et al. (1995a).

Table 2.15. Desmutagenic effects of mutagen-induced cells against 4-nitroquinoline-1-oxide

Number of passages in the presence of mutagen	Number of revertants per plate*	Inhibition, %
0	291 ± 14	63
1	226 ± 15	72
2	25 ± 1	100

*Spontaneous background was 23 ±revertants per plate.
From Vorobjeva et al. (1995a).

This observation was not due to an accumulation of mutants, insofar as growing the cells in a medium without mutagen resulted in the same antimutagenic effects both of culture liquid and of the cells as in the control cells (not induced). As suggested above, one of the possible mechanisms of desmutagenicity may be related to the interaction of electrophilic mutagens and glutathione. It has been suggested (Ketterer, 1986) that glutathione (GSH) and GSH transferase is the most likely system for the prevention of electrophilic mutagenicity. The GSH transferase activity of the induced cells was 50% higher than that of the control cells, while the content of sulfhydryl compounds was almost identical (Table 2.16).

Table 2.16. Glutathione transferase activity and the content of sulfhydryl compounds in mutagen-induced and control cells of *P. shermanii*

Cells	GSH transferase activity, nmol/min/mg	Sulfhydryl compounds, nmol/mg protein	
		free	total
Control	4.0	81.0	101.25
Induced	6.0	80.0	100.0

From Vorobjeva et al. (1995a).

Thus, propionic acid bacteria (and perhaps some other microorganisms) possess two levels of defense: (i) external, mediated by the secreted compounds and cell surface structures acting as desmutagens, and (ii) intracellular, operating by metabolic modulation, DNA reparation and replication, and inducible detoxification. The finding of antimutagenicity and its induction in propionic acid bacteria by mutagens reveals an additional useful feature of these organisms, which can be utilized through the consumption of food enriched in these bacteria.

Various industries involving propionic acid bacteria use lactate (in cheese making), glucose (in vitamin B_{12} production), or whey lactose (in obtaining biomass to prepare food and leavening additives) as the fermented carbon source. Antimutagenic activity of the culture liquid was detected with any of the three carbon sources used. The activity was highest when bacteria were grown on lactate and lower in media containing glucose or lactose (Vorobjeva et al., 1993b). The demonstration of desmutagenic action of the cultural liquid of propionic acid bacteria has additional and special perspectives connected with the statement, that desmutagens found in bacterial systems often show the ability to inhibit carcinogenesis in laboratory animals, although this could not be ascertained with bioantimutagens (Kada et al., 1986).

In addition to the well-studied *P. acnes*, there have been reports on the oncolytic and radioprotective activity of clostridial spores (Blanca et al., 1989), lactobacilli (Axelsson et al., 1989), and *Pseudomonas jananensis*, a new species of *Pseudomonas* showing sarcomostatic activity (Cai et al., 1989). It may be recalled, however, that the antitumor activity of lactic acid bacteria had been noticed quite a long time ago. So, Bogdanof et al. (1962) reported an antitumor action of the medium in which *Lactobacillus bulgaricus* was grown; according to the report by Reddi et al. (1973), yogurt prepared with *Lb. bulgaricus* and *Streptococcus thermophilus* inhibited proliferation of Ehrlich ascites tumour cells in mice.

Recently, Lankaputhra and Shah (1998) showed that both live and dead cells of probiotic bacteria, including 6 strains of *Lactobacillus acidophilus* and 9 strains of bifidobacteria, express antimutagenic activity against the following mutagens and promutagens: N-methyl-N'-nitro-N-nitroso-guanidine, aflatoxin B, 2-amino-3-methyl-3H-imidazoquinoline, 2-amino-3-

methyl-9H-pyrido-(3,3-6)indole, 2-nitrofluorene, 4-nitro-O-phenylene-diamine, and 4-nitroquinoline-N-oxide (NQO). The authors found that live cells bound or inactivated the mutagens permanently, whereas dead cells released the mutagens upon extraction with dimethylsulfoxide. When organic acids produced by these bacteria: acetic, butyric, lactic and pyruvic acid were tested for antimutagenic activity, butyric acid and, to a lesser extent, acetic acid were found to be the most effective. Lactic and pyruvic acids exhibited lower activities against all the mutagens studied except NQO. Thus, organic acids, the products of bacterial metabolism, contribute to the antimutagenic activities of probiotic bacteria. These findings emphasize once again the importance of maintaining viable populations of probiotic bacteria in the gut, where various mutagens may often be found.

In 1989, the Japanese firm Morinaga reported that a bacterium known as K-303 and found only in Japan, can synthesize substances with antiaflatoxin activity. Aflatoxin formed by moulds is a poison possessing mutagenic and carcinogenic activities. The substance was named antiaflatoxin, and its production was planned for 1991. Keeping in mind the well-known advantages of the cultivation of microorganisms in comparison with the cultivation of plants and animals, a high rate of microbial multiplication, independence from weather and climate etc., one can predict good prospects for antimutagens of microbial origin.

2.5   Reactivation of Cells Inactivated by UV light and Other Stress Factors

In addition to the antimutagenic activity against chemical mutagens such as that demonstrated in our experiments (Vorobjeva et al., 1991), cell extracts of propionibacteria displayed a reactivative action on UV-inactivated *E. coli* AB 1157. It is interesting that the reactivative action of the wild-type strain of *P. shermanii* was stronger than that of its mutant strains, superproducers of vitamin B_{12}, or *P. rubrum* producing a red pigment (according to the literature, pigmented strains are better protected from UV light than apigmented strains), or even *Deinococcus radiodurans* known for its exclusive resistance to X-rays and UV light. Therefore, the protective action of cell extracts does not necessarily correlates with self-resistance of bacteria to UV-rays.

Among the species of propionibacteria tested, i.e., *P. rubrum, P. raffinosaceum, P. acidipropionici, P. shermanii, P. coccoides, P. acnes* the highest effect was found in *P. shermanii* and *P. acidipropionici*, while *P. coccoides* and *P. acnes* did not exert any protective effects in UV-inactivated *E. coli* cells (Vorobjeva et al., 1996b). The protective substances in the extract are dialyzable and have molecular weights higher than 12 kDa.

Moreover, nondialyzed extracts showed lower protective actions than the dialyzed ones. Maximal protective action of the extract was revealed when it was applied at concentrations of 20-25 μg of protein per ml. One can assume that at high concentrations some components of the extract may inhibit the binding of the active agent(s). Preincubation of the non-irradiated test culture with the dialysate for 5, 30, and 60 min did not change the number of grown colonies, i.e., we observed a true protective action of the extract. It was also established, that the protective effect of the extract did not depend on the carbon source utilized by propionibacteria during growth, since extracts prepared from cells grown in media with lactate or glucose showed approximately the same protective effect.

A similar protective action was found both in the case of pre- (5 min) and postincubation (30 min) of the extract with irradiated cells. Thus a screening action of the extract can be largely excluded by these observations. Heating the dialysate for 10 min at 50°C and 70°C or for 1 min at 92°C caused some reduction of its activity, which still remained relatively high, but upon heating for 12 min at 92°C the activity was reduced three-fold. Reactivative activity of the extract was retained after storage up to 1.5 months at –20°C and more than one year after lyophilization. Lipase treatment had no effect on the reactivative activity of the extract; RNAse and especially DNAse treatment lowered it significantly, while proteinase K treatment alone or followed by trypsin treatment reduced the activity 5.5-fold or completely, respectively.

Previously (Adler et al., 1966), only filamentous strains of *E. coli* have been shown to be reactivated by an extract of *E. coli*, and reactivation was related to the mechanism of the initiation of cell division. We have for the first time demonstrated reactivation of a nonfilamentous strain of *E. coli* AB1157 by extracts of propionic acid bacteria. Possible mechanisms of reactivation include stimulation of reparation systems, antimutagenic action and/or stabilization of DNA. The protective effect was very high, reaching a factor of 25 (calculated as the ratio of the number of colonies grown in the presence of extract, to the number of colonies grown in the absence of extract), although the complete recovery of all the cells was never reached. This phenomenon may perhaps be explained by the fact that the maximal effect was observed at a very low viability of the injured cells (0.006%).

To investigate the protective mechanism of the dialysate, the latter was tested on isogenic *E. coli* strains with various impairments of the DNA repair systems. The sensivity to UV light increased in the following order: *E. coli* AB1157, PolA, UvrA, and RecA. In the mutant PolA-1 the systems for rapid DNA repair and excision repair by short fragments are blocked due to the lack of DNA polymerase 1. The mutant has the system of excision repair, though slower than the wild type, and the remaining open gaps are equivalent to unexcised pyrimidine dimers triggering the SOS system.

The sensitivity of *S. typhimurium* TA100 (UvrB⁻) was intermediate between the sensitivities of UvrA and RecA strains. The dialysate exerted a protective effect on all tested strains, subjected to UV light. Survival of the protected suspensions of the PolA strain was 5 and 25 times higher compared with the survival rates of unprotected suspensions of 0.18 and 0.04%, respectively. Survival of the protected suspension of UvrA strain (deficient in the excision repair system) was 8 and 2 times higher compared with the survival rates of unprotected suspensions of 0.014 and 1.14%, respectively. The recombination and SOS repair are impaired in the RecA strain, the activity of the excision repair system being intact. The protective dialysate increased the survival of the RecA strain three-fold at the survival rates of unprotected cells of 0.08-0.017%. If the viability of unprotected cells was higher than 0.14%, the protection was weak. So the protective effect of dialysate was more pronounced in the cases of wild type and PolA type than with RecA and UvrA strains, which are more sensitive to UV light than the former two strains (Vorobjeva et al., 1995c).

As we have seen, the efficiency of reactivation was inversely proportional to the rate of cell survival. It was also shown (Vorobjeva et al., 1995c) that the two types of reactivation of UV-injured *E. coli* cells, photoreactivation (PhR) and reactivation by the dialysate, are different but completely additive. Maximal reactivation can be attained by PhR followed by the protective action of the dialysate. It was found (Fraikin et al., 1995; Vorobjeva et al., 1996b) that the dialysate reduced lethal effects of UV-C and UV-B radiation not only in prokaryotes, but eukaryotes as well, i.e., the yeasts *Candida guilliermondii* and *Saccharomyces cerevisiae*. The reactivation occurred when the dialysate was added to irradiated suspensions immediately after irradiation or 15 min later. The dialysate exerted its protective and reactivative action not only in UV-inactivated *E. coli*, but in Gram-positive and Gram-negative bacteria including *E. faecalis*, *P. shermanii* itself and other strains of different species of propionibacteria, *S. typhimurium* and even in a representative of archaea, *Halobacterium salinarium*. But the dialysate failed to reactivate those cells that were irradiated with visible light (400-600 nm) or light of the entire optical range (above 290 nm).

It is known that the inactivation caused by UV-B and UV-C light is determined primarily by photochemical DNA damage. Therefore, it may be suggested that the dialysate triggers some protective mechanisms for the repair of such DNA damage. The fact that the dialysate fails in visible light can be explained (Fraikin et al., 1995) by the light-induced damage to a hypothetical receptor localized in the plasma membrane, which mediates the reactivating effect of the dialysate.

Both bacteria and yeasts were protected from injuries and reactivated not only by the whole dialysate, but also by two of its fractions (active factors, AF), obtained by treating the dialysate with ammonium sulfate to 20-40% (AF-1) and 60-80% (AF-2) saturation. Active proteinaceous factors protect cells from different stress factors. The stress systems we studied were the heat shock regulon (htpR-controlled), the oxidation stress regulon (oxyR-controlled), and the SOS regulon (lexA-controlled). These regulons can be induced separately, but some agents induce more than one regulon (for example, $CdCl_2$) and such agents may generate multiple signals in the affected cell.

Heat shock proteins (HSPs) in *E. coli* are synthesized when cells are exposed to H_2O_2 (Van Bogelen et al., 1987), ethanol (Travers et al., 1982), UV light (Krueger and Walker, 1984) and heavy metals (Linquist, 1988). HSPs defend the cell from toxic effects and other stresses. The genes controlling the response against stress factors are regulated both separately and coordinately. Cellular responses to external stressors are often mediated by the heat-shock proteins possessing chaperone functions. Therefore, we suggested that a protein factor (AF) may exert an universal action, showing protective and reactive effects in organisms injured by different and unrelated stressors, and our findings support this (Vorobjeva et al., 1997).

Heat shock and ethanol treatment. Reactivative action of the dialysate was demonstrated in *E. coli, S. cerevisiae* and *C. guilliermondii* subjected to heat shock (Table 2.17). The efficiency of the reactivation was inversely proportional to the viability of the inactivated cells, thus replicating the regularity detected for UV-irradiated bacteria and yeasts. Moreover, the two dialysate fractions that showed reactivative and protective activity in UV-irradiated *E. coli* also showed reactivative effects in bacterial cells inactivated by heating.

It is known (Van Bogelen, 1987) that a shift to 42°C or addition of 4% ethanol preferentially induces the heat shock regulon in *E. coli* cells. Both ethanol and heat shock induce similar lesions in the cells (Piper, 1995): they increase the permeability of the membrane and have negative effects on membrane-linked processes, and cells respond to the two stresses by the induction of synthesis of heat shock proteins. Therefore, it was not surprising that the AF afforded protection and reactivation to the ethanol-injured cells of *E. coli*. The survival of cells preincubated with AF followed by ethanol treatment (4% solution, 15 min) was six times higher than of control cells. The reactivative effect of the AF alone was much weaker.

Table 2.17. Reactivation of *E. coli*, *S. cerevisiae* and *C. guilliermondii* cells inactivated by heating

Heating conditions	Viability, %			
	no dialysate	+ dialysate	fraction 1	fraction 2
Escherichia coli				
45°C, 30 min	74	91	–	–
45°C, 90 min	23	85	–	–
42°C, 60 min	58	–	90	88
Saccharomyces cerevisiae				
42°C, 30 min	66	91	–	–
42°C, 60 min	5.9	9.5	–	–
42°C, 90 min	4.5	8.4	–	–
Candida guilliermondii				
42°C, 60 min	33.8	32.9	–	–
42°C, 90 min	20.5	28.0	–	–

From Fraikin et al. (1995).
Cells were post-incubated for 15 min with the active factors from *P. shermanii*

Heavy metals. The following salts of heavy metals were used as stress factors: $CdCl_2$, $ZnSO_4$, $CoCl_2$ and $CuSO_4$. It was shown in *E. coli* (Van Bogelen et al., 1987) that $CdCl_2$ induced 17 proteins and evoked SOS-, heat-shock and oxidation stress responses, the latter two being predominant. $CdCl_2$ was found (Lee et al., 1983a, b) to be the most potent agent in inducing the synthesis of adenylated nucleotides, accumulation of which is consistent with a role of alarmones in the oxyR-mediated response. The dialysate of *P. shermanii* exerted protective and reactivative action in *E. coli* injured by the exposure to the above mentioned heavy metals (Table 2.18). Also in Table 2.18 it is seen that the viability of cells pre- or postincubated with the AF in the case of exposition to $CdCl_2$ increased 7.7- and 7.6-fold, to $CuSO_4$, 5.2- and 5.1-fold, to $ZnSO_4$, 5.6- and 5.3-fold, and to $CoCl_2$, 3.5- and 3.4-fold, respectively.

Nalidixic acid. In *E. coli* the SOS regulon is the primary responder to nalidixic acid, and the heat shock regulon is the secondary one (Van Bogelen et al., 1987). Nalidixic acid was applied at a concentration of 600 μM for 60 min. In this case the cell survival was 11%. Viability of the cells preincubated with AF increased up to 14%, and of the postincubated cells up to 16%.

Hydrogen peroxide. Hydrogen peroxide has been shown to damage DNA *in vitro* (Demple, Linn, 1982). H_2O_2 causes a complete but transient inhibition; induces primarily the oxidative stress response and also the SOS response (Van Bogelen et al., 1987). The AF exerted a marked protective action in *E.*

coli inactivated by H_2O_2, and the effect was again (see above) inversely proportional to the rate of cell survival (Table 2.19).

Table 2.18. Protection and reactivation of *E. coli* cells inactivated by Cd, Cu, Zn and Co ions by the active factor (AF) from *P. shermanii*

Conditions of treatment	Cell number, ×10⁶/ml	Viability, %	Cell division index**
Control (untreated)	121 ± 7.5	100	
CdCl₂, 2 mM, 15 min			
Cd-salt alone	9.4 ± 4.6	7.8	
AF*, then Cd-salt	72 ± 9.8	59.5	7.7
Cd-salt, then AF	71 ± 9.8	58.7	7.6
CuSO₄, 2mM, 15 min			
Cu-salt alone	11.5 ± 1.38	9.5	
AF, then Cu-salt	59 ± 4.2	48.8	5.1
Cu-salt, then AF	60 ± 10.5	49.6	5.2
ZnSO₄, 2mM, 15 min			
Zn-salt alone	11.0 ± 9.0	9.1	
AF, then Zn-salt	58.0 ± 4.0	48.0	5.3
Zn-salt, then AF	62.0 ± 4.9	51.2	5.6
CoCl₂, 1 mM, 15 min			
Co-salt alone	21.1 ± 1.0	25.1	
AF, then Co-salt	74.0 ± 3.0	88	3.5
Co-salt, then AF	71 ± 3.0	84.5	3.4

*AF, active factor. **Cell division index is the ratio of colony-forming units in the presence and absence of the active factor. Protection and reactivation were assayed by pre- and post-incubation, respectively, with the AF (20 μg protein per ml suspension), both incubations lasting for 10 min. From Vorobjeva et al. (1997).

Table 2.19. Protection and reactivation by the active factor (AF) from *P. shermanii* of *E. coli* cells inactivated by treatment with hydrogen peroxide,

Treatment	Cell number, ×10⁶/ml	Viability, %	Cell division index**
Control (untreated)	34 ± 3.5	100	
H₂O₂, 5 mM, 15 min			
H₂O₂ alone	1.83 ± 0.7	5.3	
AF*, then H₂O₂	5.7 ± 0.6	16.7	3.16
H₂O₂, then AF	5.0 ± 0.8	15.3	3.0
H₂O₂, 10 mM, 15 min			
H₂O₂ alone	0.3 ± 0.1	0.6	
AF, then H₂O₂	3.1 ± 0.9	6.7	11.1
H₂O₂, then AF	2.5 ± 0.3	5.5	9.1

*AF, active factor. **Cell division index is the ratio of colony-forming units in the presence and absence of the active factor. Protection and reactivation were determined as shown in Table 2.18. From Vorobjeva et al. (1997).

Keeping in mind the above-mentioned observations, it can be assumed that the AF performs better in those cases when cellular repair systems are insufficient. When survival of the inactivated cells was 5.3%, preincubation with the AF increased it up to 16.7%, but when survival was 0.6%, then preincubation with the AF increased it up to 6.7%, i.e., more than 10-fold.

Isolation of a protector protein from P. shermanii. The finding of reactivative effects of the bacterial AF in a wide spectrum of organisms inactivated by different stress factors undoubtedly represents theoretical and practical interest. At the same time, this finding inevitably and urgently poses a question of how such an action is realized, in other words, what is the mechanism of reactivation at the cellular and molecular level. To elucidate the mechanism, it was necessary to isolate an individual substance responsible for the observed effects. Without the isolation and characterization of the individual active compound(s) the above-mentioned suggestions about the type of interactions between the AF and the cell remained only hypothetical.

An initial purification of the AF protein was performed by salt fractionation of the soluble proteins of *P. shermanii*. Active substances were found in protein fractions precipitated by ammonium sulfate at 20-40% (AF-1) and 60-80% (AF-2) saturation (Fraikin et al., 1994).

By ion-exchange chromatography on DEAE-Sepharose FF, the two ammonium sulfate fractions (primary fractions) were separated into subfractions (Fig. 2.8). Seven subfractions showed protective activity in *E. coli* inactivated by UV light. When the subfractions were analyzed by SDS polyacrylamide gel electrophoresis and HPLC, it was found that the subfraction AF2-2.5 contained mostly a single protein, whereas the other subfractions were composed of mixtures of proteins and peptides of different molecular weights. The molecular weight of subfraction AF2-2.5 was determined at 44 ± 2 kDa (Zinchenko et al., 1998). The protective activity of subfraction AF2-2.5 in UV-inactivated *E. coli* was dependent on its concentration in the range of 10-60 μg of protein per ml.

The isolated protector also showed reactivative and protective effects in *E. coli* subjected to heating, ethanol treatment and $CdCl_2$ (2 mM). The isolation of the individual compound, responsible for the protective and reactivative properties of the AF, opens new ways for understanding the mechanism of the antistress action of this protein preparation.

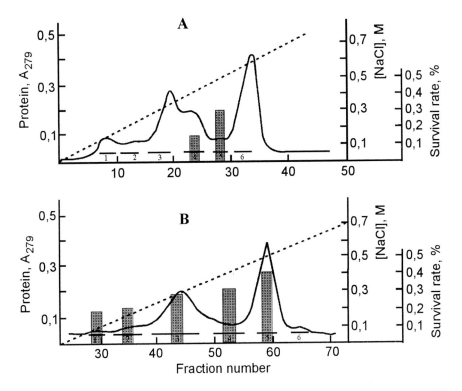

Figure 2.8. Refractionation of cell-protective proteins in primary active fractions (AFs) from *P. shermanii* by anion-exchange chromatography on DEAE-Sepharose. The primary active fractions were AF1-2 in A and AF2-2 in B. Shaded columns show cell viability as percent in the presence of respective fractions. Further details in Zinchenko et al. (1998).

Chapter 3

Transformations of Energy

3.1   Fermentation

3.1.1   Biochemistry of fermentation

Propionic acid fermentation is of major significance for the energetics of propionic acid bacteria. The main fermentation products are propionic and acetic acids and CO_2 (Mashur et al., 1971; Foschino et al., 1988). Formic and succinic acids (Mashur et al., 1971) as well as acetoin and diacetyl (Tomka, 1949; Antila, 1956/57; Lee et al., 1969, 1970) are also produced, but in smaller amounts. Other volatile aromatic substances are dimethylsulfide, acetaldehyde, propionic aldehyde, ethanol and propanol (Keenan and Bills, 1968; Dykstra et al., 1971). Propionic acid fermentation differs from the other types of fermentation by the high ATP yield and by some unique enzymes and reactions.

Propionic acid fermentation is not limited to propionibacteria; it functions in vertebrates, in many species of arthropods, in some invertebrates under anaerobic conditions (Halanker and Blomquist, 1989). In eukaryotes the propionic acid fermentation operates in reverse, providing a pathway for the catabolism of propionate formed via β-oxidation of odd-numbered fatty acids, by degradation of branched-chain amino acids (valine, isoleucine) and also produced from the carbon backbones of methionine, threonine, thymine and cholesterol (Rosenberg, 1983). The key reaction of propionic acid fermentation is the transformation of L-methylmalonyl-CoA(b) to succinyl-CoA, which requires coenzyme B_{12} (AdoCbl). In humans vitamin B_{12} deficit provokes a disease called pernicious anemia, which is accompanied by the accumulation of propionic and methylmalonic acids in blood and urine (for details, see Chapter 7).

In vertebrates, in addition to the main pathway of propionate catabolism (conversion into succinate), alternative pathways may function under conditions when the main pathway is blocked. In the termite *Zootermopsis angusticollis* a high B_{12} content apparently is due to the presence of microorganisms in the stomach (Wakayama et al., 1984). There is evidence that vitamin B_{12} is used by termites for the conversion of succinate to methylmalonate and incorporation of the latter instead of malonyl-CoA into methyl-branched hydrocarbons. Therefore, in termites the direction of carbon flow is the same as in bacteria (from succinate to propionate), but opposite to that found in vertebrates. From the aforesaid it is clear that studying propionic acid fermentation is important not only for understanding the biochemistry of propionic acid bacteria, but of many other organisms, including humans.

Propionic acid fermentation is found (Wegner et al., 1968; Haase et al., 1984) in various bacteria of the genera *Propionibacterium*, *Rhodospirillum*, *Micrococcus*, *Rhizobium*, *Mycobacterium*, as well as in a protozoan, *Ochromonas malhamensis*. Only in propionibacteria it serves as the main pathway of energy generation, while in other species it is just another way of living. Evidently, this is the reason why in propionibacteria some essential and most vulnerable reactions (see below) are duplicated by different enzymes, and reactions used to regenerate reducing equivalents are enclosed in a cycle, while in eukaryotes and other organisms the cycle is open.

It is known that cycling of a process provides a greater degree of autonomy, independence from the environment; that is why such processes as the tricarboxylic acid cycle, CO_2 fixation in autothrophs, pentose monophosphate pathway and some others operate as cyclic processes, supplying the organism with vital substances. The chemistry of propionic acid fermentation, schematically represented in Fig. 3.1, has been thoroughly investigated and reviewed (Stjernholm and Wood, 1963; Hettinga and Reinbold, 1972b; Vorobjeva, 1976).

Isolation and identification of the intermediates of glucose fermentation, verification of the expected end products (Wood et al., 1937) and analysis of the distribution of labeled products (Wood and Leaver, 1953) showed that glycolysis is the main pathway of glucose utilization by propionic acid bacteria. Glucose is phosphorylated, forming hexose monophosphate (van Niel, 1928) and hexose diphosphate (Pett and Wynne, 1933). Transcarboxylase, hexosephosphate isomerase (Wood et al., 1963), fructose-diphosphate aldolase (Sibley and Lehninger, 1949; Wood et al., 1963) and triosephosphate dehydrogenase activities are found in propionibacterial cells (van Demark and Fukui, 1956).

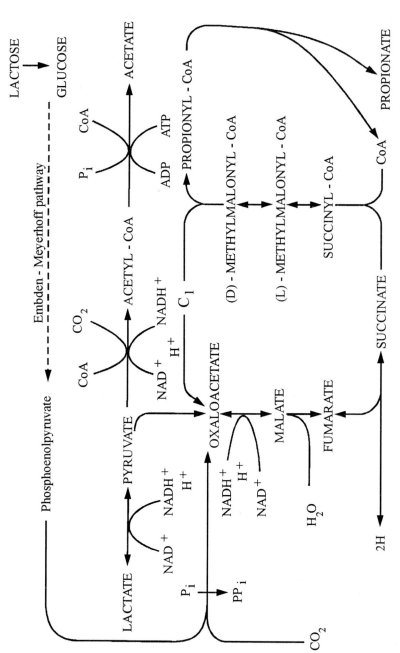

Figure 3.1. Metabolic pathways of propionic acid bacteria.

Acetone-dried cells of *P. pentosaceum* can phosphorylate glucose, arabinose, glycerol and some other substrates at the expense of the phosphate group of ATP (Stone et al., 1937; Barker and Lipmann, 1949). However, the degradation of substrates is relatively insensitive to the inhibitor of glycolysis NaF, and analysis of the labeled metabolic products showed (Wood et al., 1937; Volk, 1954; Leaver et al., 1955) that the hexose monophosphate pathway also contributed to the degradation of glucose.

Propionic acid bacteria can utilize gluconate (Fukui, 1952), and in cell free extracts of *P. pentosaceum* the key enzymes of hexose monophosphate pathway, transketolase and transaldolase, are found (van Demark and Fukui, 1956). Among the various intermediates of pentose fermentation, D-arabinose-5-P and the enzyme catalyzing the isomerization of D-arabinose-5-P to D-ribulose-5-P are found (Volk, 1954). An essential observation is that ribose-1-C^{14} as well as gluconate labeled in different positions are converted to acetate, propionate and succinate by the cell extract of *P. shermanii* (Stjernholm and Flanders, 1962). Distribution of the labeled products led to the conclusion (Swick and Wood, 1960) that the substrates entered first the hexose monophosphate pathway forming fructose-6-P, which then was used in the glycolytic pathway and transcarboxylation cycle. Pyruvic acid, an obligate intermediate in fermentation, was isolated from the culture liquid of propionic acid bacteria (Wood and Werkman, 1934). Pyruvate can be converted to propionate in several ways:

Pyruvate $\longrightarrow$ acrylate $\longrightarrow$ propionate;
Pyruvate $\longrightarrow$ lactate $\longrightarrow$ propionate;
Pyruvate $+ C_1 \rightarrow$ succinate $\rightarrow$ methylmalonate $\rightarrow$ propionate.

The first two options are not realized in propionic acid bacteria, so that propionate is formed in the third reaction (Werkman and Wood, 1942). This is confirmed experimentally by the ability of these bacteria to convert succinic acid into propionic acid (Wood and Werkman, 1940). In animal tissues propionate and C_4-acids (Flavin et al., 1955) are interconnected by the following reaction:

$$\text{Propionyl-CoA} + CO_2 + \text{ATP} \xleftrightarrow{\text{propionyl carboxylase}} \text{methylmalonyl-CoA(a)} + \text{ADP} + P_i$$

Propionate is activated first by forming propionyl-CoA, then it is carboxylated in a process requiring ATP and producing methylmalonyl-CoA. It was suggested that propionic acid bacteria form propionate by reversing the reaction shown above (Kaziro and Ochoa, 1964). However, the rate of the reverse reaction is very slow and does not correlate with the high rates of propionate production. But propionic acid bacteria have other means to synthesize C_4-dicarboxylic acids from C_3- and C_1-compounds.

One of these is by heterotrophic CO_2 fixation, first discovered in propionic acid bacteria by Wood and Werkman (1936, 1938). The CO_2 fixation is especially high when the bacteria are grown on glycerol. Both the fixation of carbon dioxide and formation of succinate by propionibacteria are inhibited by NaF (Wood and Werkman, 1940). Another pathway of the condensation of C_3- and C_1-compounds was discovered by Swick and Wood (1960), who showed that these bacteria contain a transcarboxylase that has a role in producing propionate from methylmalonyl-CoA. This was preceded by an observation (Wood and Leaver, 1953) that propionic acid bacteria have two mechanisms for CO_2 fixation and only one of these is inhibited by NaF. The enzyme, discovered by Swick and Wood, catalyzed a new type of biochemical reactions—transcarboxylation between a carboxyl donor and an acceptor:

Methylmalonyl-CoA + pyruvate $\rightleftharpoons$ propionyl-CoA + oxaloacetate.

The discovery by the American scientists was confirmed in our investigations (Vorobjeva, 1958b, 1976) by estimating the carbon balance in the end products of propionic acid fermentation. Reactions leading to the production of propionic acid in propionibacteria can be represented by the following sequence (Delwiche, 1948; Johns, 1951):

$$\text{Pyruvate + methylmalonyl-CoA(a)} \xrightleftharpoons{\text{transcarbo xylase}} \text{oxaloacetate +}$$
$$\text{propionyl-CoA} \quad (3.1)$$

$$\text{Oxaloacetate} + 4H^+ \xrightarrow{\text{citric acid cycle}} \text{succinate} \quad (3.2)$$

$$\text{Succinate + propionyl-CoA} \xrightleftharpoons{\text{CoA trans ferase}} \text{succinyl-CoA +}$$
$$\text{propionate} \quad (3.3)$$

$$\text{Succinyl-CoA} \xrightleftharpoons{\text{methylmalonyl isomerase}} \text{methylmalonyl-CoA(b)} \quad (3.4)$$

$$\text{Methylmalonyl-CoA(b)} \xrightleftharpoons{\text{methylmalonyl racemase}}$$
$$\text{methylmalonyl-CoA(a)} \quad (3.5)$$

$$\text{(Summary)} \quad \text{Pyruvate} + 4H^+ \longrightarrow \text{propionate} \quad (3.6)$$

This reaction sequence, which is the reverse of the propionate metabolism in animal cells, is ultimately equivalent to the reduction of pyruvate to propionate by NADH formed in glycolysis or in triose oxidation to acetate and CO_2.

The conversion of oxaloacetate to succinate is catalyzed by enzymes of the citric acid cycle: malate dehydrogenase, fumarase and succinate dehydrogenase. These enzymes were isolated from the cells of *P. shermanii*

(Allen et al., 1964), where they are known to catalyze the conversion of oxaloacetate to succinate at a high rate (Krebs and Egglestone, 1941). In propionibacteria, the succinate dehydrogenase is not inhibited by malonate, in contrast with succinate dehydrogenases of the Krebs cycle (Ichikawa, 1955).

Acetic acid, another product of propionic acid fermentation, is formed as a result of the oxidative decarboxylation of pyruvate:

$$\text{Pyruvate} + \text{NAD}^+ + \text{CoA} \xleftrightarrow{\text{pyruvate dehydrogenase}} \text{acetyl-CoA} + \text{H}^+ + \text{NADH} + \text{CO}_2 \quad (3.7)$$

$$\text{Acetyl-CoA} + \text{P}_i \xleftrightarrow{\text{phosphotransacetylase}} \text{acetyl-P} + \text{CoA} \quad (3.8)$$

$$\text{Acetyl-P} + \text{ADP} \xleftrightarrow{\text{acetylkinase}} \text{acetate} + \text{ATP} \quad (3.9)$$

The reductive steps in the formation of propionate, including the reduction of oxaloacetate and fumarate, are coupled with reoxidation of NADH supplied by the reactions of phosphoglyceraldehyde oxidation and oxidative decarboxylation of pyruvate. There is evidence (Sone, 1972) that succinate dehydrogenase of *P. arabinosum* is in fact fumarate reductase that catalyzes the oxidation of NADH, L-lactate, L-glycerol-3-phosphate (L-GP) coupled with the reduction of fumarate. Fumarate reductase is localized mainly in the particulate fraction, while GP-dehydrogenases are localized both in the particulate and soluble fractions. Cytochrome b, nonheme Fe, flavoproteins and lipids are found in the particulate fraction. These compounds, together with the respective dehydrogenases, comprise a specific type of electron transport system that oxidizes NADH, L-GP and lactate. In the presence of these substrates oxygen consumption is strongly inhibited by fumarate.

The addition of fumarate to the suspension of *P. freudenreichii* oxidizing lactate was shown to result in a partial oxidation of cytochromes b and a_2 (de Vries et al., 1977). In the presence of 2H-heptyl-4-hydroxyquinoline-N-oxide, an inhibitor of cytochrome b, lactate dehydrogenase activity was reduced by 85% if fumarate served as an electron acceptor, but only by 25% in the presence of methylene blue (in the latter case H_2 is transferred to the dye without participation of cytochrome b). Keeping in mind the evidence presented above, the propionic acid fermentation can be viewed as a harmonious combination of the soluble and membrane-bound redox enzymes. The involvement of cytochrome b in anaerobic electron transport to fumarate confirms the suggestion (Bauchop and Elsden, 1960) that there is an oxidative phosphorylation site between fumarate and succinate in propionibacteria.

3.1.2   The enzymes

Reaction 3.1, the key reaction of propionic acid fermentation, is catalyzed by pyruvate carboxytransphosphorylase, a unique biotin-dependent transcarboxylase (see below). There are other reactions of carboxyl group transfer catalyzed by phosphoenolpyruvate (PEP) carboxytransphosphorylase and phosphoenolpyruvate carboxykinase, but these (i) do not require biotin and (ii) use CO_2 as the source of carboxyl groups. The actual species involved may be HCO_3^- (or H_2CO_3) rather than free CO_2 (Cooper et al., 1968), since free CO_2 is not evolved in the PEP carboxytransphosphorylase reaction (Swick and Wood, 1960). Propionic acid bacteria are able to decarboxylate succinate, producing CO_2 in a biotin-dependent reaction (Delwiche,1948; Lichstein, 1958). If succinate is accumulated as the end product, then the cycle (see Fig. 3.1) is broken, and oxaloacetic acid is not supplied by reaction 3.1, but is formed primarily by CO_2 fixation onto PEP catalyzed by PEP carboxytransphosphorylase (PEP-CTP).

PEP carboxytransphosphorylase. The enzyme catalyzes the first reaction in the CO_2-fixation sequence, the major CO_2-fixing mechanism in propionibacteria:

$$PEP + CO_2 + P_i \xleftrightarrow{\quad PEP-CTP \quad} \text{oxaloacetate} + PP_i \qquad\qquad (3.10)$$

It can be seen that the reaction requires an energy-rich compound, phosphoenolpyruvate, and inorganic orthophosphate (P_i), that it is reversible and stimulated when pyrophosphatase is added (Siu and Wood, 1962). The rate of the forward reaction is seven times higher than the reverse reaction. In the absence of CO_2 the enzyme catalyzes an irreversible conversion of PEP and inorganic phosphate to pyruvate and PP_i (Lochmüller et al., 1966):

$$PEP + P_i \longrightarrow \text{pyruvate} + PP_i \qquad\qquad (3.11)$$

It is assumed that both the reactions are catalyzed by the complex of PEP-P_i-enzyme. CO_2 competes for the complex, driving the reaction towards oxaloacetate and thus decreasing the rate of pyruvate production. Reaction 3.11 is irreversible under experimental conditions, but reaction 3.10 is reversible and interesting in that PP_i can be used to form PEP from pyruvate (Davis and Wood, 1966). Therefore, the PP_i derived from ATP can be reutilized, thus acting as a control mechanism for PEP preservation. And since PP_i strongly inhibits the PEP carboxytransphosphorylase reaction, PEP can be diverted to the Krebs cycle (Frings and Schlegel, 1970).

Under conditions unfavorable for transcarboxylation, CO_2 fixation onto PEP becomes vitally important for bacteria, but the availability of PEP can limit the production of C_4-compounds. Nevertheless, for this situation Nature supplied propionibacteria with the enzyme pyruvate-phosphate dikinase (Evans and Wood, 1968). The enzyme is induced by growth on lactate, it generates pyrophosphate and catalyzes the conversion of pyruvate to phosphoenolpyruvate *in vivo*.

The reaction of CO_2 fixation onto phosphoenolpyruvic acid by PEP carboxytransphosphorylase is considered (O'Brien and Wood, 1974) as a control mechanism of propionic acid fermentation. They observed a conversion of the enzymatically active tetrameric form of PEP carboxytransphosphorylase isolated from *P. shermanii* into a less active dimeric form induced by oxalate, malate and fumarate. Therefore, the loss of activity by enzyme dissociation, accompanied by increased proteolysis, is an effective means of controlling the level of intermediates in propionic acid fermentation. Differential abilities of propionibacteria to fix CO_2 could be associated (Wood and Leaver, 1953) with their abilities to carry out the reaction $CO_2 \rightarrow C_1$ and to form sulfhydryl complexes with C_1.

Pyruvate-phosphate dikinase catalyzes the interconversions of pyruvate, ATP and P_i, on one side, and PEP, AMP and PP_i, on the other:

$$\text{Enzyme} + \text{ATP} \rightleftharpoons \text{enzyme-}PP_i + \text{AMP} \tag{3.12}$$

$$\text{Enzyme-}PP_i + P_i \rightleftharpoons \text{enzyme-}P + PP_i \tag{3.13}$$

$$\text{Enzyme-}P_i + \text{pyruvate} \rightleftharpoons \text{enzyme} + \text{PEP} \tag{3.14}$$

$$\text{Pyruvate} + \text{ATP} + P_i \rightleftharpoons \text{PEP} + \text{AMP} + PP_i \tag{3.15}$$

A phosphorylated form of the enzyme was isolated (Milner and Wood, 1972) after an incubation of the enzyme with ^{32}P-PEP. Evans and Wood (1968) isolated a pyrophosphorylated form of the enzyme and demonstrated experimentally a 'ping-pong' mechanism of the dikinase reaction. It has been suggested that various dikinases play an important role in gluconeogenesis (Slack and Hatch, 1967; Reeves et al., 1968), since they are found in parasitic amebae and *Bacteroides symbiosis*, which lack the ubiquitous pyruvate kinase (Reeves et al., 1968). The dikinase reaction is shifted to the right by the utilization of PEP for biosynthesis and by the hydrolysis of PP_i formed in the reaction by pyrophosphatases.

Pyruvate carboxytransphosphorylase (PCTP). The enzyme, frequently referred to as the **transcarboxylase**, catalyzes the carboxyl transfer reaction 3.1 without an intermediary of CO_2 or any significant expenditure of energy:

$$\text{Methylmalonyl-CoA} + \text{pyruvate} \rightleftharpoons \text{propionyl-CoA} + \text{oxaloacetate}$$

Pyruvate carboxytransphosphorylase contains about 1.5 μg of biotin per mg of protein. Under low ionic strength and alkaline conditions the enzyme dissociates spontaneously into the inactive subunits. There are two species of subunits, one contains biotin, and the other contains metals, Co and Zn (Wood et al., 1963; Northrop and Wood, 1969; Gerwin et al., 1969). The transcarboxylase activity with succinyl-CoA is 10-15 times lower than with methylmalonyl-CoA. Specificity to carboxyl donors is fairly broad, with the K_M values for propionyl-CoA of 1.0 mM; for acetyl-CoA, 0.5 mM; for butyryl-CoA, 0.1 mM; and for acetoacetyl-CoA, 0.025 mM. The broad specificity to the CoA esters enables the carboxyl group of oxaloacetate to enter the citric acid cycle or to be used in the synthesis of fatty acids.

The transfer of carboxyl groups by transcarboxylation is advantageous for cell economy, since in this way a relatively slow propionylcarboxylase reaction is bypassed and wasteful use of energy for CO_2 activation by ATP is avoided. It is possible that a similar carboxyl group transfer takes place in the synthesis of fatty acids, since in this case no formation and activation of CO_2 is evident, which ensures favorable kinetics of the process. The transcarboxylase has a broad pH optimum of 5.5-7.8 and is not sensitive to SH-group reagents. The enzyme is stereospecific to the isomer (a) of methylmalonyl-CoA, which is produced enzymatically by a racemase.

Malate dehydrogenase. The enzyme catalyzes the reduction of oxaloacetate to malate:

$$\text{Oxaloacetate} + \text{NADH} + H^+ \rightleftharpoons \text{malate} + \text{NAD}^+ \tag{3.16}$$

Fumarase. A "powerful" fumarase is found in the cells of *P. shermanii* (Krebs and Egglestone, 1941). The enzyme catalyzes interconversions between fumarate and malate:

$$\text{Malate} \underset{+H_2O}{\overset{-H_2O}{\rightleftharpoons}} \text{fumarate} \tag{3.17}$$

The fumarase from *P. pentosaceum* is inhibited by thiol reagents and mercurials (Ayres et al., 1962). Enzymatic activity is nearly constant in the range of pH 5-7.

Fumarate reductase (succinate dehydrogenase). The enzyme catalyzes the reduction of fumarate to succinate. With various dyes used as hydrogen acceptors, the enzyme has optimum activity between pH 7.4 and 7.8. The K_M of the enzyme ranges from $2.2 \ 10^{-3}$ to $7 \ 10^{-4}$ M (Lara, 1959).

Pyruvate dehydrogenase complex. Allen et al.(1964) and later de Vries et al.(1973) postulated that this complex in *P. shermanii* includes pyruvate dehydrogenase that uses thiamine diphosphate as coenzyme, dihydrolipoyl transacetylase containing lipoic acid, and dihydrolipoyl dehydrogenase containing NAD and FAD. The complex catalyzes the following reaction:

$$\text{Pyruvate} + \text{NAD} + \text{CoA} \rightleftharpoons \text{acetyl-CoA} + \text{NADH} + \text{H}^+ + CO_2 \quad (3.18)$$

Pyruvate dehydrogenase of propionibacteria differs from that of aerobic bacteria in that it does not depend on lipoate and has a similarity with pyruvate cytochrome *b* oxidoreductase. Castberg and Morris (1978) reported the isolation from cells of *P. shermanii* of pyruvate oxidase (reducing 2,6-dichlorophenolindophenol (DCIP)) and pyruvate dehydrogenase system (reduces NAD). The pyruvate oxidase could not use NAD as an electron acceptor, and the NAD-dependent enzyme did not transport electrons to DCIP. Unlike the pyruvate oxidase of *E. coli*, the enzyme from *P. shermanii* was not activated by phosphatydylcholine in the presence of SDS, and the presence of thiamine diphosphate and Mg^{2+} was not required for the activity of purified preparations.

Castberg and Morris (1978) concluded that in *P. shermanii* growing anaerobically on lactate the classical pyruvate dehydrogenase complex is not expressed. They suggested that under anaerobic conditions the pyruvate oxidase could participate in electron transport from pyruvate to fumarate via cytochrome *b*.

Methylmalonyl-CoA mutase (isomerase). Methylmalonyl-CoA mutase catalyzes the interconversion of succinyl-CoA and methylmalonyl-CoA(b), the reaction (reaction 4) that plays an important role in the transformation of pyruvate to propionate in propionibacteria. The isomerization reaction consists in splitting the C_1-C_2 bond and an intramolecular transfer of the whole thioester group to C_3 with a simultaneous transfer of H in the opposite direction (from C_3 to C_2):

$$\overset{4}{\underset{3 \quad 2 \quad 1}{CH_3-\underset{|}{\overset{COOH}{CH}}-CO-CoA}} \rightleftharpoons \underset{4 \ 2 \quad 3 \quad 1}{HOOC-CH_2-CH_2-CO-CoA}$$

An AdoCbl-dependent methylmalonyl-CoA mutase was isolated from the cells of *P. shermanii* and purified to homogeneity (Francalani et al., 1986). Its molecular weight was reported at 124,000 (Zagalak et al., 1974), which is close to the molecular weight of the enzyme from mammalian sources (165,000). The enzyme is composed of two non-identical subunits with molecular weights of 79,000 and 67,000. The enzyme does not require K^+ or NH_4^+. In mutases of animal sources the cobamide coenzyme is tightly bound to the apoenzyme, but in bacterial mutases it is easily separated. The apoenzyme from animal tissues only binds coenzyme B_{12} or benzimidazole coenzymes; the bacterial mutase is less specific. The apoenzyme of *P. shermanii* is activated by vitamin B_{12}, oxycobalamin or even the anucleotidic factor B (Overath et al., 1962). Deaminated cobalamins are not, however, active. Thus, modifications in the nucleoside part of the coenzyme result in a complete loss of enzymatic activity, but some variations in the nucleotide part are permitted.

Methylmalonyl-CoA racemase. The enzyme catalyzes the transformation of the isomer (b) of methylmalonyl-CoA to the isomer (a), which can be used at subsequent steps of propionic acid fermentation:

$$\text{Methylmalonyl-CoA(b)} \xrightarrow{\text{racemase}} \text{methylmalonyl-CoA(a)} \qquad (3.19)$$

The mechanism of racemization consists in the α-hydrogen atom being ionized and split off. Racemase activity is difficult to separate from transcarboxylase, propionyl-CoA transferase and methylmalonyl-CoA mutase (Allen et al., 1963). The racemase was isolated and purified from *P. shermanii* (Allen et al., 1964). The molecular weight of the enzyme is 27,000-29,000. The enzyme is unusually resistant to heat treatment: 50% of the activity is retained upon heating for 5 min at 100°C.

CoA transferase. CoA transferase catalyzes a reversible transfer of coenzyme A from propionyl-CoA or acetyl-CoA to succinate, resulting in the formation of propionic acid. The reaction can be described as follows:

$$\text{Propionyl-CoA} + \text{succinate} \rightleftharpoons \text{succinyl-CoA} + \text{propionate} \qquad (3.20)$$

The transferase purified from *P. shermanii* had a much higher activity than those from *P. pentosaceum* or *Micrococcus lactilicus* (Allen et al., 1964). The pH optimum ranged from 6.5 to 7.8; the apparent K_M for succinyl-CoA in the transfer to acetate was $1.3 \cdot 10^{-4}$ M and for the transfer to propionate it was $6.8 \cdot 10^{-5}$ M. The enzyme is very stable as evidenced by a very small loss of activity on storage at $-10°C$ in 60% ammonium sulfate for two years.

Acetyl kinase. In contrast with the acetyl kinase of *E. coli*, the enzyme of *P. shermanii* is equally active with propionate and acetate, but in the fermentation it most likely catalyzes the formation of acetate (Pawelkiewicz and Legocki, 1963). The K_M for propionate and acetate is 0.1 and 0.14 mM, respectively.

The exact mechanism of the formation of acetate by propionic acid bacteria is not known. Acetylkinase catalyzes the following reactions:

$$\text{Acetyl-P} + \text{ADP} \rightleftharpoons \text{acetate} + \text{ATP} \tag{3.21}$$

$$\text{Propionate} + \text{ATP} \rightleftharpoons \text{propionyl-P} + \text{ADP} \tag{3.22}$$

The optimal temperature for this kinase is $50°C$ under standard conditions, two maxima of the enzymatic activity are observed at pH 6.7 and 8.1. Mg^{2+} is essential for activity, but N-ethylmaleimide and iodoacetate at concentrations of 10^{-2} M are not inhibitory.

Lactate dehydrogenase. *P. pentosaceum* contains the enzymes capable of oxidizing lactate in the presence of fumarate (Barker and Lipmann, 1944). Lactate dehydrogenase was isolated and purified from propionibacteria (Serzedello et al., 1969). Molinari and Lara (1960) found that the enzyme (concentrated 15-fold) had an infinite affinity for methylene blue, DCIP and 1,2-naphthoquinone-4-sulphonate. The optimum pH for lactate oxidation was found to be 7.7 at $30°C$. Purified preparations were thermolabile, stimulated by NH_4^+ and Mg^{2+}, and inhibited by pyruvate, oxalate, thiol reagents, quinacrine, chloroquine, quinine, dicoumarol, vitamin K_1, pentachlorophenol, thyroxine and hydrazine. FAD was the only flavin found in the enzyme. *P. shermanii*, *P. pentosaceum* and *P. technicum* contain lactate dehydrogenases, which are much more active with L-lactate than with D-lactate.

3.1.3   Energy yield

The uniqueness of propionic acid fermentation is due to the participation of PEP carboxytransphosphorylase, the enzyme not found in the other organisms that synthesize propionate. Due to the presence of this enzyme the propionic acid fermentation functions as a cyclic process (for the significance of cycling, see above). Another peculiarity of this fermentation is related to the way propionate is formed, which is coupled with the reduction of fumarate to succinate and the oxidation of pyruvate to acetate and CO_2. The electron transport accompanying these reactions is coupled with oxidative phosphorylation and ATP synthesis. The third unique characteristic of this fermentation is a large ATP yield, superior to the other known fermentations.

Three moles of ATP are derived from 1.5 moles of glucose in the course of glycolysis. Then one triose (pyruvate) is oxidized to CO_2 and acetyl-P, and acetate and ATP are formed from the latter in the presence of ADP and acetyl kinase. Two other molecules of pyruvate enter the cycle. Cytochrome *b* is involved in the anaerobic transport of two electrons from NADH to fumarate (Cox et al., 1970), and 2 moles of ATP are formed by oxidative phosphorylation (Van Gent-Ruijters et al., 1975). If two electrons are transported from lactate or glycerol-3-phosphate to fumarate, the formation of 1 mole of ATP is postulated.

Apart from the fermentation, other processes are known, which may contribute to the apparent high ATP yields in propionibacteria. For instance, translocation of a metabolite, if coupled with transmembrane proton movements, can significantly augment the protonmotive force that drives the ATP synthesis, e.g., lactate efflux in a symport with protons in lactic acid bacteria.

But a more important mechanism for some anaerobic species is the generation of the electrochemical potential of Na^+ by membrane-bound decarboxylases, such as methylmalonyl CoA-decarboxylase (Dimroth, 1988). In these biotin-dependent enzymes (Dose, 1989) decarboxylation is coupled with the translocation of sodium ions out of the cell, so they act as primary electrogenic sodium pumps (Dimroth, 1988). The electrochemical sodium gradients generated by these decarboxylases can drive directly the ATP synthesis (Fig. 3.2A) via the Na^+-dependent ATP synthase, first described in a marine bacterium, *Propionigenium modestum* (Laubinger and Dimroth, 1987). Alternatively, the Na^+ gradient can be converted by the ubiquitous Na^+-H^+-exchanger into the protonmotive force and utilized by the H+-dependent ATP synthase, as shown in Fig. 3.2B.

The mechanisms discussed above may also operate in the propionibacteria, which are capable of decarboxylating methylmalonyl-CoA and are highly tolerant to NaCl. Theoretically, the decarboxylation of 2 moles of methylmalonyl-CoA will yield 2/3 mole of ATP. Therefore, 1.5 moles of glucose can yield as much as 6 moles of ATP, which is consistent with the ATP yields calculated on the basis of the molar Y_{ATP} and Y_S coefficients. In *P. freudenreichii* the Y_{ATP} coefficient varies in the range of 15.5-16.9 g dry weight per mol ATP under anaerobic conditions (De Vries et al., 1973; Pritchard et al., 1977).

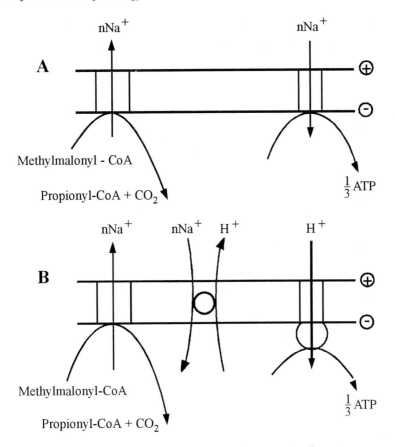

Figure 3.2. Coupling of methylmalonyl-CoA decarboxylation and generation of the electro-chemical potential of sodium ions across the cytoplasmic membrane. The sodium gradient may drive ATP synthesis according to two possible mechanisms (A and B) discussed in the text. Reproduced from Dimroth (1988), with permission.

A direct determination of the ATP content in *P. shermanii* at different growth phases (Gaitan and Vorobjeva, 1981) was carried out using the luciferin-luciferase method (Strehler and McElroy, 1957) with immobilized luciferase (Brovko et al., 1978). In cells growing on glucose the ATP level increased continuously up to 5 nmol per mg dry weight (Fig. 3.3). The ATP level dropped toward the end of the log-phase, but remained above 4 nmol/mg. The decrease was most prominent in the mid-log phase; in the stationary phase the ATP level stabilized. The decline in ATP levels was accompanied by the accumulation of polyphosphates (Gaitan and Vorobjeva, 1981).

The ATP level in *P. shermanii* depended on the substrate used to grow the cells, but was independent of the specific growth rate. In cells growing on lactate the ATP level was lower, reaching a maximum of 3 nmol per mg dry weight (Fig. 3.4). In addition, the ATP level remained essentially unchanged when the growth was inhibited by excess lactate (Fig. 3.5).

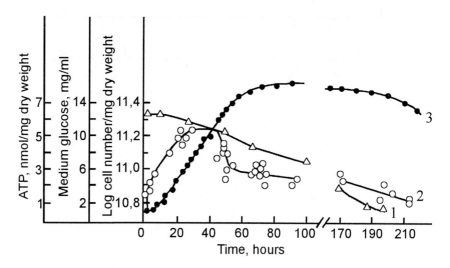

Figure 3.3. Changes in cellular ATP content of *P. shermanii* growing on glucose. Curve 1 - cellular ATP content; 2 - medium glucose concentration; 3 - growth curve. From Gaitan and Vorobjeva (1981).

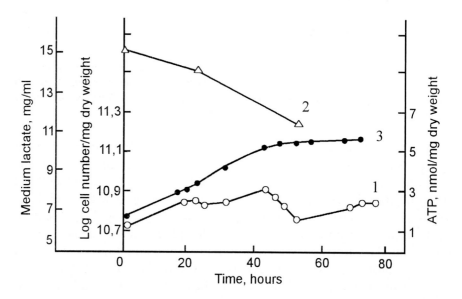

Figure 3.4. Cellular ATP content of *P. shermanii* growing at optimal lactate concentrations. Curve 1 - cellular ATP content; 2 - medium lactate concentration; 3 - growth curve. From Gaitan and Vorobjeva (1981).

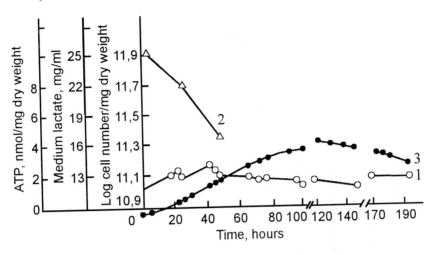

Figure 3.5. Cellular ATP content of *P. shermanii* during growth inhibition by excess lactate. Curve 1 - cellular ATP content; 2 - medium lactate concentration; 3 - growth curve. From Gaitan and Vorobjeva (1981).

Under conditions of the growth arrest caused by nitrogen starvation, cellular ATP was maintained at a constant but low level; fermentation continued, supplying the energy for endogenous metabolism. But starving cultures required higher ATP levels to start growing; the cells could not divide until a certain threshold was reached. In lactate cultures that resumed growing the increase in the ATP pool was accompanied by the utilization of the alkaline- and salt-soluble fractions of polyphosphates, suggesting that polyphosphates are involved in maintaining cellular ATP levels and supplying energy for other processes (Gaitan and Vorobjeva, 1981). The nearly constant ATP pool in cells growing with excess lactate or in non-growing cells (deprived of nitrogen) indicates the existence of a strict control of the ATP level in propionibacteria.

3.1.4 Fermentable substrates and the ratio of products

Typical for the propionic acid fermentation is the formation of identical fermentation products from C_6-, C_5-, C_4- and C_3-compounds. The ratio of products may differ, depending to a large extent on the degree of oxidation of the utilized carbon source. The ratio of propionic to acetic acid in glycerol medium was found to be 2:1, in lactate medium it was 1:1.5, and in pyruvate medium it was 1:2 (Vorobjeva, 1958a). Utilization of pyruvate resulted in the production of acetic acid with a constant rate (Vorobjeva, 1958b). The yield of propionic acid was twofold lower, and after four days it was almost unchanged (Fig. 3.6). No equimolarity between the acetate and CO_2

formation was observed, which may be due to the active CO_2 fixation into organic compounds, producing C_4-compounds.

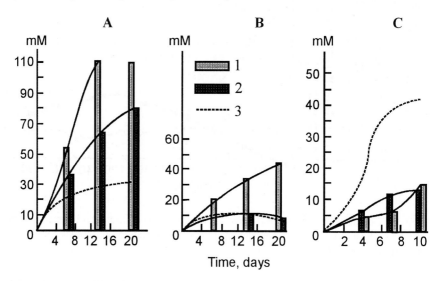

Figure 3.6. Growth curve of *P. shermanii* in synthetic media. A: with sodium lactate; B: with sodium pyruvate; C: on glycerol supplemented with 0.2% aspartic acid. Curve 1 - acetic acid; 2 - propionic acid; 3 - cellular carbon. From Vorobjeva (1958a).

The propionic acid fermentation has a different and 'reverse' character in synthetic media with glycerol as the sole carbon source. Under strictly anaerobic conditions glycerol is fermented in the presence of fumarate as an electron acceptor (Vorobjeva, 1958b). Obviously, the anaerobic metabolism of this substrate requires its oxidation approximately to the level of carbohydrates. This explains why the first phase of the fermentation of glycerol is accompanied by the formation of a more reduced product, propionic acid, rather than oxidized products (as in the case of lactate and pyruvate) (Fig. 3.6). The degree of reduction of the carbon source has a significant effect upon the ratio of fermentation products (which changes during bacterial growth) and upon the yield of biomass. With lactate and pyruvate only 5 and 7%, respectively, of the fermented substrate was found in the cells. With glycerol 30% of the fermented carbon source was retained in the cell (Vorobjeva, 1958a, b). Therefore, glycerol is the most 'profitable' substrate for the biomass and propionic acid production by propionibacteria.

Evidently, C_3-compounds are utilized via the dikinase reaction (at least with lactate and pyruvate as substrates), gluconeogenetic reactions, hexose monophosphate and diphosphate pathways. It is suggested that glycerol degradation proceeds through the formation of glycerol-3-phosphate and dihydroxyacetone (Stjernholm and Wood, 1963). In cells growing on glycerol catalytic quantities of oxaloacetate are required (Wood, 1961). Fumaric acid and amino acids, present in yeast extract, also initiate growth by acting as electron acceptors (Vorobjeva, 1958a).

Lactate can be oxidized directly to pyruvate by FAD-dependent lactate dehydrogenases, and pyruvate can be further metabolized to acetate and CO_2. However, propionate is formed as a result of the reduction of lactate or pyruvate via the methylmalonyl-CoA pathway.

P. pentosaceum is one of the rare microorganisms that are able to grow anaerobically in media containing C_4-alcohols, e.g. erythritol. Fermentation of erythritol by this microorganism results in the formation of erythritol-phosphate, erythrulose-phosphate, glycerol-3-phosphate, phosphoglyceric acid, ribose-5-phosphate, xylulose-5-phosphate, glucose-6-phosphate, fructose-6-phosphate, mannose-6-phosphate, fructose-1,6-diphosphate and sedoheptu-lose-7-phosphate (Wawskiewicz and Barker, 1968). The authors explain the presence of formaldehyde in this fermentation by the splitting of erythritol phosphate between the C_3 and C_4 atoms. They suggest that the degradation of erythritol provides phosphorylated C_3-, C_5-, C_6- and C_7-compounds to the cell.

Quantitatively, the product ratios are similar when glycerol, erythritol, adonitol or mannitol are fermented (Wood and Leaver, 1953). The same type of radioactive labeling (by CO_2) is found in products arising from the fermentation of C_3- and C_6-substrates. These data show that there are no separate mechanisms for C_4- and C_5-compounds.

Increasing the proton concentration in the medium causes a change in the ratio of the main end products: the formation of acetic acid is increased ($0.77 \cdot 10^{-2}$ M at pH 6.85 and $1.33 \cdot 10^{-2}$ at pH 4.95) and of propionic acid is appreciably decreased (from $4.56 \cdot 10^{-2}$ M to $2.79 \cdot 10^{-2}$ M) (Ibragimova et al., 1971). Lowering the pH results in a reduced dissociation of organic acids; undissociated molecules are more toxic to bacteria than the ionized species, and undissociated propionic acid is more toxic than acetic acid. The authors (Ibragimova et al., 1971) suggest that unfavorable effects of a low pH on the cells are counteracted by shifting the fermentation to a lower propionate production. Under the influence of high lactate concentrations (4.6%) and hydrogen ions (pH <4.95) fermentation of the substrate increases without a corresponding synthesis of biomass (Ibragimova and Sacharova, 1972), i.e. an uncoupling of energetic and constructive processes is observed.

The formation of reduced products increases in the presence of reducing agents (for instance, H_2) in the medium. As an electron transferring reducing intermediate was used anthraquinone-2,6-disulfonic acid and cobaltous sepalchate (Emde and Shink, 1990a, b). Propionibacteria were grown on glucose in a fixed potential three-electrode system. Under these conditions the yield of propionic acid was increased up to 90-100% (versus 73% in the control). It is interesting that the electrode system had no significant effect on the fermentation process if the bacteria were grown on lactate.

As mentioned above, propionic, acetic, succinic acids and CO_2 are the main, but not the sole fermentation products in propionic acid bacteria. Propionibacteria contain enzymatic systems responsible for the formation of formate, with variable activity levels in different species. *P. shermanii* and *P. arabinosum* produced up to 0.077 and 0.160 μM formic acid, respectively, in glucose media (Mashur et al., 1971). In *P. jensenii* the formation of formic acid was observed in neutral or weakly alkaline lactate media, but the quantity of formate decreased towards the end of the fermentation. Addition of sodium formate to lactate stimulates the growth of propionibacteria (Vorobjeva, 1958a). A small production of formate was found in glucose fermentations of *P. shermanii*, *P. pentosaceum*, *P. rubrum* and *P. petersonii* (Vorobjeva, 1972).

Formate is the usual end product in polyalcohol fermentation by propionibacteria (Wawskiewich, 1968). Many propionibacteria produce significant amounts of acetoin and diacetyl (Antila, 1956/1957; Lee et al., 1969, 1970). The yield of diacetyl at 21°C is higher than at 32°C or 37°C (Lee et al., 1969, 1970), the pH-optimum for its formation is pH 4.0-4.5. Diacetyl is reduced to acetoin and 2,3-butylenediol. Addition of citrate to the milk increases the yield of diacetyl and delays its conversion to reduced products. In small quantities acetaldehyde is also formed (Keenan and Bills, 1968). It is suggested (Antila, 1956/1957) that these products are not formed by fermentation, but by the condensation of two pyruvate molecules, forming L-acetolactic acid. Propionibacteria are able to decarboxylate L-acetolactic acid, which may explain the observed high rates of CO_2 production.

Some species of propionibacteria produce lactate from the fermented glucose. The quantity of lactate varies as a function of pH and Eh values, medium composition, conditions of aeration and nature of the strain (Ichikawa, 1955; Foschino et al., 1988). Different amounts of lactic acid are produced by different species of propionibacteria fermenting L-xylose and L-arabinose (Wood, 1961).

3.1.5   The Pasteur effect

In propionic acid bacteria the Pasteur effect, although less prominent than in the yeast, was found for the first time in *P. pentosaceum* in 1939 (Chaix and Fromageot). It is known that in yeasts growing under anaerobic conditions the effect is manifested by consuming more substrate to synthesize a unit of biomass than under aerobic conditions. In yeasts, the mechanism of the Pasteur effect may be connected with changes in the enzyme activities of the hexose monophosphate pathway or with an altered state of the cell membrane induced by the transition from aerobic to anaerobic metabolism (Doelle, 1981), although there is no unanimity on this question (Lagukas, 1987). The Pasteur effect in propionibacteria is connected with the synthesis of catabolic enzymes. Whereas fermentation in yeasts is considered (Berry, 1985) a means of survival in unfavorable environments (lack of oxygen), it is a normal way of life for propionic acid bacteria, although a transition to aerobic metabolism is also possible.

Basically, oxygen concentrations up to 5000 ppm (microaerophilic conditions) have no effect on the growth rates of three strains: *P. acidipropionici (pentosaceum)*, *P. shermanii* and *P. thoenii*. The same trend continued up to 20,000 ppm O_2 (atmospheric conditions) with *P. acidipropionici* and *P. thoenii*, while the growth of *P. shermanii* was much slower and the stationary phase was delayed (Canzy et al., 1993). When a culture of *P. pentosaceum* growing under microaerophilic conditions ($1 \mu M$ O_2) was transferred abruptly to anaerobic conditions, its growth almost stopped, but pyruvate was accumulated and acetate and propionate began to be formed. In several hours a normal course of the fermentation, growth rate and lactate consumption was restored (van Gent-Ruijters et al., 1976). It was concluded that the delay in fermentation was due to the repression by oxygen of the synthesis of cytochrome *b*, fumarate reductase and other enzymes of the complex participating in fermentation. Therefore, the Pasteur effect in propionibacteria is regulated at the level of synthesis of catabolic enzymes and not at the level of their activity.

Following an abrupt transition of *P. pentosaceum* from anaerobic to aerobic metabolism ($10 \mu M$ O_2), the growth rate increases at first, but the formation of acetate from lactate and glucose is slowed down, propionate formation is completely repressed and pyruvate is accumulated in the medium (van Gent-Ruijters et al., 1976; Schwartz et al., 1976). A decrease in the activities of the citric acid cycle enzymes: malate dehydrogenase, fumarase and NADH oxidase, lactate oxidase, NADH-dependent fumarate reductase, lactate-dependent nitrate reductase is observed upon the transition from anaerobic to aerobic ($10 \mu M$ O_2) metabolism (van Gent-Ruijters, 1975).

Placed under strictly aerobic conditions (on the surface of a plate), *P. shermanii* does not grow at all. It was found that the lack of growth is explained by the repression of cytochromes *b*, *d* and *a*, which is due to the repression by oxygen of the synthesis of ALA synthetase and ALA dehydratase (Menon, Shemin, 1967).

In liquid medium, *P. shermanii* and *P. freudenreichii* grew more slowly under aeration than *P. pentosaceum* or *P. rubrum* (de Vries et al., 1972), and the extent of inhibition of the cytochrome synthesis was higher in strains more sensitive to oxygen. A complete inhibition of the cytochrome synthesis by high concentrations of oxygen will block both the oxidative phosphorylation and growth.

However, even an anaerobic *P. shermanii* can be adapted to aerobic existence by a long period of chemostat cultivation. In this case a certain level of oxygen in the medium was found to be very important (Pritchard et al., 1977). Increasing the partial pressure of oxygen (pO_2) from 0 to 42 mm Hg resulted in a reduction of the biomass yield and activities of membrane-bound enzymes: NADH oxidase, succinate dehydrogenase and the two lactate dehydrogenases (at pO_2 of 20-42 mm Hg). The content of cytochrome *b* was reduced by increasing the pO_2 from 10 to 42 mm Hg as compared with anaerobic conditions, and superoxide dismutase (SOD) activity was raised by increasing the pO_2 from 0 to 85 mm Hg. The content of cytochromes b_{560}, a_{600} and d_{628} was at a minimum between 20 and 42 mm Hg.

A striking feature of that study (Pritchard et al., 1977) was that increasing the pO_2 further up to 330 mm Hg brought about an increase in the biomass yield compared with anaerobic conditions, in the activities of membrane bound enzymes, in the content of cytochromes and the ratio of cytochromes $a + d$ to cytochrome *b*. The level of the CO-binding cytochrome was also increased, while the ratio b_{560}/total cytochromes declined from 71% (anaerobic conditions) to 48%. The pO_2 of 330 mm Hg was the highest oxygen concentration that allowed establishing a stationary state. More than 70% of the metabolized lactate was oxidized to acetate and CO_2, but the rate of propionate formation was very low. The high molar growth yields observed under these conditions suggested that the oxidation of lactate and NADH via the cytochrome electron-transport system was coupled to the ATP synthesis (Pritchard et al., 1977). Therefore, in addition to substrate level phosphorylation, the ATP synthesis in propionibacteria can proceed by oxidative phosphorylation as well.

When the O_2 concentration exceeded the utilization capacity of the culture so that free oxygen accumulated in the medium (when the pO_2 was 520 mm Hg), the growth was suppressed, NADH oxidase and L-lactate dehydrogenase were inactivated, SOD activity was reduced, but D-lactate dehydrogenase and succinate dehydrogenase activities remained unchanged.

How can one explain the described behavior of *P. shermanii*? It is suggested (Pritchard et al., 1977) that at low partial oxygen pressures (20-42 mm Hg) the bacterium is caught in a trap: the pO_2 is sufficient for the repression of enzymes involved in anaerobic metabolism and for the degradation of redox components adapted to anaerobic existence, but not sufficient for the induction of synthesis of the electron transport enzymes needed for aerobic metabolism. If the pO_2 is in the range of 40-300 mm Hg, then such an induction (adaptation) is possible and the culture is able to conduct an energetically more profitable aerobic style of life.

The limit of aerotolerance of propionic acid bacteria is reached (Shwartz, 1973) when the oxidizing systems: NADH oxidase, L-lactate dehydrogenase and superoxide dismutase are inactivated by the soluble oxygen accumulated in the medium.

3.2   Respiration

Although van Niel in 1928 had noticed the oxidative capacity of propionic acid bacteria, and French researchers (Chaix and Fromageot, 1939; Chaix and Andemand, 1940) quantitatively described it, there were no attempts to grow these bacteria under aerobic conditions before 1958, owing to the well-entrenched view of their strict anaerobiosis. But the old views are usually less binding to the young than to the old, which is why our first steps in science were associated with studies, bold enough for those times, of aerobic metabolism of the 'anaerobic' propionic acid bacteria.

Our observations (Vorobjeva, 1959) of the metabolic consumption of molecular oxygen—even in the stationary phase—by propionic acid bacteria which, furthermore, could be adapted to grow under aerobic conditions, although initially met with skepticism by some experts, were followed by several confirmatory reports, thus superceding the old views about the lifestyles of propionic acid bacteria. By now it is well established that bacterial responses to oxygen can vary widely among the strains—of the 4 species tested: *P. acidipropionici*, *P. thoenii*, *P. jensenii* and *P. freudenreichii* only the former two could grow on the surface of solid medium in the air (Canzi et al., 1993). At the same time, the growth curves of all the strains were almost identical when cultivated in a fermenter with oxygen concentrations ranging from 30 to 20,000 (atmospheric) ppm; however, at the highest concentration the growth of *P. shermanii* was markedly suppressed. Under aerobic conditions all the strains consumed O_2, and at the atmospheric O_2 concentration the ratio of acetic to propionic acid was increased.

3.2.1   Oxidative activity

Tricarboxylic acid (TCA) cycle and glyoxylate shunt. It is known that the oxidative capacity of bacteria is the result of co-ordinated work of the enzymes of glycolysis, TCA cycle, hexose monophosphate shunt, electron transport in the respiratory chain, and flavin respiration. It was shown (Krainova and Bonarceva, 1973) that *P. petersonii* and *P. shermanii*, grown on glucose, contain all the enzymes of the TCA cycle (Fig. 3.7) and glyoxylate shunt. The highest activities of citrate synthetase, isocitrate dehydrogenase, aconitase, succinate thiokinase and fumarase were expressed in *P. shermanii*, and of α-ketoglutarate dehydrogenase, succinate dehydrogenase and malate dehydrogenase in *P. petersonii*. The rate of respiration was suppressed by $5 \cdot 10^{-3}$ M monofluoroacetate, a specific inhibitor of the TCA cycle (fumarase inhibitor), by 21 and 55% in *P. shermanii* and *P. petersonii*, respectively, showing the importance of the TCA cycle for the energy metabolism in propionibacteria.

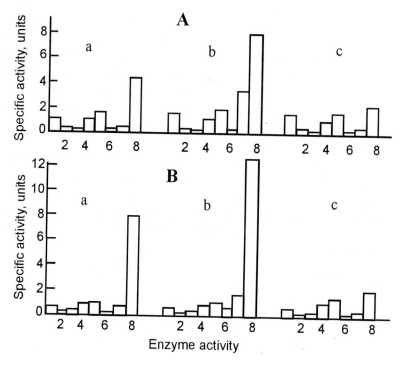

Figure 3.7. Specific activities of the tricarboxylic acid cycle enzymes in *P. shermanii* (A) and *P. petersonii* (B) grown under aerobic conditions (a), stationary conditions (b), or anaerobic conditions (c). 1: isocitrate dehydrogenase; 2: aconitase; 3: α-ketoglutarate dehydrogenase; 4: citrate synthase; 5: succinate thiokinase; 6: succinate dehydrogenase; 7: fumarase; 8: malate dehydrogenase. From Krainova and Bonarceva (1973).

It has been established (Delwiche and Carson, 1953) that propionic acid bacteria are able to oxidize the intermediate products of the TCA cycle. Under anaerobic conditions the TCA cycle is also functional, and its role may not be limited to anabolic processes. In these conditions nitrate and fumarate can act as terminal electron acceptors in propionic acid bacteria. It is well known that the TCA cycle provides microorganisms with precursors for biosynthetic reactions, and plays an essential role in both the catabolic and anabolic metabolism.

Oxidation of different substrates. We found (Vorobjeva, 1959) that glycerol can be used as a sole carbon source by *P. jensenii* only under aerobic conditions. If fumarate was added to minimal medium, then glycerol fermentation proceeded under anaerobic conditions with fumarate acting as an electron acceptor. Propionibacteria can oxidize compounds more reduced than glycerol, namely, alkanes and long-chain primary alcohols (Table 3.1). Oxidation of hydrocarbons is suppressed by the inhibitors of cytochrome oxidases, NaN_3 (10^{-4} M) and KCN (10^{-3} M), respectively, by 88 and 96%, which is similar to the degree of inhibition observed for glucose oxidation by *P. pentosaceum*.

Table 3.1. Respiration rates of propionic acid bacteria on cetyl alcohol and *n*-hexadecane

Organism	Substrate	Time, min			
		30	60	90	120
		Qo_2, μg O_2/ mg dry weight per h			
P. globosum	cetyl alcohol	38.3	43.4	42.0	45.0
	n-hexadecane	18.2	16.0	16.0	20.0
P. pentosaceum	cetyl alcohol	18.6	30.0	26.4	14.0
	n-hexadecane	11.0	13.3	12.8	14.0

From Vorobjeva et al. (1979).

Hydrocarbons, similarly to glucose, are oxidized to CO_2 by propionibacteria (Fig. 3.8). Cetyl alcohol and palmitic acid, found in the incubation medium (Table 3.2), apparently serve as intermediates in the oxidation of hexadecane (Vorobjeva et al., 1979a). Some CO-binding proteins, known as cytochromes P450, are involved in the hydroxylation of alkanes. In this context, the finding of CO-binding pigments in four species of propionic acid bacteria (de Vries et al., 1972) should be noted.

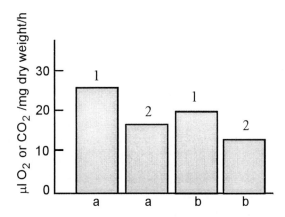

Figure 3.8. Oxygen uptake and CO_2 production from glucose (a) and a mixture of *n*-alkanes (b) by *P. petersonii*. Curve 1 - oxygen uptake; 2 - carbon dioxide. From Vorobjeva et al. (1979a).

Table 3.2. Cetyl alcohol and palmitic acid levels in propionic acid bacteria growing on *n*-hexadecane

Organism	Cetyl alcohol	Palmitic acid
	μg per mg dry weight	
P. raffinosaceum	1.5	–
P. jensenii	7.6	13.0
P. thoenii	6.7	–
P. technicum	2.0	1.0

From Vorobjeva et al. (1979).

Propionibacterial oxygen uptakes with substrates of varying oxidation levels are shown in Table 3.3 (Schwartz, 1973). The highest oxygen uptake was observed with Na-lactate and glycerol; with the other substrates the O_2 uptake was the lower, the more oxidized was the substrate. In our experiments (Bonarceva et al., 1973a) the maximal rate of respiration was found in *P. shermanii* grown under semi-aerobic (stationary) conditions with glucose in rich corn steep-based medium (glucose-CS medium); the rate was slightly lower in anaerobically grown, and 20 times as low in aerobically grown cells. The respiration activity was higher in glucose-CS medium than in semi-synthetic glucose-peptone medium; the younger was the culture, the higher was the respiration rate. Respiration rates in *P. petersonii* were approximately 10 times lower than in *P. shermanii* (in glucose-CS medium). *P. petersonii* grown for 48 h in an anaerostat consumed oxygen more actively than the cultures grown aerobically or semi-aerobically (Bonarceva et al., 1973b). Similar observations were made by Schwartz (1973), who showed that the preincubation of bacterial suspensions under nitrogen activated the O_2 uptake (Table 3.3).

Table 3.3. Oxygen uptake by propionic acid bacteria respiring on different substrates

Substrate	P. shermanii PZ3			P. petersonii 62		
	Fresh cells	Starved in:		Fresh cells	Starved in:	
		air	nitrogen		air	nitrogen
	μmol O_2/g wet weight per min					
None	3.1	0.6	0.8	3.0	0.7	0.4
D-glucose	4.0	2.5	2.6	3.5	1.5	1.4
DL-lactate	9.0	5.9	10.1	10.5	4.3	6.7
Glycerol	5.5	2.4	5.3	3.7	0.9	1.4
Pyruvate	3.7	1.9	1.7	3.1	1.6	1.9
Succinate	3.9	1.4	1.4	2.8	1.4	1.7
Propionate	4.4	1.2	5.7	4.9	0.8	1.5
Acetate	3.4	0.4	1.0	3.2	0.7	0.5

Reproduced from Schwartz (1973), with permission.

The higher respiration rates in anaerobically as compared with aerobically grown cells (Bonarceva et al., 1973b) may be due to the repression of flavin enzymes by oxygen (Table 3.4). Evidently, the flavin-based respiration is the major route of oxygen uptake in such species as *P. shermanii*. A high demand for oxygen is a typical feature of the flavin respiration, as well as low turnover numbers of flavoprotein oxidases.

Table 3.4. Content of flavins in the cells of *P. shermanii* and *P. petersonii* at different conditions of aeration

Flavin	P. shermanii			P. petersonii		
	Anaerobic	Semi-aerobic	Aerobic	Anaerobic	Semi-aerobic	Aerobic
			μg per g			
FAD	151	123	96	48	40	46
FMN	39	52	20	28	16	15
FAD+FMN	190	175	116	76	56	61

From Bonarceva et al. (1973b).

Cultures grown aerobically were more sensitive to inhibitors (e.g., to atebrine) than those grown anaerobically. This finding may be explained by the repression of the glycolytic system under aerobic conditions, when energy generation as well as reoxidation of NADH depend greatly on the activity of flavoproteins. At the same time, the activities of flavin oxidases, NADH oxidase and especially pyruvate oxidase, were lower in aerobically than in anaerobically grown cultures. As shown by de Vries et al. (1972), not only the flavoprotein synthesis, but also the cytochrome synthesis is strongly repressed under aerobic conditions in *P. shermanii* and *P. freudenreichii*. Obviously, this is why these strains grow slowly in the presence of oxygen, in contrast with *P. rubrum* and *P. pentosaceum* (Table 3.5).

Table 3.5. Influence of aeration on growth and cytochrome synthesis by propionic acid bacteria

Organism	Growth conditions	Growth rate (ΔE_{660}/h)	Differential rate of cytochrome synthesis (μmol/g dry weight)
P. freudenreichii	anaerobic	0.050	0.23
	aerobic	0.013	0.01
P. shermanii	anaerobic	0.050	0.22
	aerobic	0.013	0.01
P. rubrum	anaerobic	0.043	0.16
	aerobic	0.050	0.06
P. pentosaceum	anaerobic	0.044	0.25
	aerobic	0.067	0.05

Reproduced from de Vries et al. (1972), with permission.

Following the transition of *P. freudenreichii* from anaerobic to aerobic growth in lactate medium (de Vries et al., 1972), the production of propionate, acetate and succinate was blocked, and an accumulation of pyruvate was observed (Fig. 3.9). These observations showed that under aerobic conditions the bacterium gained energy by oxidative phosphorylation rather than by substrate-level phosphorylation. A mean value of 15 was determined for the Y_O (dry weight/g·atom O_2 consumed) in these experiments. The Y_{ATP} values (dry weight/mol ATP) in different microorganisms vary from 8 to 20 (Forrest and Walker, 1964). From these data the P/O quotient in *P. freudenreichii* was calculated to be between 1 and 2 (de Vries et al., 1972).

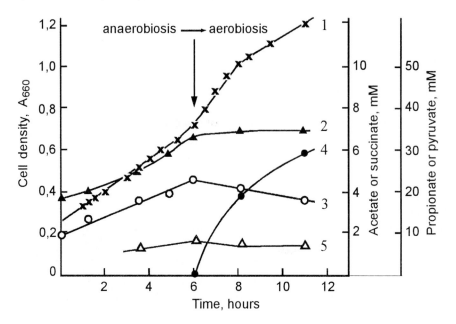

Figure 3.9. The effect of transition from anaerobic to aerobic conditions on growth and lactate metabolism of *P. freudenreichii*. Curve 1 - growth; 2 - acetate; 3 - propionate; 4 - pyruvate; 5 - succinate. Reproduced with permission from W. De Vries et al., *J. Gen. Microbiol.* 71: 515-524, 1972 © Society for General Microbiology.

3.2.2   Actions of inhibitors

One of the methods for studying the composition of a respiratory chain is by using inhibitors, although bacterial respiratory chains differ from the mitochondrial chain by non-uniform actions of the inhibitors and uncouplers. Respiratory chains of many bacteria are insensitive to azide, dinitrophenol, antimycin A, and oligomycin. We found (Vorobjeva et al., 1982) that NaN_3 (1mM) exerts essentially no inhibitory action and KCN (2mM) displays a marked (33%) inhibitory effect on the O_2 uptake by *P. globosum*, but a smaller (12%) effect in the case of *P. petersonii*; in the other bacteria studied the inhibitory effects of KCN and NaN_3 were in the range of 3–12% (Table 3.6). Such effects of azide and cyanide may indicate the absence of a typical cytochrome oxidase or the presence of a cytochrome oxidase with a low sensitivity to KCN.

In another study, cyanide inhibited the oxidation of D- and L-lactate and NADH oxidase activity in membrane particles of *P. shermanii* (Pritchard and Asmundson, 1980). The oxygen uptake insensitive to the inhibition by 100 μM KCN amounted to 30, 15, 80 and 10% of the total oxygen uptake with L-lactate, D-lactate, succinate and NADH as substrates, respectively. Cyanide-resistant respiration is widespread among living organisms. It is attributed to flavoprotein respiration with participation of oxidases resistant to KCN (Beevers, 1961); it is also explained by the presence of a *b*-shunt, i.e., a by-pass to oxygen from cytochrome *b* (see below).

Table 3.6. Effects of respiration inhibitors on oxygen uptake by propionic acid bacteria

Organism	Oxygen uptake, nmol O_2/mg dry weight per min	Inhibition, %	
		KCN, 2mM	NaN₃, 1mM
P. coccoides	131.70	6	8
P. globosum	80.30	33	10
P. shermanii	70.10	9	0
P. pentosaceum	75.80	7	5
P. raffinosaceum	73.00	9	3
P. petersonii	70.40	12	5

From Bonarceva et al. (1973a).

The respiration resistant to KCN, NaN₃ and CO may be related to the functions of oxygenases, enzymes that add molecular oxygen directly onto substrates. In the presence of KCN (2.5-5.0 mM) the rate of oxygen uptake by membrane particles of *P. pentosaceum* was reduced and the total amount of oxygen consumed per mole of substrate (NADH, D-, L-lactate) was increased (Pritchard and Asmundson, 1980). The increase was higher at higher KCN concentrations, and was accompanied by the production of H_2O_2. 30% of the oxidizable D-lactate was oxidized with the concomitant production of H_2O_2. There was no H_2O_2 production in the absence of KCN. Oxidation of NADH and L-lactate was also accompanied by H_2O_2 production, although in smaller amounts. The authors concluded (Pritchard and Asmundson, 1980) that in the presence of KCN above 20 mM the O_2 consumption proceeds, at least partly, via a pathway associated with H_2O_2 production. Apparently, the pathway in question is the electron transport to flavoproteins and FeS-proteins in the dehydrogenase system of the CN-resistant respiration.

Since only 30-50% of the CN-resistant respiration results in H_2O_2 formation, the existence of an alternative system of CN-resistant oxidases and peroxidases was suggested. Peroxidase activity is found in propioni-bacteria (Vorobjeva et al., 1986). Since CN-resistant oxidases have a low affinity to oxygen, it was suggested (Pritchard and Asmundson, 1980) that under microaerophilic conditions they do not play any essential role.

Other oxidase inhibitors, such as CO and NaN_3, only slightly affected the respiratory activity of the cultures (Bonarceva et al., 1973a, b). The data mentioned above show that propionic acid bacteria do not have a typical cytochrome oxidase sensitive to CO and KCN. The peculiarity of the propionibacterial respiration is also expressed (Chaix and Fromageot, 1939) in that it is stimulated by H_2S, similar to chlorella, but unlike the respiration of other microorganisms.

2,4-Dinitrophenol and gramicidin, uncouplers of the oxidative phosphorylation in mitochondria, also suppressed the respiration in propionibacteria. Especially sensitive to the action of uncouplers were cells of *P. petersonii*, in which the respiration was completely blocked by gramicidin at 10 µg/ml, while in *P. shermanii* it was inhibited by 93% at 100 µg/ml gramicidin. Dinitrophenol at a concentration of $1 \cdot 10^{-3}$ M decreased the rate of respiration by 40% in *P. shermanii*, but blocked it completely in *P. petersonii*.

As expected, the inhibitors of flavoproteins (atebrine, amytal and rotenone) suppressed the respiration of propionic acid bacteria. Rotenone ($2 \cdot 10^{-4}$ M) decreased the rate of respiration of *P. shermanii* and *P. petersonii* by 38 and 46%, respectively. Rotenone and atebrine blocked the respiration of *P. shermanii* to a smaller extent than of *P. petersonii*; and cells grown under aerobic conditions were more sensitive to the inhibitor than grown under anaerobic conditions. This observation can be explained by the fact that the glycolytic system is considerably repressed under aerobic conditions, and energy production as well as reoxidation of NADH depend greatly on the activity of flavin-dependent enzymes. Amytal ($1 \cdot 10^{-2}$ M), a specific inhibitor of the initial component of the respiratory chain, NADH-flavoprotein, inhibited the respiration of *P. shermanii* and *P. petersonii* by 42% and 50%, respectively.

These data show, therefore, that the propionibacterial respiration is suppressed by those inhibitors, which are specific to flavoprotein oxidases. Specifically, in *P. shermanii* and *P. petersonii* an NADH-oxidase activity was found (Bonarceva et al., 1973b) that was several times higher in the fraction of membrane fragments than in the cell extract, and was higher in *P. shermanii* than in *P. petersonii*. Dicoumarol ($1 \cdot 10^{-3}$ M) inhibited the NADH-oxidase activity by 65 and 67% in *P. shermanii* and *P. petersonii*, respectively. Cyanide ($1 \cdot 10^{-3}$ M) completely blocked the NADH-oxidase activity in *P. petersonii*, but showed no significant effect in *P. shermanii*; antimycin had no effect on the NADH-oxidase activity.

P. shermanii and *P. petersonii* contain pyruvate oxidase activity related to flavoproteins. The enzyme activity was almost two times higher in cells grown under anaerobic than aerobic conditions. The presence of pyruvate oxidase shows that the propionibacteria are able to oxidize a substrate (pyruvate) directly by oxygen (see above); at the same time, evidently, the flavin-dependent respiration can also function anaerobically using electron acceptors other than oxygen.

3.2.3   Electron transport chain

The membrane-bound electron transport chains (ETC) catalyzes the transfer of reducing equivalents (protons and electrons) from NADH, glycerol-3-phosphate and lactate to oxygen, nitrate or fumarate (Sone, 1972; de Vries et al., 1972, 1977). *P. pentosaceum* contains a constitutive nitrate reductase and can reduce nitrate as a terminal electron acceptor during lactate utilization. Nitrate respiration is linked with the ATP synthesis (van Gent-Ruijters et al., 1975).

In *P. shermanii* the activities of NADH- and NADPH-dehydrogenases were almost 10 times higher, in *P. petersonii*, NADH-dehydrogenase activity was 3.5 times higher and NADPH-dehydrogenase slightly higher in the fraction of membrane fragments than in the cell extract (Bonarceva et al., 1973b). Dicoumarol ($3 \cdot 10^{-4}$ M) inhibited by 60 and 65% the activity of NADH-dehydrogenases and by 82 and 90% the activity of NADPH-dehydrogenases in *P. shermanii* and *P. petersonii*, respectively. This indicates that the dehydrogenases do not bind menadione in the respiratory chain of propionic acid bacteria.

NADH-dehydrogenase purified 30-fold has a molecular mass of 215,000 and an absorption maximum at 408 nm, which indicates the presence of Fe-flavin (van Gent-Ruijters et al., 1975). The best electron acceptor, according to the authors, is ferricyanide. If the specific activity of the enzyme with ferricyanide is set arbitrarily to 100 units, then it amounts to 1.2 units with dichlorophenol indophenol, to 0.6 units with menaquinone, to 0.5 units with cytochrome *c*, and to 0.001 units with oxygen. Both soluble and particulate NADH-dehydrogenases are present.

In addition, *P. shermanii* and *P. petersonii* contain glucose dehydro-genases and succinate dehydrogenases, tightly bound in the respiratory chain (Table 3.7). Atebrine ($1 \cdot 10^{-3}$ M) inhibited the succinate dehydrogenase activity by 50% and 40% in *P. shermanii* and *P. petersonii*, respectively; this, together with the ability of the succinate dehydrogenase to use a typical electron acceptor of flavin-dependent enzymes, 2,6-dichlorophenol indophenol (DCIP), shows that it is a flavoprotein. At the same time, dicoumarol ($1 \cdot 10^{-3}$ M) had no effect on the succinate dehydrogenase activity (Bonarceva et al., 1973b), suggesting that the enzyme is not associated with vitamin K in propionic acid bacteria.

Table 3.7. Specific activities of respiratory chain dehydrogenases in *P. shermanii* and *P. petersonii*

Enzyme	Strain	Particulate fraction	Cell extract
NADH-dehydrogenase	*P. shermanii*	8.50	0.85
	P. petersonii	1.43	0.67
NADPH-dehydrogenase	*P. shermanii*	10.0	1.05
	P. petersonii	2.66	2.38
Succinate dehydrogenase	*P. shermanii*	2.48	1.43
	P. petersonii	0.14	3.10
Glucose dehydrogenase	*P. shermanii*	13.3	0.95
	P. petersonii	2.30	0.81

From Bonarceva et al. (1973a).

In another propionibacterium, *P. arabinosum*, glycerol 1-phosphate dehydrogenase is found in the particulate fraction (Sone, 1972); prosthetic groups of the enzyme include non-heme iron and flavins. DCIP, phenazine methosulfate, and ferricyanide all were good electron acceptors for the enzyme; molecular oxygen was a poor electron acceptor, and H_2O_2, produced during interaction with oxygen, inactivated the enzyme. Amytal, *p*-chloromercuribenzoate (PCMB), dipirydyl, thinoyltrifluoroacetone, dicoumarol and UV-light (360 nm) inactivated the glycerolphosphate dehydrogenase.

It has been suggested that menadiones are components of the electron transport chain (ETC). Indeed, we (Bonarceva et al., 1973b) found naphthoquinones in propionibacteria (Table 3.8). Cox et al. (1970) showed that during the oxidation of NADH, lactate and malate in *E. coli* ubiquinones may act at two sites of the ETC: before and after cytochrome *b*. At the same time, there is evidence for quinone-independent pathways of electron transport.

Table 3.8. Cellular content of quinones in propionic acid bacteria

Organism	Growth conditions	Naphthoquinones, μg/g dry weight
P. shermanii	Anaerobic	43.2
	Semianaerobic	45.6
	Aerobic	50.4
P. petersonii	Anaerobic	42.2
	Semianaerobic	43.2
	Aerobic	43.2

From Bonarceva et al. (1973b).

A possible component of the ETC is non-heme iron that reacts with oxygen at a site close to flavoproteins and thus may serve as an 'alternative oxidase' resistant to cyanide (Allen and Beard, 1965). The fraction of membrane fragments from *P. shermanii* and *P. petersonii* contains non-heme iron (Bonarceva et al., 1973b). In the membrane fraction of *P. petersonii*, a more aerobic species, the content of non-heme iron is higher than in *P. shermanii*. In aerobically grown cells of *P. petersonii* the content of non-heme iron was four times higher than in anaerobically grown cells (Table 3.9). Memranes of *P. shermanii* contain two different cytochrome *b* components: b_{556} and b_{562}.

Table 3.9. Non-heme iron in particulate fractions of *P. shermanii* and *P. petersonii*

Organism	Growth conditions	Non-heme iron, μmol/mg protein
P. shermanii	Aerobic	2.3
	Anaerobic	2.9
P. petersonii	Aerobic	11.3
	Anaerobic	3.7

From Bonarceva et al. (1973b).

Membrane suspensions of *P. pentosaceum*, grown on lactate in the presence of nitrate, reduced cytochrome *b* in the presence of L-lactate and NADH. The reduced cytochrome *b* was oxidized when nitrate or fumarate was added to anaerobic suspensions (van Gent-Ruijters et al., 1976), or when suspensions were bubbled with oxygen (Pritchard and Asmundson, 1980). Reduction of the cytochrome was more rapid when suspensions were oxidized by oxygen in the presence of NADH than L-lactate. The finding that NADH is a more effective electron donor to oxygen than lactate is explained by the higher activity of NADH oxidase compared with lactate oxidase in *P. pentosaceum* (van Gent-Ruijters et al., 1976). In *P. shermanii* KCN (10 mM) inhibited reoxidation of the reduced cytochrome by oxygen, but had no effect on its reoxidation by fumarate (Pritchard and Asmundson, 1980). These observations show that cytochrome *b* may be involved in different pathways of electron transport in propionic acid bacteria.

On the basis of spectroscopic investigations cytochrome *d* (d_{627}) has been considered (Pritchard and Asmundson, 1980) as the best candidate for the role of a terminal oxidase, although small peaks of cytochrome *a* and a CO-binding cytochrome, presumably cytochrome *o*, have also been found in propionic acid bacteria (de Vries et al., 1972). But the relatively high concentration of cyanide needed to inhibit oxidase activities is more typical for cytochrome *d* than for cytochromes *a* or *o* (Arima and Oka, 1965). In contrast with oxygen respiration, fumarate respiration of propionic acid bacteria is insensitive to cyanide (10mM) when NADH, D- or L-lactate is being oxidized. By increasing the concentration of KCN ten times a strong inhibition of fumarate reductase activity was observed, but the inhibition was lower with L-lactate as an electron donor than with the two other substrates. It means that at the higher concentration KCN inhibited a site other than the terminal oxidase, and the site itself may not always be the same with different substrates (Pritchard and Asmundson, 1980).

Fumarate exhibits a strong inhibitory action on oxygen uptake (Sone, 1972). Conversely, NAD-fumarate reductase activity was significantly lowered when cells were grown in the presence of nitrate, and even more so—in the presence of oxygen. Nitrate is a weak inhibitor of fumarate reductase activity. Since nitrate reductase activity is found in *P. pentosaceum* grown anaerobically in the absence of nitrate, nitrate reductase is a constitutive enzyme. The highest nitrate reductase activity was found in the cells of *P. pentosaceum* grown in the presence of nitrate (twice as high as without it), and the lowest was in the cells grown in the presence of a low (10 μM) concentration of oxygen (van Gent-Ruijters et al., 1976); therefore, oxygen acts as a potent repressor of fumarate and nitrate reductases in propionibacteria.

On the basis of these investigations the ETC of propionic acid bacteria can be schematically represented as shown in Fig. 3.10.

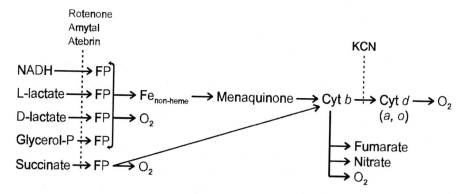

Figure 3.10. Respiratory pathways in propionibacteria. Continuous lines indicate pathways of electron transport, dashed lines show inhibitor sites. FP, flavoprotein dehydrogenases. Electrons can be transported to oxygen directly from cytochrome *b*. Nitrate and fumarate can also serve as terminal acceptors.

Electron transport in the respiratory chain of propionic acid bacteria shown above can be accompanied by ATP synthesis. The following sites of coupling between oxidation and phosphorylation may be tentatively identified by analogy with the mitochondrial ETC: NAD—FP, naphtho(ubi)quinones—cytochromes *b-c*, terminal oxidase—O_2. In most of the bacterial species investigated the P/O quotient appeared to be lower than unity (Gelman et al., 1972).

3.2.4 Oxidative phosphorylation

From the discussions above it follows that up to the 1980s sufficient evidence had accumulated, indicating a potential equipment of propionibacteria for aerobic life. One of the essential steps was the discovery of oxidative phosphorylation in propionic acid bacteria (Bruchatcheva et al., 1975).

In these experiments ^{32}P-labelled phosphate was used to evaluate the activity of oxidative phosphorylation. The method is based on the analysis of radioactive phosphorus incorporated into the organic phase by phosphorylating membrane particles within short periods of time. The incubation mixture had the following composition (mM): K-phosphate buffer pH 7.4, 1.0; MgCl$_2$, 1.0; glucose, 25.0; ADP, 0.1; NADH, 1.0; hexokinase, 43 units; inorganic ^{32}P-phosphate free of pyrophosphate, 720 cpm; membrane particles, 2.5 mg protein; in a total volume of 2.0 ml. Radioactive phosphorus in organic phase (ATP, glucose-6-phosphate) was counted and the amount of phosphate esters was calculated according to the formula:

$$\frac{(\text{cpm } P_{\text{labile}})_{\text{exp}} - (\text{cpm } P)_{\text{contr}}}{\text{cpm } per \text{ } \mu\text{g·atom } P_i \text{ standard}} = \mu\text{g·atom terminal group ATP.} \quad (3.23)$$

Oxygen uptake was measured by following NADH oxidation spectrophotometrically at 340 nm. From these measurements the ratio P/NADH was determined (the ratio of phosphate esterified to the quantity of NADH oxidized). It was shown that the electron transport by the respiratory chain of aerobically grown P. shermanii and P. petersonii was accompanied by oxidative phosphorylation with P/NADH ratios of 0.34 and 0.49, respectively. P. shermanii is more sensitive to the toxic action of oxygen than P. petersonii, so when grown anaerobically the P/NADH ratio was reduced to 0.02 in P. shermanii, but only to 0.26 in P. petersonii, being 17 and 2 times lower, respectively, than in aerobically grown cells.

Oxidative phosphorylation was also demonstrated by an indirect method, by estimating how much substrate was used for the synthesis of one unit of biomass (Y_S). As shown in Table 3.10, the two species display maximal growth yield coefficients (Y_S) under aerobic conditions of growth, therefore, in the presence of oxygen the substrate is utilized more effectively, especially by P. petersonii, which produces 2.5 times more ATP per unit substrate under aerobic conditions. In P. shermanii the difference is smaller, indicating that metabolic changes induced by the transition from anaerobic to aerobic conditions are less profound.

Table 3.10. Glucose utilization and biomass accumulation by *P. shermanii* and *P. petersonii* in different conditions of aeration

Organism	Growth* conditions	Glucose consumed, g	Biomass produced, g/l	Y_S, g dry weight/ mol substrate	Mol ATP per mol substrate consumed
P. shermanii	Aerobic	6.10	1.58	46.7	4.67
	Anaerobic	10.90	2.12	35.0	3.50
P. petersonii	Aerobic	7.60	1.98	47.0	4.70
	Anaerobic	4.12	0.47	20.3	2.03

*Cultures were grown for 24 h.   From Bruchatcheva et al. (1975).

On the basis of similar calculations (Y_S and Y_{ATP}) it has been assumed that anaerobic electron transport to NO_3^- in *P. pentosaceum* is also linked with oxidative phosphorylation (van Gent-Ruijters et al., 1976). The authors concluded that one mole of ATP is formed when two electrons are transferred from lactate or glycerol 1-phosphate to nitrate ($P/NO_3^- = 1$), and two moles of ATP are formed when $2e^-$ are transferred from NADH to nitrate ($P/NO_3^- = 2$).

Oxygen uptake by the whole cells, extracts and membrane fraction of *P. shermanii* is considerably higher, but the P/NADH ratio is lower than in *P. petersonii*. It indicates that in *P. shermanii* only a fraction of the O_2 consumed is used for the respiration-driven ATP synthesis. Another, larger fraction of the O_2 consumption is not linked with ATP production. One reason for the low effectiveness of oxidative phosphorylation in bacteria can be the presence of a non-phosphorylating bypass resulting in 'free oxidation'. Apparently, such bypasses are present in propionic acid bacteria, which is suggested by the finding that NADH is oxidized not only by the particulate fraction, but also by the soluble non-phosphorylating fraction (Bonarceva et al., 1973b). The soluble (cytoplasmic) fraction of propionic acid bacteria is unable to carry out oxidative phosphorylation, accordingly, NADH oxidation is not sensitive to KCN ($1 \cdot 10^{-3}$ M), an inhibitor of cytochrome oxidases (Bruchatcheva et al., 1975).

So, although both membrane and soluble fractions of *P. shermanii* and *P. petersonii* can oxidize NADH, only the oxidation by the particulate fraction is linked with the accumulation of energy.

The oxidative phosphorylation by the respiratory chains of *P. shermanii* and *P. petersonii* is sensitive to uncouplers and inhibitors: 2,4-DNP ($1 \cdot 10^{-4}$ M) reduced the P/NADH ratio in both cultures by 20-40%; gramicidin ($1 \cdot 10^{-6}$ M) by 60%; *o*-phenanthroline ($1 \cdot 10^{-4}$ M) by 20% in *P. shermanii* and by 80% in *P. petersonii*. A marked inhibition of the oxidative phosphorylation by the iron chelator *o*-phenanthroline in *P. petersonii* as compared with *P. shermanii* may indicate an essential role of non-heme iron in the oxidative phosphorylation of the former. Pretreatment of the particulate fraction of propionibacterial cells by UV-light led to a decrease in P/NADH ratio by

40% and 85% in *P. shermanii* and *P. petersonii*, respectively. These data indicate a possible (direct or indirect) participation of such respiratory chain components as vitamin K_2, cytochromes, non-heme iron in the oxidative phosphorylation of propionic acid bacteria.

In addition, such terminal electron acceptors as oxygen, nitrate and fumarate, natural to propionic acid bacteria, can be substituted during growth by an entirely foreign compound, ferricyanide, widely used *in vitro* as an artificial electron acceptor. Ferricyanide was supplemented to batch cultures of *P. freudenreichii* DSM 20271, growing in a bioreactor equipped with a three-electrode amperometric system, designed to regulate the redox potential of the medium (Emde and Schink, 1990a, b). When present at 0.25 mM, ferricyanide was completely reduced in the process of utilization of lactate or glycerol as substrates. However, when the concentration of ferricyanide was raised to 8 mM, it resulted in a considerably increased growth yield and a shift towards greater acetate production. In experiments with a fixed potential glycerol, lactate and propionate were oxidized to acetate and CO_2, and electrons were quantitatively transferred to the working electrode. The yield of biomass in these conditions amounted to 29.0, 13.4, and 14.2 g per mole of glycerol, lactate or propionate, respectively, as oxidized substrates. The results were interpreted (Emde and Schink, 1990a) as indicating a role for oxidative (in addition to substrate-level) phosphorylation in ATP production.

The potential of electrons transferred to ferricyanide in this system was estimated at −80 to −140 mV. Assuming that a minimum ΔE_0 of about 140 mV is needed to translocate one proton across the cytoplasmic membrane, the ATP yield by oxidative phosphorylation in the case of glycerol, lactate and propionate oxidation to acetate must be 1.0, 0.66 and 0.66, respectively. The yields of biomass observed experimentally were in accord with the calculated ATP yields. The authors also showed that the reduction of ferricyanide ($E_0 = +430$ mV) occurs close to the potential of menaquinone (Emde and Schink, 1990a).

Cells of *P. thoenii* P127 grown at pH 7.0 with or without lactate had a proton motive force (PMF) in the log-phase of 146 mV, an average ΔpH of 0.93, and $\Delta \Psi$ of −91 mV (Rehberger and Glatz, 1995). When the medium pH fell to 4.75 in the stationary phase, the cells maintained an internal pH of 6.1, while the PMF was increased to −175 mV.

3.3 Antioxidative Defense

In biological systems molecular oxygen can be reduced by any of the three possible mechanisms:

$$O_2 + 4e^- + 4H^+ \longrightarrow 2H_2O \tag{3.24}$$

$$O_2 + 2e^- + 2H^+ \longrightarrow H_2O_2 \tag{3.25}$$

$$O_2 + 1e^- \longrightarrow O_2^- \tag{3.26}$$

Reduction to H_2O takes place when electron transfer in the respiratory chain is completed by terminal oxidases. Reduction to H_2O_2 is associated with flavoproteins in flavin respiration, and superoxide radicals (O_2^-) arise constantly in the respiratory chain and some other biological systems. Hydroperoxide is toxic to the cell and is degraded by catalase and/or peroxidase. Superoxide radicals are highly toxic to the cell, in addition, they promote the formation of more toxic products, namely, singlet oxygen and hydroxyl radicals:

$$O_2^- + H_2O_2 \longrightarrow H_2O_2 + O_2 + OH^{\bullet} \tag{3.27}$$

$$O_2^- + OH^{\bullet} \longrightarrow O_2^{\bullet} + OH^- \tag{3.28}$$

This is why almost all aerobic and aerotolerant organisms are equipped with the enzyme superoxide dismutase (SOD) that captures and, in conjunction with catalase, destroys these radicals:

$$O_2^- + O_2^- + H^+ \xrightarrow{\text{SOD}} H_2O_2 + O_2 \tag{3.29}$$

$$H_2O_2 \xrightarrow{\text{catalase}} H_2O + \frac{1}{2}O_2 \tag{3.30}$$

The enzymes SOD, catalase and peroxidase are components of the antioxidative defense of the cell. As follows from the above, the CN-resistant respiration of propionic acid bacteria is responsible for most of the oxygen consumed by the cell with the attendant production of H_2O_2. The same type of respiration is the main source of superoxide radicals. It was shown (Vorobjeva and Kraeva, 1982) that NADH oxidation by membrane fractions of three strains, representing three different species, is accompanied by the formation of superoxide radicals (Table 3.11). Succinate oxidation, however, was not accompanied by a noticeable production of radicals. The highest rate of superoxide production was found in *P. globosum*, followed by *P. coccoides* and *P. shermanii*. Antimycin inhibited NADH oxidase activity in all the strains and simultaneously increased superoxide production by 32, 36 and 15%, respectively, in *P. shermanii*, *P. globosum* and *P. coccoides*. This showed that superoxide radical formation by propionic acid bacteria occurs in that part of the respiration chain that precedes the site of antimycin action.

Table 3.11. NADH oxidase activity and NADH-dependent superoxide
radical generation in propionic acid bacteria

Organism	NADH oxidase, nmol O_2/mg/min	Superoxide production, nmol/mg/min
P. globosum	420	35.2 ± 3.2
P. coccoides	330	13.6 ± 1.5
P. shermanii	280	8.4 ± 1.4

From Vorobjeva and Kraeva (1982).

Superoxide dismutase (SOD). Oxygen radicals are removed by SOD, which has been found first in *P. shermanii* (Schwartz and Sporkenbach, 1975; Pritchard et al., 1977; Kraeva and Vorobjeva, 1981b). By electrophoretic separation of proteins in polyacrylamide gel (PAGE), followed by activity staining, we (Vorobjeva and Kraeva, 1982) identified SODs in *P. globosum, P. shermanii, P. coccoides, P. petersonii, P. pentosaceum* and *P. raffinosaceum* (Fig. 3.11), which were represented by a single isoenzyme. The SOD of *P. globosum* contained iron and was resistant to KCN (10mM), but was inhibited by NaN_3 (1mM) by 25-30% and inactivated by H_2O_2 (5 mM). This profile of sensitivities to inhibitors and H_2O_2 is typical for iron-containing SOD enzymes.

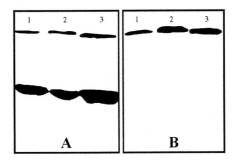

Figure 3.11. Detection by polyacrylamide gel-electrophoresis of superoxide dismutase (A) and peroxidase (B) activities in crude extracts of *P. coccoides* (1), *P. shermanii* (2) and *P. globosum* (3). A: gel stained for superoxide dismutase activity according to Beauchamp and Fridovich (1971). Stain solutions contained no KCN. B: gel stained for peroxidase activity according to Gregory and Fridovich (1974). From Kraeva and Vorobjeva (1981b).

According to our data (Vorobjeva and Kraeva, 1982), the molecular weight of the non-denatured SOD is $45,000 \pm 5,000$, but is reduced to $22,000 \pm 2,000$ in denaturing conditions; therefore, the SOD of *P. globosum* is composed of two identical non-covalently linked subunits. The enzyme is quite thermostable: upon heating at 70°C for 10 min at pH 7.2 was unchanged, and heating at 90°C for 5 min decreased the activity by 84%. Boiling for 5 min completely inactivated the enzyme. The enzyme has a pH-optimum at pH 6.1-7.5. The SOD of *P. globosum* is not membrane-bound and is localized in the soluble fraction (Kraeva and Vorobjeva, 1981a). SOD activity varied little with growth conditions of propionibacteria, being only slightly higher under aerobic than anaerobic conditions (Table 3.12).

Table 3.12. Effects of aeration on superoxide dismutase activity (units/mg protein) in propionic acid bacteria

Organism	Growth conditions	
	Anaerobic	Aerobic
P. coccoides	20	26
P. shermanii	11	17
P. globosum	22	29
P. pentosaceum	11	14
P. petersonii	8	14
P. raffinosaceum	9	12

From Vorobjeva and Kraeva (1982).

In chemostat cultures of *P. shermanii* an increase in SOD activity was observed (Pritchard et al., 1977) upon raising the pO_2 from 0 to 85 mm Hg, but increasing the pO_2 still further resulted in a decrease, although SOD activity was evident even at toxic concentrations of oxygen. The higher SOD (and catalase) activities correspond with the higher superoxide production of the strains (compare Tables 3.11 and 3.12), giving evidence of the importance of these enzymes in protecting the cell from superoxide radicals.

Catalase. Propionic acid bacteria synthesize considerable amounts of catalase (Vorobjeva et al., 1968; Kraeva and Vorobjeva, 1981b), which distinguishes them from other anaerobic microorganisms. Aeration has no significant effect on the enzyme synthesis (Table 3.13), so that under anaerobic conditions these bacteria show high catalase activities, the highest being in *P. coccoides* and the lowest in *P. petersonii*. In iron-deficient media catalase is not produced and H_2O_2 accumulates in the medium up to 1.74 mM. In cultures supplemented with iron H_2O_2 is undetectable.

Table 3.13. Effects of aeration on catalase activity (units/mg protein) in propionic acid bacteria

Organism	Incubation conditions		
	Anaerobic	Microaerophilic	Aerobic
P. globosum	1800	1900	1900
P. shermanii	1500	1500	1700
P. pentosaceum	1100	1100	1200
P. petersonii	600	800	800
P. coccoides	2800	2800	2900

From Kraeva and Vorobjeva (1981b).

KCN and NaN$_3$ inhibited the catalase activity by 83% and 90%, respectively (Kraeva and Vorobjeva, 1981b), which can be taken as an indirect indication of the hemin nature of the enzyme. The highest catalase activity (as well as superoxide dismutase) was found in the cytoplasm of *P. globosum, P. shermanii* and *P. coccoides* (Kraeva and Vorobjeva, 1981b; Vorobjeva and Kraeva, 1982). Simultaneous presence of SOD and catalase in the cell makes it possible to destroy superoxide and peroxide radicals formed in oxidative reactions.

Peroxidase. Peroxidase activity and the corresponding enzyme have been found in propionic acid bacteria for the first time in our laboratory (cf. Fig. 3.11) (Kraeva and Vorobjeva, 1981b). The specific activity was 0.076 units/mg of protein (Vorobjeva et al., 1986). It was shown that the maximal rate of *o*-dianizidine (peroxidase substrate used in these experiments) oxidation is attained at an H$_2$O$_2$ concentration of 0.03 mM, and the K$_M$ for H$_2$O$_2$ is 15 μM. Like other peroxidases of microbial origin, peroxidases of propionic acid bacteria display a very high specificity to H$_2$O$_2$. Maximal stability of the enzyme was observed in the temperature range of 20-30°C. Heating at 100°C for 10 min caused a complete inactivation of the enzyme, confirming its protein nature. The pH optimum is in the range of 6.8-7.0, while other peroxidases usually have pH-optima in the range of pH 4.5-5.5. The inhibition of peroxidase activity by Na$_2$S, NaN$_3$ and KCN (Table 3.14) suggests that the peroxidase of *P. shermanii* is related to heme-containing enzymes.

Table 3.14. Effects of inhibitors on peroxidase activity of *P. shermanii*

Inhibitor	Concentration, M	Inhibition, %
Na$_2$S	1×10^{-5}	64
	1×10^{-4}	75
NaN$_3$	1×10^{-2}	3
	1×10^{-1}	60
KCN	1×10^{-2}	50
	1×10^{-1}	95

From Vorobjeva et al. (1986).

Chapter 4

Biosynthetic Processes and Physiologically Active Compounds

4.1 Nutrition

4.1.1 Nitrogen sources

Ammonium salts and amino acids. Propionic acid bacteria can synthesize all amino acids by assimilating nitrogen from ammonium salts (Kucheras and Gebhardt, 1972). These bacteria also contain peptidases capable of supplying the cell with all essential amino acids. As have been recently shown (Langsrud et al., 1995), the proteolytic system of propioni-bacteria consists of at least two weak proteinases, one cell wall-associated and another intracellular or membrane-bound. In 7 strains of propioni-bacteria found in Swiss cheese at the stage of ripening the proteolytic activity was maximal at 15, 30, 37 and 45°C (Perez-Chaia et al., 1988). At 21°C no strain displayed any proteolytic activity. Evidently, such a complex pattern of temperature dependence is due to the presence of several enzyme systems. The pH-optima of the proteases of these strains were in the range of pH 5.7-6.1, except for *P. acidipropionici*, whose activity was weakly dependent on the pH between 5.1-6.4.

Growth of propionibacteria in Emmental and Gruyere cheeses is accompanied by proline accumulation, with which the flavor of hard rennet cheese is associated (Antila and Antila, 1968). Amounts of proline are greater in Swiss-type cheese than in any other. Proline formation during cheese ripening is due to the action of propionibacterial peptidases rather than to biosynthesis (Corre et al., 1995). In *P. freudenreichii* subsp. *shermanii* both prolidase and prolinase were found (Corre et al., 1995). The addition of glutamate, ornithine, citrulline and especially arginine to

128

$(NH_4)_2SO_4$-containing medium resulted in a significant stimulation of growth (Joseph and Wixom, 1972), suggesting the presence in propionibacteria of the ornithine–urease cycle associated with interconversions of 5-carbon amino acids: glutamic acid → glutamic acid semialdehyde → ornithine → citrulline → arginine → ornithine. Proline was poorly utilized (Joseph and Wixom, 1972). Histidine, proline, phenylalanine, leucine, glycine and aspartic acid could substitute for $(NH_4)_2SO_4$ with no change in the growth rates. Threonine, serine, ornithine and valine were utilized less efficiently. Still, the best yield of biomass was obtained with $(NH_4)_2SO_4$ (Kucheras and Gebhardt, 1972).

Resting cells of *P. freudenreichii* under both aerobic and anaerobic conditions rapidly degraded aspartic acid, glycine and alanine, and more slowly other amino acids. But large strain and species variations were observed, and certain specificities in the metabolism of different amino acids were found (Langsrud et al., 1995).

In the presence of all amino acids in the medium bacterial growth is facilitated, since amino acids are incorporated into proteins in a readily available form. Preparation methods of casein hydrolysate were found to have important effects on propionibacterial growth. Mixtures of amino acids produced by alkaline, acid and trypsin hydrolysis were tested (Zodrow et al., 1963a) and the best result was achieved with the trypsin hydrolysate. It was suggested that in this case tryptophane is preserved and some peptides are formed which have specific stimulatory effects on the bacterial growth. Nitrate inhibits the deamination of amino acids in cheese by propionic acid bacteria (Peltola and Antila, 1953).

Nitrate reduction. According to Kaspar (1982), *P. acidipropionici, P. freudenreichii, P. jensenii, P. shermanii* and *P. thoenii* can reduce nitrate to nitrite and further to N_2O. Formation of N_2O from nitrite in prokaryotes may represent a mechanism of detoxification rather than transformation of energy. N_2O was not further reduced; oxygen inhibited nitrate reduction by *P. acidipropionici* and *P. thoenii*. The enzymes of nitrate and nitrite reduction were either constitutive or derepressed in anaerobiosis; only *P. pentosaceum* contained a constitutive nitrate reductase. Nitrate stimulated the synthesis of nitrate reductase in *P. acidipropionici*; specific growth rates and biomass yields were increased by the addition of nitrate. Nitrite at a concentration of 10 mM was not inhibitory.

The ability to reduce nitrate was investigated by Swart et al. (1995) and was shown to vary from species to species and from strain to strain, being strongly affected by environmental factors. *P. acidipropionici, P. acnes* and *P. freudenreichii* subsp. *freudenreichii* were able to reduce nitrate, in contrast with *P. jensenii* and *P. thoenii* that were unable to do so in yeast extract-based media containing KNO_3 at pH 8.0. No nitrate reduction was observed in the presence of glucose in these media. *P. acidipropionici* and *P. acnes* were able to further reduce the nitrite formed to nitrogen, in contrast with *P. freudenreichii* subsp. *freudenreichii* (Swart et al., 1995). Kaspar (1982) came to the conclusion that some propionibacteria are able to perform a dissimilative nitrate reduction down to NH_3.

Nitrogen fixation. Keeping in mind a rather wide distribution of nitrogen fixation among prokaryotes (Postgate, 1982) and phylogenetic links of the propionibacteria with a group of actinomycetes and their relatives, some of which are capable of nitrogen fixation, we decided to investigate this process in propionibacteria. Two cultures of *P. shermanii* and *P. petersonii* differing in intracellular corrinoid levels and sensitivity to oxygen were investigated. *P. petersonii* is more aerotolerant and contains less vitamin B_{12} than *P. shermanii*. It could not be ruled out that some complex compounds of cobalt (corrinoids represent such complexes) can participate in the N_2-activation by chemosorption on metal-protein complexes. It has been reported (Shpokauskas et al., 1965) that the strains of *Azotobacter* capable of corrinoid formation are the most active fixators of molecular nitrogen.

Our post-graduate students (Baranova and Gogotov, 1974) found that the two propionibacterial species can grow with no nitrogen source added under conditions favorable for nitrogen fixation. In these conditions the cells incorporated ^{15}N from $^{15}N_2$ using pyruvate as carbon and energy source. Cells of *P. shermanii* displayed a higher nitrogen fixation activity than *P. petersonii*: it constituted 32 nmol ^{15}N/mg protein and 6 nmol ^{15}N/mg protein in 15 h, respectively. Nitrogen fixation activity was dependent on the nature of carbon source present in growth media and in the assay mix (Table 4.1). Cells of *P. shermanii* grown on pyruvate or fumarate reduced acetylene more actively (we used the acetylene method to assay N_2 fixation) than cells grown on glucose. Nitrogen fixation occurred in the presence of NADH and NADPH as hydrogen donors (Table 4.2), the latter supporting the highest nitrogenase activity.

Table 4.1. In vitro reduction of acetylene by *P. shermanii* cells grown on different carbon sources

Test substrate	Carbon source in growth medium		
	Pyruvate	Glucose	Fumarate
	nmol C_2H_4 per mg protein in 5 h		
None	10.6	1.8	8.6
Pyruvate	47.5	7.8	nd*
Glucose	nd	6.8	nd
Fumarate	nd	nd	27.0

*nd, not determined.   From Baranova and Gogotov (1974).

Table 4.2. Nitrogen fixation by propionibacterial suspensions in the presence of different reductants

Compound	P. shermanii		P. petersonii	
	^{15}N-NH$_3$, µg	^{15}N fractional abundance, %	^{15}N-NH$_3$, µg	^{15}N fractional abundance, %
None	1.9	0.031	2.1	0.033
Pyruvate	22.0	0.340	16.1	0.320
Glucose	16.1	0.310	20.6	0.350
NADH	39.6	0.340	72.3	0.280
NADPH	28.9	0.355	28.9	0.340

From Baranova and Gogotov (1974).

It is known that in nitrogen-fixing organisms nitrogenase activity is accompanied by hydrogenase activity. We were able to demonstrate the hydrogenase activity in *P. shermanii* and *P. petersonii*, which amounted to 17.2 and 7.8 µl H$_2$ /h per mg protein, respectively, in the presence of methylviologen as hydrogen acceptor. Like the nitrogenase activity, the hydrogenase activity in *P. shermanii* grown on pyruvate was considerably higher than in the cells grown on glucose: 1510 and 30 µl H$_2$ /h per mg protein, respectively.

4.1.2   Vitamins

Thiamine, biotin, pantothenic acid, riboflavin and vitamin B$_{12}$ are involved in propionic acid fermentation. Biotin forms the prosthetic group of methyl-malonyl-CoA transcarboxylase; pantothenate is a constituent of CoA; thiamine is not the coenzyme (co-carboxylase) of the enzyme carboxylase like in other organisms, for acetaldehyde has not been detected in propionibacteria (although traces were recently found), but it may function as a component of dehydrogenases in oxidative phosphorylation of α-keto acids. Riboflavin is a constituent of FAD and FMN. Propionibacteria can synthesize vitamins B$_2$ and B$_{12}$ in considerable amounts (see below), but the other three vitamins must be supplied. Some strains can grow in synthetic media without thiamine (Silverman and Werkman, 1939; Delwiche, 1949), in some other strains thiamine can be replaced by *p*-aminobenzoic acid.

There is no obligatory requirement for riboflavin, although it stimulates the growth. Coenzyme A can replace pantothenic acid for *P. freudenreichii* (Moat and Delwiche, 1950), and desthiobiotin (sulfur-free precursor of biotin) is more active in *P. pentosaceum* than biotin (Lichstein, 1955). Stjernholm and Wood (1963) were unable to obtain cultures of *P. shermanii* deficient in biotin, since in the absence of added biotin the bacteria started to synthesize it. Our experiments (Konovalova and Vorobvjeva, 1969) showed that *P. shermanii* and *P. technicum* are able to synthesize thiamine, pyridoxine, folic acid, riboflavin, nicotinic acid and, naturally, cobamides.

The vitamins synthesized were found in cells and in culture liquid. If the medium contained vitamins in sufficient amounts, no additional synthesis occurred.

Potato extract, orange juice and yeast extract all stimulate the fermentation of glucose and acid production by propionic acid bacteria (Tatum et al., 1936). Stimulation by potato extract is associated with some essential growth factors. If synthetic medium is supplemented with yeast extract, then the addition of individual vitamins (biotin, pantothenate, thiamine or *p*-aminobenzoic acid) is unnecessary (El-Hagarawy, 1957). In connection with the ability of *P. shermanii* to synthesize vitamins Karlin (1966) suggested to include these bacteria into dairy products. For example, kefir enriched with *P. shermanii* contained increased amounts of vitamin B_1, B_2, B_6, PP, B_{12}, pantothenate, folic and folinic acid as compared with control samples. Especially high increases in the latter four vitamins were observed.

4.1.3   Sulfur assimilation (peculiarities)

It is known that sulfur-containing amino acids stimulate the growth of propionibacteria, which are tolerant to high concentrations of H_2S in the medium (Fromageot and Chaix, 1937; Chaix and Fromageot, 1939). These bacteria can form dimethylsulfide (having an antimutagenic action) from sulfur-containing amino acids present in milk peptides; they also can form sulfitocobalamin (Wagner and Bernhauer, 1964), a corrinoid compound, in which the cobalt atom is linked with SO_3.

We (Vorobjeva and Charakhchjan, 1983) showed that propionic acid bacteria can utilize any source of sulfur, from the most oxidized (sulfate) to the most reduced (sulfide) (Table 4.3). Good growth was observed in the presence of 6 mM sulfide, a concentration ten times higher than the purple nonsulfur bacterium *Rhodopseudomonas globiformis* can tolerate.

Table 4.3. Growth yields of *P. shermanii* in media with different sulfur sources

Sulfur source	Concentration, mM	Biomass* after 144 h, mg/ml
None	–	0.23
SO_4^{-2}	30	0.82
SO_4^{-2}	2.0	0.88
SO_3^{-2}	37	0.12
SO_3^{-2}	2.5	0.83
S^{-2}	6.0	0.67
S^{-2}	30	0.10
$S_2O_3^{-2}$	5.5-13.5	0.70
S^0 (elemental)	6.0	0.52

* Mean values of replicate experiments.
From Vorobjeva and Charakhchjan (1983).

Good growth of *P. shermanii* was supported by thiosulfate and elemental sulfur. In the latter case the culture released sulfide. The reduction of sulfur to H_2S can proceed by enzymatic and nonenzymatic pathways, for example, via the formation of sulfide as in *E. coli* (Okada et al., 1982). The ability of *P. shermanii* to utilize thiosulfate, sulfite, elemental sulfur and sulfide shows that these compounds serve as intermediates of the assimilatory sulfate reduction in propionic acid bacteria.

An unusual pattern of sulfate utilization was observed in growing cultures of *P. shermanii*. Uptake of $^{35}SO_4^{2-}$ by the cells had an oscillatory pattern, alternating between the periods of sulfate utilization and release (Fig. 4.1). The accumulation of SO_4^{2-} in the medium was maximal in the exponential phase of growth (24-28 h).

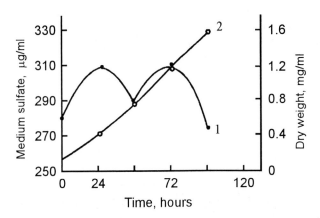

Figure 4.1. Sulfate utilization and biomass accumulation by *P. shermanii*. Curve 1 - medium sulfate concentration; 2 - dry biomass. From Vorobjeva and Charakhchjan (1983).

It was also found that the direction of sulfate fluxes in or out of the cell depended on its intracellular pool. When sulfate-depleted cells were transferred to a new medium, they actively accumulated sulfate first, but then started to release it (Fig. 4.2). Cells grown first in the medium high in sulfate, when transferred to a fresh medium, started to release sulfate at once. Interestingly, such an oscillatory pattern of sulfate utilization was not a common property of all propionic acid bacteria. For instance, *P. petersonii* steadily took up sulfate from the medium, while the two closely related strains, *P. shermanii* and *P. freudenreichii*, showed an oscillatory pattern of sulfate consumption, like the distantly related *E. coli* (Fig. 4.3).

In resting cells of *P. shermanii* a sharp increase in radioactivity (^{35}S) was observed in the first 2 min of incubation, followed by a decrease due to the release of sulfate into the medium, and a new increase in radioactivity 6 min later (Fig. 4.4). When the sulfate concentration in the medium was less than 50 µM, no sulfate was released if the cells were first grown with no sulfur source. If growing cells were not limited in sulfate, then the oscillatory character of sulfate uptake was independent of the presence of sulfate in the medium. Sulfate release into the medium also showed an oscillatory character (Fig. 4.5). Sulfate transport into the cell could be described by Michaelis-Menten kinetics with a K_t of 13.3 µM.

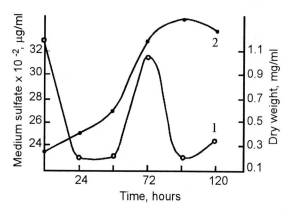

Figure 4.2. Sulfate utilization and biomass accumulation by *P. shermanii* depleted of sulfur. Curve 1 - medium sulfate; 2 - dry biomass. From Vorobjeva and Charakhchjan (1983).

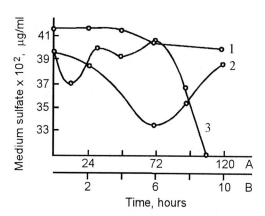

Figure 4.3. Sulfate utilization by bacterial cultures: *P. petersonii* (1), *P. freudenreichii* (2) and *E. coli*-52 (3). Time axis B refers to *E. coli*. From Vorobjeva and Charakhchjan (1983).

Sulfate transport is an energy dependent process. The cells deprived of any sources of energy following a starvation period of 30 min accumulated 3.5 times less sulfate than did the cells preincubated for 15 min in the medium containing an energy source, lactate. In the absence of an energy source the initial rate of sulfate uptake was decreased 5 times as compared with the control. Temperature optimum for SO_4^{2-} uptake was at 28-33°C

(Fig. 4.6). Thus, the results show that in propionic acid bacteria sulfate enters the cell by way of active transport, like in other microorganisms studied so far.

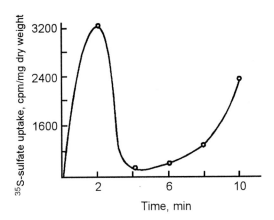

Figure 4.4. ^{35}S-Sulfate uptake by cell suspension of *P. shermanii*. From Charakhchjan and Vorobjeva (1984).

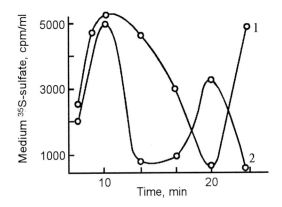

Figure 4.5. Release of ^{35}S-sulfate by *P. shermanii* into the medium containing 40 mM Na$_2$SO$_4$ (1) or no sulfate (2). From Charakhchjan and Vorobjeva (1984).

The most active sulfate uptake was observed in the range of pH 5.5-7.0; it rapidly decreased above pH 7.0 (Fig. 4.7). This may be due to the high affinity of protons to the system of SO$_4^{2-}$ transport; in yeasts it was shown (Borst-Powels, 1981) that the sulfate transport occurs in a symport with protons. Sulfate, and especially thiosulfate, considerably inhibited the transport of the labeled sulfate, in contrast to cysteine, which showed no significant effect on sulfate uptake, so that the cells grown with cysteine utilized SO$_4^{2-}$ to the same extent as those grown with sulfate. It means that cysteine has no role in the regulation of the sulfate transport in propionibacteria, while other intermediates of the pathway may be involved in this regulation.

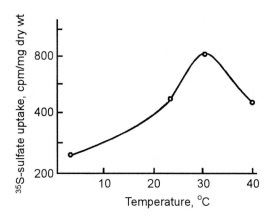

Figure 4.6. Temperature dependence of sulfate uptake by *P. shermanii*. Sulfate uptake was estimated by incubating the cells with $Na_2{}^{35}SO_4$ for 4 min. From Charakhchjan and Vorobjeva (1984).

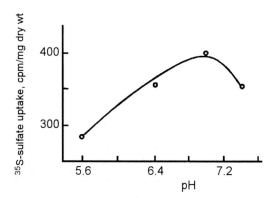

Figure 4.7. pH-dependence of sulfate uptake by *P. shermanii*. Cells were incubated with $Na_2{}^{35}SO_4$ for 3 min. From Charakhchjan and Vorobjeva (1984).

We suggest that in some propionibacteria sulfate release into the medium, which is dependent on the intracellular sulfate pool, can serve as a regulatory mechanism for the latter. It cannot be excluded, however, that alternations between sulfate uptake and release are the result of a number of interdependent processes in the cell, forming circadian rhythms. In particular, sulfate can be produced in the reaction catalyzed by APS-ammonium adenylyl transferase, which represents an exchange of the sulfo group of adenosine 5-phosphosulfate for an amino group; in this reaction sulfate and adenosine 5-phosphoamidate (APA) are formed. It has been suggested (Schift and Fankhaueser, 1981) that APA mediates regulatory links between the nitrogen and sulfur metabolism.

4.1.4   Carbon sources

Lactate. Hard rennet cheeses represent a selective habitat for propionic acid bacteria, since they contain lactate formed as the end product of lactose fermentation by lactic acid bacteria. Unlike many other bacteria, propionibacteria can utilize lactate efficiently, which is the reason why propionibacteria are so abundant in hard, ripened cheeses. Propionic acid bacteria use lactate best in the presence of yeast extract (Antila, 1954), but even higher stimulatory effect is exerted by cell-free extracts of lactic acid bacteria, *Streptococcus thermophilus* and *Lactobacillus* spp. (Hietaranta and Antila, 1953). Lactate as a carbon source supports higher growth rates of propionic acid bacteria than lactose (El-Hagarawy et al., 1954).

Growth yields and growth rates of *P. freudenreichii* on DL-lactate, glucose and galactose were similar and higher than those on lactose (Piveteau et al., 1995a). D-lactate utilization started only after all L-lactate had been consumed. The authors showed that in mixtures of DL-lactate and individual sugars lactate was utilized before the sugars.

Carbohydrates and alcohols. Glucose is the preferred carbon source for biosynthetic processes, although propionibacteria grow equally well on lactose, lactate, pyruvate and glycerol, and can ferment a variety of carbohydrates (Table 4.4). The growth is promoted, if glucose is sterilized dissolved in basal medium instead of being added aseptically to the medium sterilized separately (Field and Lichstein, 1958a). The authors suggested that upon heating glucose could react with phosphates and amino acids, forming a factor that replaces CO_2, which is necessary for the initiation of growth. An effect similar to the combined sterilization of glucose in basal medium is obtained by supplementing the medium with N-D-glycosyl-glycine, glycyl-L-asparagine, asparagine and some dicarboxylic acids, or by acid-hydrolyzed casein (Field and Lichstein, 1958b).

Both glucose and 2-deoxy-D-glucose are actively taken up by anaerobically grown cells of *P. shermanii* by an energy-dependent transport process (Warnecke et al., 1982). The uptake and formation of D-glucose-6-phosphate from glucose by cell-free extracts is dependent on ATP, not on phosphoenolpyruvate (PEP). An inhibitor of membrane ATPases, dicyclohexyl carbodiimide, at a concentration of 1 µM inhibited the sugar transport by 67%. Sugar uptake was not dependent on the membrane potential or on the proton gradient. An uncoupler of respiration, FCCP, caused a complete depolarization of the membrane potential at concentrations below 100 μM, but even at 200 μM it inhibited the sugar transport only by 30%. Collapsing the membrane potential with valinomycin, while inducing K^+ loss from the cell, had no effect on sugar transport.

Propionibacteria are capable of utilizing glucose from such an unusual medium as egg white, which is highly viscous (up to 50 centipoises). Pretreatment of liquid egg white with ammonium sulfate (0.1%) and sodium phosphate (0.2%) promoted the bacterial growth and utilization of carbohydrates from egg white (Stoyanova et al., 1979). These investigations were performed in connection with studying possible ways of using propionibacteria to remove the sugars present in egg white (see Chapter 7).

Propionic acid bacteria can use citrate as a carbon source. In the presence of lactate the utilization of citrate is suppressed (Hietaranta and Antila, 1954). Propionibacteria can oxidize acetate and propionate. The oxidation is intensified in the presence of thiamine, Mg^{2+} and K^+ ions displaying synergistic actions (Quastel and Webley, 1942). In the absence of thiamine Mg^{2+} and K^+ stimulate the oxidation of succinate, fumarate, lactate, ethanol, propanol, and glucose. It has been suggested (Quastel and Webley, 1942) that K^+ enhanced the effect of Mg^{2+} by increasing the permeability to the latter.

Under anaerobic conditions *P. pentosaceum* can grow on erythritol (Wood and Liever, 1953), which only few microorganisms can utilize. Propionic, acetic, formic and succinic acids are produced from erythtritol. Metabolic pathways of erythritol (and other carbon sources) are shown above, in Chapter 3.

The cutaneous bacteria *P. acnes* and *P. avidum* can use hydrocarbons, lipids, and amino acids as sources of carbon and energy. *P. granulosum*, which lacks active proteases, cannot use peptides. *P. avidum* is the strain most adaptive to nutritive conditions: it grows well in complex media without glucose, but at the same time responds with a higher yield of biomass to adding glucose to the medium as compared with two other strains (Greenman et al., 1981). Lipase-secreting cutaneous bacteria can split triglycerides of skin fats *in vivo* and utilize glycerol.

Alkanes. We were first (Vorobjeva et al., 1979a) to demonstrate the ability of propionic acid bacteria to oxidize and to grow on *n*-alkanes. Habituation of these bacteria to dairy products, limitations with regard to carbon sources, which for many strains are limited to 3 or 4 carbohydrates and lactate, a tradition of cultivating propionic acid bacteria only under anaerobic conditions precluded any idea of a possible use of *n*-alkanes by these bacteria. Theoretically, however, this possibility could not be excluded, based on the following known facts:

Table 4.4. Fermentaton of different carbohydrates by the species of *Propionibacterium*

Acid produced from:	P. *frfr.*	P. *frgl.*	P. *frsh.*	P. *th.*	P. *acp.*	P. *jn.*	P. *av.*	P. *acn.*	P. *lym.*	P. *gr.*
Adonitol	v	·	$+^w$	$+^w$	w^+	$+^w$	v	$-^+$	+	−
Amygdalin	$-^w$	−	−	$+^w$	w^-	w^-	$-^w$	−	w^-	$-^w$
Arabinose	$+^w$	+	$+^w$	$-^+$	+	v	v	$-^w$	−	$-^w$
Cellobiose	w^-	w	$-^+$	w^-	$+^w$	d	$-^w$	$-^+$	−	−
Dulcitol	−	−	−	−	−	−	−	−	−	−
Erythritol	$+^-$	+	$+^w$	+	$+^w$	$+^w$	$+^w$	$-^+$	+	−
Esculin	−	−	−	+	w^-	$-^w$	−	−	−	−
Esculin hydr.	+	+	+	+	+	+	+	−	−	−
Fructose	+	w	$+^w$	+	+	+	+	$+^-$	$+^-$	+
Galactose	$+^w$	+	+	+	+	+	+	$+^-$	$-^w$	w^-
Glycerol	$+^w$	+	+	+	+	+	+	$+^-$	−	+
Glycogen	w^-	−	−	$-^+$	v	$-^w$	−	−	−	−
Inositol	v	w	$+^-$	$+^-$	+	$+^-$	$-^+$	$-^w$	v	−
Inulin	−	−	−	−	$-^w$	$-^w$	−	−	−	−
Lactose	−	+	+	d	+	v	v	−	−	−
Maltose	w^-	w	w	$+^w$	+	$+^-$	+	$-^w$	+	$+^-$
Mannitol	−	−	−	−	+	+	$-^+$	d	−	$-^+$
Mannose	$+^w$	+	+	+	+	+	+	$+^-$	−	$+^w$
Melezitose	−	w	w^-	v	$+^w$	$+^w$	$+^-$	−	−	v
Melibiose	−	w	d	$+^-$	v	$+^w$	$-^+$	−	−	v
Raffinose	−	−	−	$-^+$	d	$+^w$	$-^+$	−	−	d
Rhamnose	−	−	−	−	$+^w$	−	−	−	−	−
Ribose	$+^w$	+	v	+	$+^w$	$+^w$	$+^w$	$-^+$	+	v
Salicin	$+^w$	−	$-^w$	$+^w$	+	$+^-$	v	−	−	$-^w$
Sorbitol	−	−	$-^+$	v	+	$-^+$	$-^+$	$-^+$	−	−
Sorbose	−	w	w^-	−	$+^-$	$-^w$	$-^w$	$-^w$	−	$-^w$
Starch	−	w	w^-	$+^w$	$+^w$	v	$-^+$	$-^w$	w^-	$-^w$
Starch hydr.	−	·	−	$-^+$	−	−	−	−	d	−
Sucrose	w^-	−	$-^+$	$+^w$	+	$+^w$	+	$-^+$	d	$+^w$
Trehalose	−	−	−	+	$+^w$	$+^w$	$+^w$	$-^+$	−	w^-
Xylose	$-^w$	w	w^-	$-^w$	$+^w$	v	$-^+$	$-^w$	−	$-^w$

*P. frfr., *freudenreichii* ss. *freudenreichii*; frgl., *freudenreichii* ss. *globosum*; frsh., *freudenreichii* ss. *shermanii*; th., *thoenii*; acp., *acidipropionici*; jn., *jensenii*; av., *avidum*; acn., *acnes*; lym., *lymphophilum*; gr., *granulosum*.

+, strong acid, i.e. pH 5.5 or below in 90-100% of strains; w, weak acid, i.e. pH 5.5-6.0 in 90-100% of strains; −, acid not produced, i.e. pH 6.0 in 90-100% of strains; d, reaction positive in 40-50% of strains; v, reaction variable within a strain; ·, not tested. When any of the above used as *superscripts*, reaction of 10-40% of strains. Uninoculated xylose and arabinose, under CO_2 (prereduced media), are often pH 5.9. For cultures in these media, pH 5.4-5.7 was considered weak acid and pH below 5.4 strong acid.

Adapted from Cummins and Johnson (1986), with permission.

(i) evolutionary relatedness of propionic acid bacteria, mycobacteria and corynebacteria, many of which are capable of using hydrocarbons; (ii) propionic acid bacteria contain large amounts of phosphathidylinositol (Brenan and Balou, 1968) and trehalose (Winder et al., 1967), which are found in large quantities in microorganisms growing on hydrocarbons (Jwanny et al., 1974); (iii) propionic acid bacteria contain enzymes of the TCA cycle, the glyoxylate shunt, components of respiratory chain, oxidases, including phenol oxidase (Barksdale, 1970), which determine the oxidative properties of these bacteria and the use of oxygen in their metabolism; (iv) among the propionibacteria there are inhabitants of the intestinal tract of ruminants, where methane is formed in large amounts.

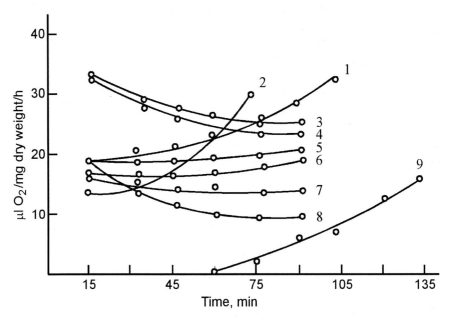

Figure 4.8. Oxidation of a mixture of *n*-alkanes by different strains of propionic acid bacteria. *P. globosum* (1); *P. thoenii* (2); *P. technicum* (3); *P. globosum* Cob (4); *P. freudenreichii* (5); *P. pentosaceum* CCM (6); *P. raffinosaceum* (7); *P. pentosaceum* Cob (8); *P. coccoides* (9). From Vorobjeva et al. (1979a).

We found (Vorobjeva et al., 1979a) that propionibacteria of different species were able to oxidize a mixture of *n*-alkanes of the following composition (%): dodecane, 0.4; tridecane, 4.2; tetradecane, 15.7; pentadecane, 26.9; hexadecane, 26.8; heptadecane, 17.9; octadecane, 6.2; and eicosane, 0.5. They also oxidized individual *n*-alkanes with chain lengths from 8 to 18 carbon atoms as well as cetyl alcohol. The results of manometric investigations showed that the bacteria (except cocci) oxidized *n*-alkanes in the mixture without a noticeable lag-period (Fig. 4.8). The absence of a latent period can be associated with a synergism of alkanes, when the oxidation of one alkane is stimulated by the presence of another, so that the lag-period in fact is smoothed over. The same result can arise as a consequence of additive effects of the low oxidation activities towards individual alkanes that form the mixture.

In propionic acid cocci the active oxygen uptake was higher than in the other strains, and had a latent period of 1-2 h. The intensity of respiration by *P. technicum, P. globosum, P. freudenreichii, P. pentosaceum, P. raffinosaceum* changed little with time. In *P. globosum* and *P. technicum* the rate of oxygen uptake increased with time. In *P. pentosaceum* and *P. acnes* the enzymes of alkanes oxidation probably are constitutive. The intensity of respiration, regardless of whether the cells were grown on glucose or alkanes, was approximately equal (Table 4.5). Chloramphenicol, a protein synthesis inhibitor, could not prevent the oxidation of alkanes in *P. pentosaceum*, but completely inhibited it in propionic acid cocci (Fig. 4.9), indicating that the corresponding enzymes are inducible in the latter.

Table 4.5. Oxygen uptake in a mixture of *n*-alkanes by propionic acid bacteria grown on glucose or hydrocarbons

Organism	Time, min		
	30	60	90
P. pentosaceum		Q_{O_2}	
grown on glucose	17.5	17.6	18.0
grown on hydrocarbons	20.1	18.0	17.6
P. acnes			
grown on glucose	16.0	16.7	18.7
grown on hydrocarbons	14.7	14.9	15.3

From Vorobjeva et al. (1979a).

Table 4.6. Oxidation of individual *n*-alkanes by suspensions of *P. pentosaceum*

Alkane	Time, min						
	30	60	90	120	150	180	210
	μl O_2 per mg dry biomass						
Dodecane	0	0	0	1.2	1.3	1.7	1.9
Tridecane	0	0	0	1.8	2.0	2.0	2.0
Tetradecane	0	0	2.1	3.0	3.2	3.3	3.3
Pentadecane	0	1.2	4.1	5.4	7.2	10.0	12.5
Hexadecane	0	0.8	2.2	4.2	5.6	7.3	9.4

From Vorobjeva et al. (1979a).

Individual hydrocarbons with chain lengths of C_8, C_9, and C_{10} were not oxidized by *P. pentosaceum*, probably because of their toxicity. Among the *n*-alkanes tested pentadecane and hexadecane were oxidized better than the others. In this respect it should be noted that the main fatty acid in *P. pentosaceum* is C_{16}-acid (Moss et al., 1969). C_{12}, C_{13} and C_{14} alkanes were oxidized much slower, the slower the shorter chain they had. No individual alkane tested was oxidized by *P. pentosaceum* during the first 30 min. In the next 30 min the oxidation of hexadecane and pentadecane started, but the maximal Qo_2 was not reached even 3 h later. In the case of the other alkanes the lag-phase took 60 to 90 min and the Qo_2 became constant 30 min after the onset of respiration (Table 4.6).

These results, coupled with observations on cell morphology shown in Figs. 4.10-4.13, allowed us to conclude that *n*-alkanes are not unnatural substrates for propionic acid bacteria. The cells of cutaneous and dairy propionic acid bacteria grown in media containing kerosine, diesel fuel, hexadecane, cetyl alcohol, were not different in morphology from the cells grown on glucose. In thin sections of the cells of *P. petersonii* (Fig. 4.13) inclusions, apparently of lipid nature, were found.

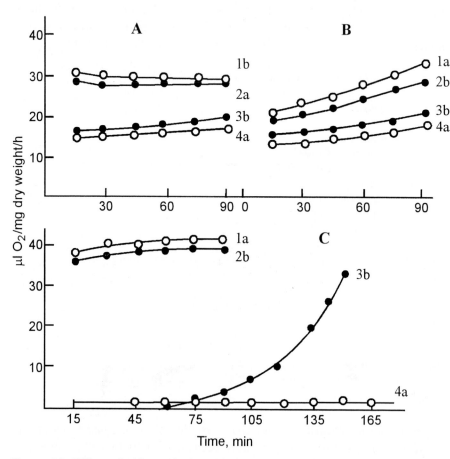

Figure 4.9. Effects of chloramphenicol on the oxidation of *n*-alkanes and glucose by propionic acid bacteria. Strains: *P. pentosaceum* CCM (A, B); *P.coccoides* (C). Oxygen uptake was assayed in the presence (a) or absence (b) of chloramphenicol at 100 μg/ml. Substrates: glucose (1, 2); *n*-alkanes (3, 4). From Vorobjeva et al. (1979a).

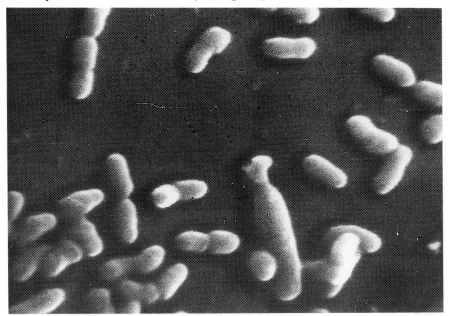

Figure 4.10. Scanning electron micrograph of *P. acnes* showing cells grown with kerosine. × 12,660. From Vorobjeva et al. (1979a).

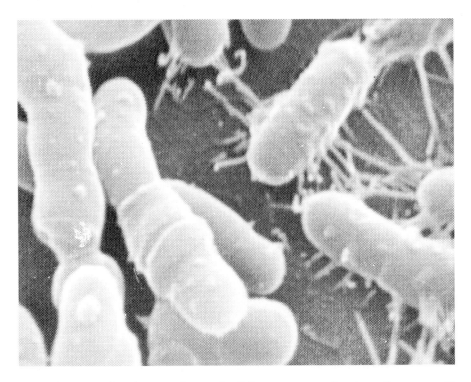

Figure 4.11. Scanning electron micrograph of *P. pentosaceum* showing cells grown with diesel fuel. × 50,000. From Vorobjeva et al. (1979a).

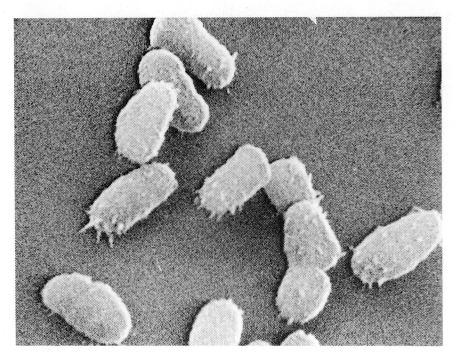

Figure 4.12. Scanning electron micrograph of *P. technicum* showing cells grown with diesel fuel. × 21,100. From Vorobjeva et al. (1979a).

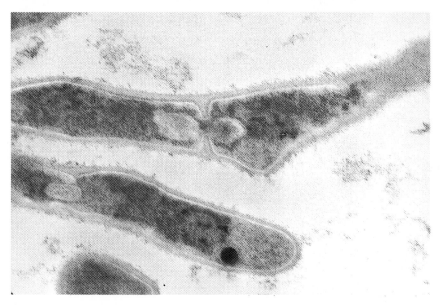

Figure 4.13. Transmission electron micrograph of a thin section of *P. petersonii* cells grown with hydrocarbons. × 60,000. From Vorobjeva et al. (1979a).

4.2   Quantitative Characteristics of Growth

In batch culture the growth of propionibacteria can be described by the curve shown in Fig. 4.14 for *P. shermanii*. When the bacteria are grown in glucose-peptone medium using a 10% (v/v) inoculum, a lag-phase of 3.5 to 4 h is observed. The specific growth rate (μ) is 0.21-0.22 h^{-1} in the exponential phase (18-24 h) and about 0.01 h^{-1} at the beginning of the stationary phase (72-73 h). In continuous cultures growing on glucose the maximal growth rate of *P. shermanii* was 0.07 h^{-1} and the generation (cell division) time was 9.9 h (Wagner et al., 1968).

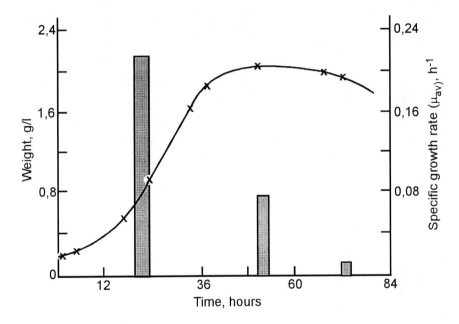

Figure 4.14. Growth curve and specific growth rates of *P. shermanii* growing with glucose. Mean specific rates are shown by shaded columns. From Vorobjeva (1976).

In batch culture the kinetics of growth is described by the Monod equation:

$$\mu = \mu_{max} \frac{S}{K_S + S},$$
(4.1)

where μ_{max} is the maximal specific growth rate, S is the substrate concentration, and K_S is the substrate concentration at which $\mu = \frac{1}{2} \cdot \mu_{max}$.

This formula applies only when the concentration of lactate used as a substrate is below 1.2%. When the lactate concentration is higher, the growth rate is limited by the end products formed during fermentation, first of all, by propionic acid. Using the method of short-term continuous cultures, the dependence of the growth rate on propionate concentration was shown (Ibragimova and Sacharova, 1974) to obey the Ierusalimsky equation, which is similar to the equation describing non-competitive inhibition:

$$\mu = \mu_0 \frac{K_P}{K_P + P},$$ (4.2)

where μ_0 is the specific growth rate in the absence of inhibitory product, P is the concentration of growth-inhibiting product, and K_P is the concentration of growth-inhibiting product at which $\mu = \frac{1}{2}\cdot\mu_0$.

The dependence of the specific growth rate on initial substrate concentrations is shown in Table 4.7. Increasing the glucose concentration from 1.2 to 2.2% had little effect, while increasing the lactate concentration from 1.3 to 2.5% depressed the specific growth rate of *P. shermanii* almost twofold (Gaitan and Vorobjeva, 1981).

Table 4.7. Dependence of growth rates of *P. shermanii* in log phase on carbon source concentration

Carbon source, %	Specific growth rate (μ), h^{-1}
Glucose	
1.2	0.033
1.3	0.026
1.5	0.035
2.2	0.038
Lactate	
1.3	0.023
2.5	0.014

From Gaitan and Vorobjeva (1981).

Slowing down of the growth rate of propionibacteria by hydroxyl and hydrogen ions is also described by the Ierusalimsky equation. For hydroxyl ions this formula is valid only for cultures in the phase of accelerating growth, while for hydrogen ions in the late phases as well. In continuous culture the growth rate is limited by hydrogen ions in a narrow pH range from 4.95 to 5.2 (Ibragimova et al., 1971). It should be kept in mind that growth is inhibited both by hydrogen ions and ions of organic acids, as well as by undissociated molecules of propionic and acetic acids. At the same time, propionibacteria are able to regulate the pH by shifting it in the direction favorable for growth, if medium pH in the chemostat is in the range of 4.4 to 7.45 (Ibragimova and Sacharova, 1974).

Growth yield coefficients, Y_G, were measured by Babuchowski et al. (1987) in three species of propionibacteria: *P. jensenii*, *P. acidipropionici* and *P. thoenii*, growing in complex medium with peptone and yeast extract.

The dependence of Y_G on initial sugar concentrations was described by two linear regression equations for the low (less than 7 mM maltose) and high (up to 22.2 mM glucose and 11.7 mM maltose) concentrations. The corresponding curves did not pass through the origin of coordinates, suggesting that components of the medium, other than sugars, were also utilized. Thus, sugar metabolism is different at low and high concentrations (Babuchowski et al., 1987).

The effects of glucose concentration on biomass yields, specific growth rates and the activities of hydrolytic enzymes in three species of cutaneous propionibacteria were also studied (Greenman et al., 1981). The biomass yield (Y_S) of *P. acnes*, *P. avidum* and *P. granulosum*, growing in semi-synthetic medium in chemostat culture, was increased by raising the glucose concentration from 0.2% (limiting concentration) to 0.3-0.4%, and remained unchanged on raising it still further. In the absence of glucose *P. granulosum* had a low yield and a low growth rate (μ_{max} of 0.02 h^{-1}), apparently being a sugarolytic strain, whereas *P. acnes* and *P. avidum*, in contrast, showed maximal values of Y_G and μ_{max} (0.16 and 0.18 h^{-1}, respectively) in the absence of glucose. At glucose concentrations of 0.2 and 0.5% the μ_{max}s were 0.08 and 0.10 h^{-1} in *P. acnes* and *P. avidum*, respectively, corresponding to only half of those observed in glucose-free medium. Obviously, the cutaneous bacteria *P. acnes* and *P. avidum* show a clear preference for peptides and amino acids, present in rich medium, over glucose as sources of carbon and energy.

Growing these strains at the maximal glucose concentration tested (0.5%) resulted in the lowest activities of exogenous lipases, hyalouronidases and acid phosphatases in the medium (for more details about extracellular enzymes, see below). This shows that the synthesis of enzymes involved in the utilization of amino acids can be repressed by glucose (Greenman et al., 1981).

Growth-inhibiting factors (low pH, high concentrations of carbon sources such as lactate, end products of metabolism) had no effect on the content of RNA, DNA, polysaccharides and lipids, but decreased protein synthesis (Ibragimova and Sacharova, 1974), leading to a lowered protein content of the cells, i.e., uncoupling of energetic and biosynthetic processes. The most sensitive to the unfavorable factors were cells at the beginning of the exponential phase.

The quantitative relationship between protein synthesis and substrate utilized can be described as $X = C_1/S_U$, where X is the level of protein synthesis, S_U is the quantity of substrate utilized per unit biomass weight, and C_1 is a proportionality coefficient.

The specific growth rate is linearly related to protein synthesis (Ibragimova and Sacharova, 1972) and can be expressed as $\mu = C_2 \cdot X + B$, where C_2 is a proportionality coefficient, and B is the specific growth rate in the absence of protein synthesis.

Finally, by combining the two expressions derived from theoretical considerations and substantiated experimentally, we obtain that the specific growth rate, μ, is directly proportional to the level of protein synthesis and inversely proportional to the quantity of substrate utilized per unit biomass weight:

$$\mu = \frac{C_1 \times C_2}{S_U} + B. \qquad (4.3)$$

What is the molecular basis of the inhibition of protein synthesis observed under unfavorable conditions? The decrease in protein content may be caused by a direct action of inhibitors on the activity (synthesis) of enzymes, involved in protein synthesis. It is known, for example, that high lactate reduces the activity of the Krebs cycle enzymes in different cell systems (Gershanowich, 1965). If high lactate concentrations have a similar effect in the case of propionibacteria, the formation of α-ketoglutarate (an intermediate in the Krebs cycle) will be decreased. The amination of α-ketoglutarate to glutamate is almost the only means of fixing ammonia nitrogen into organic compounds in propionibacteria (Vorobjeva, 1976). Therefore, low levels of α-ketoglutarate may represent a limiting factor for protein synthesis in general.

Propionic acid bacteria are characterized by high growth yield coefficients:

$$Y_S = \frac{\text{grams of biomass produced}}{\text{moles of substrate used}}, \qquad (4.4)$$

showing what fraction of the utilized substrate is retained in the form of cell constituents, and the extent of coupling between catabolic and anabolic processes.

From the data shown in Table 4.8 it follows that *P. shermanii* and *P. petersonii* use substrates more effectively under aerobic than anaerobic conditions, which is an indirect indication of the functional oxidative phosphorylation in these bacteria. In *P. petersonii* under aerobic conditions the Y_S is more than twofold higher than under anaerobic conditions. In *P. shermanii* such a strong shift in metabolism was not observed with respect to aeration, but the effectiveness of glucose utilization was 1.5 times higher than in *P. petersonii* (Y_S of 35 and 20.3, respectively). It has been assumed (de Vries et al., 1972) that the oxidative phosphorylation is the sole means of energy generation under aerobic conditions in the following species: *P. rubrum*, *P. pentosaceum*, *P. shermanii* and *P. freudenreichii*.

Table 4.8. Effects of aeration on glucose utilization and growth yields of propionic acid bacteria

Organism	Growth conditions	Glucose consumed, g	Biomass produced, g/l	Y_S, g dry weight/ mol substrate
P. shermanii	anaerobic	10.90	2.12	35.0
	microaerophilic	10.45	2.12	36.5
	aerobic	6.10	1.58	46.7
P. petersonii	anaerobic	4.12	0.47	20.3
	microaerophilic	5.05	0.58	20.7
	aerobic	7.60	1.98	47.0

From Bruchatcheva et al. (1975).

4.3 Growth Conditions

The propionibacteria grow in a temperature range of 15 to 40°C, although there is evidence (Park et al., 1967) that the growth is possible also at lower temperatures of 3 to 7°C, and this should be kept in mind when preparing hard cheeses. The relative thermotolerance of propionibacteria has a special significance for cheese making, especially for those Swiss-type cheeses that are manufactured with the second heating of the cheese mass at 55°C for 30-40 min. The classical propionic acid bacteria are cultivated routinely at 28-30°C, and the cutaneous at 37°C. Maximal production of propionic acid by *P. shermanii* occurred after growing the culture for 16 days at 24°C, and maximal corrinoids synthesis, at 18-27°C (Zodrov and Stefaniak, 1963a).

The optimal pH for growth is 6.5-7.0; at pH 5.0 the growth virtually ceases, and at still lower pHs the viability is severely compromised (Kurtz et al., 1958). In whey medium acidified to pH 5.0 with HCl *P. jensenii* produced more propionic acid from lactose than at pH 7.0 by decreasing the acetic acid production, thus maintaining the same level of acidity (pH 5.0) (Foschino et al., 1988). Under the same conditions *P. freudenreichii* subsp. *globosum* produced only CO_2 and D-lactate as the sole products of fermentation, while the pH was reduced to 4.4. If the medium was adjusted to pH 5.0 with L-lactic acid, *P. freudenreichii* subsp. *globosum* utilized L-lactate and produced D-lactate and CO_2 (the pH was not changed significantly), while *P. jensenii* utilized L-lactate and, in small amounts, lactose, and accumulated acetate, propionate and CO_2 in the medium (the pH dropped to 4.65). Thus, by changing the ratio of metabolic products, propionibacteria can regulate the pH, provided that it does not fall below a certain threshold, and *P. jensenii* in this respect is more adaptable than *P. freudenreichii*.

At high concentrations propionate inhibits growth to a greater extent than acetate (Neronova et al., 1967), and the sodium salt of propionate is more inhibitory than calcium propionate (Antila and Hietaranta, 1953). In general, propionibacteria are halotolerant. Normal growth and fermentation were observed in lactate medium containing 4% NaCl (Peltola, 1940). Propionic acid cocci can tolerate up to 6.5% NaCl in the medium (Vorobjeva et al., 1990). In rapidly growing strains the growth is inhibited by NaCl concentrations of 6.0 and 3.0% at pH 7.0 and 5.2, respectively. Conversely, slowly growing strains tolerate higher concentrations of NaCl at pH 5.2 than pH 7.0 (Rollman and Sjostrom, 1946).

The effects of aeration on some physiological characteristics of *P. shermanii* strain VKM-103, growing on lactate with yeast extract, in short-term experiments is shown in Table 4.9. When air saturation of the medium was raised from 0.5 to 12% the growth rate, growth yield and oxidative activity of the cells were enhanced (Ibragimova and Shulgovskaya, 1979). At 12% saturation with air propionic and acetic acids were formed in very low quantities, indicating a metabolic transition from fermentation to respiration. The growth yield was 1.9 g/l under anaerobic conditions, 1.25 g/l at 0.5% saturation with air, and 0.9 g/l at 1.7% saturation. This may indicate that raising the air saturation to 1.7% led to an increase in the affinity of substrate (lactate) to the corresponding enzyme (lactate dehydrogenase).

Table 4.9. Changes in physiological properties of *P. shermanii* VKM 103 induced by varying aeration conditions

Air saturation, %	μ, h^{-1}	Y_s, g/g substrate	Qo_2, $\mu l/mg/h$	Volatile acids, g/g	
				propionic	acetic
0.5	0.095	0.065	24.25	7.18	2.50
1.7	0.112	0.083	25.65	3.76	1.99
12.0	0.166	0.100	26.23	0.87	0.58

From Ibragimova and Shulgovskaya (1979).

Oxygen produces different effects on different strains of propionic acid bacteria. The growth of *P. freudenreichii* subsp. *globosum* in air-saturated lactate medium with yeast extract was inhibited, but that of *P. jensenii* was not (Foschino et al., 1988). Under aerobic conditions in whey medium both strains produced CO_2, acetate, L- and D-lactate (but not propionate) from lactose. When *P. jensenii* was grown anaerobically in the medium with the redox potential artificially raised to between +250 and +300 mV, a growth pattern similar to that found with aerobically grown *P. jensenii* was observed. In these conditions *P. freudenreichii* subsp. *globosum* expressed a very weak enzymatic activity, but CO_2 production and the change in redox potential were almost the same as under anaerobic conditions (with initial redox potential of 100-150 mV). The authors (Foschino et al., 1988) came to the conclusion that *P. jensenii* is more sensitive to the level of redox potential, while *P. freudenreichii* to oxygen.

4.4   Anabolic Processes

For the biosynthesis of cell components a microorganism must be supplied with appropriate low molecular weight compounds such as sugars, organic acids, amino acids etc. Many of 2-, 3-, 4- and 5-carbon compounds are formed in catabolic reactions. In propionic acid bacteria these reactions comprise the propionic acid fermentation, TCA cycle and hexose monophosphate shunt. The latter supplies the cell with erythrose-phosphate, ribose-5-phosphate and reducing equivalents (NADPH) needed for many syntheses. Erythrose-4-phosphate is used in the formation of aromatic amino acids: phenylalanine, tryptophane, tyrosine. Ribose-5-phosphate is incorporated into nucleic acids. The pentose cycle and propionic acid fermentation, as mentioned before, have a number of common precursors and enzymes. The inclusion of common precursors into one or another pathway is regulated by the level of ATP (Labory, 1970), and this regulation in fact determines the ratio of catabolic and anabolic processes in the cell.

4.4.1   Amino acids

Ammonium ion is assimilated into organic compounds by three main reactions. These reactions are catalyzed by glutamate dehydrogenase, alanine dehydrogenase and aspartate-ammonia lyase. *P. petersonii* and *P. shermanii* contain NADP-dependent glutamate dehydrogenase activity. Apparently, it is the only enzyme that transfers amino groups to amino acids, since alanine dehydrogenase and aspartate-ammonia lyase were not found. Pathways for different groups of amino acids have not been extensively studied, but some data are available. It was established (Wixom et al., 1971) that *P. shermanii* and *P. arabinosum* contain an isomeroreductase of α-acetohydroxy acids, catalyzing the transformation of α-acetolactate to α,β-dihydroxyisovalerate. The same enzyme catalyzes the synthesis of α,β-dioxy-β-methylvaleric acid. Propionibacteria also contain dehydratase whose specific substrate, DL-dihydroxyvalerate, is transformed into α-ketoisovaleric acid. *P. pentosaceum* contains L-threonine dehydratase which is inhibited by L-isoleucine (end product inhibition) (Joseph and Wixom, 1972). The presence in propionic acid bacteria of these enzymes suggests a functional valine-isoleucine pathway, like that in *E. coli*.

Protein synthesis in propionibacteria is accompanied by the generation of a pool of 16 amino acids: cysteine, histidine, arginine, aspartic and glutamic acids, glycine, serine, threonine, alanine, tyrosine, valine, methionine, proline, phenylalanine and leucine. In aging cultures cellular levels of methionine, glutamic acid and valine are reduced, and serine is increased. At the same time in the cells and in the culture liquid of *P. shermanii* pyruvate and α-ketoglutarate were found (Vorobjeva, 1976). The level of ketoacids steadily decreased because of their consumption for syntheses.

4.4.2   Lipids

Propionibacteria synthesize large amounts of fatty acids, lipids and phospholipids (Konovalova and Vorobjeva, 1972), whose composition has taxonomic significance. The principal fatty acid in *P. shermanii* and *P. freudenreichii* is 12-methyldecanoic (*anteiso*-C_{15}); in addition, 13-methyldecanoic (*iso*-C_{15}), pentadecanoic, hexadecanoic and C_{11}-saturated fatty acids were found. Some strains of *P. shermanii* can also synthesize acids with C_{20},- C_{21}-, C_{22}- and C_{23}-carbon chains. In five other propionibacterial species: *P. arabinosum, P. jensenii, P. pentosaceum, P. thoenii* and *P. zeae* the principal fatty acid is C_{16}; in addition, small quantities of other acids are also synthesized (Moss et al., 1969). Phospholipids amount to about 10% of the total lipids (Lancelle and Asselinau, 1968).

In our laboratory Konovalova (1970) found that the addition of lipids extracted from *P. shermanii*, as well as Tween-80 and glycerol, to the medium resulted in an increased lipid production. These observations were confirmed and extended by others (Otherholm et al., 1970; Melnikova, 1987; El Soda, 1995), showing that propionic acid bacteria contain esterases and lipases capable of splitting various lipids. These authors detected high specific activities of lipases and glycerol ester hydrolases, which are presumably responsible for the high content of fatty acids in cheese. As a general rule, substrates containing short-chain fatty acids are hydrolyzed at a faster rate as compared with substrates containing fatty acids with longer chains; and the spectrum of esterase activities varies in different species and strains (El Soda, 1995).

The lipolytic activity of propionibacteria was shown to be higher than that of lactic acid bacteria (Depuis et al., 1993). Propionibacterial lipids apparently are not only structural components of the cell, but may also have a protective role against the action of some antibiotics. In the presence of polymyxin M (Konovalova, 1970) the production of lipids and phospholipids was increased (Fig. 4.15).

4.4.3   Polyphosphates

Propionibacteria contain large amounts of polyphosphates. High-molecular-weight (acid-insoluble) compounds containing from 70 to 500 residues of phosphoric acid represent the principal type of polyphosphates; the content of shorter acid-soluble polyphosphates is very low (less than 20 µg/g of biomass). The large polyphosphates are associated with the cell membrane where they might assist in sugar transport; short-chain polyphosphates are found free in the cytoplasm.

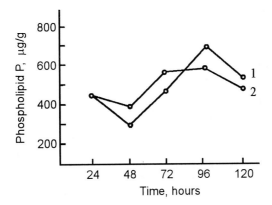

Figure 4.15. Changes in cell phospholipid content in growing cultures of *P. shermanii*. Polymyxin M was added (1) or not (2) at 50 µg/ml. From Konovalova and Vorobjeva (1972).

Salt-soluble polyphosphates may be involved in nucleic acid synthesis. The content of salt-soluble polyphosphates in *P. shermanii* reached a maximum at 72 h (3 mg/g of biomass), and subsequently decreased (Fig. 4.16). The content of polyphosphates soluble in alkali, after an insignificant decrease in the first days of incubation, increased constantly in culture and reached 4 mg/g of biomass at 120 h. The main species is represented by high-molecular-weight polyphosphates, soluble in hot 1N perchloric acid; their cellular content gradually increased up to 11 mg/g of biomass at 96 h (Fig. 4.16) (Konovalova and Vorobjeva, 1972).

Kulaev (1979) showed that such evolutionary conserved organisms as *Micrococcus lysodeicticus* and *P. shermanii* possess a glycolysis-dependent polyphosphate synthesis system in addition to the ATP-dependent one. It has been established (Bobic, 1971; Kulaev et al., 1973) that the biosynthesis of polyphosphates in propionibacteria can proceed both via the terminal phosphate of ATP as well as that of 1,3-diphosphoglyceric acid (1,3-DPGA):

$$\text{ATP} + \text{poly-P}_n \xleftrightarrow{\text{polyphosphate kinase}} \text{ADP} + \text{poly-P}_{n+1} \qquad (4.5)$$

$$\text{glucose} + \text{poly-P}_n \xleftrightarrow{\text{polyphosphate glucokinase}} \text{glucose-6-P} + \text{poly-P}_{n-1}$$

$$1,3\text{-DPGA} + \text{poly-P}_n \xleftrightarrow{\text{polyphosphate 3-phosphoglycerate kinase}}$$
$$3\text{-phosphoglycerate} + \text{poly-P}_{n+1}$$

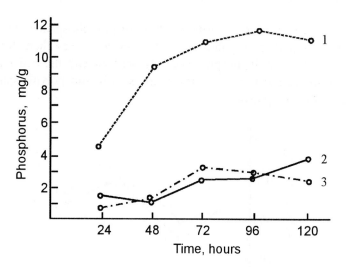

Figure 4.16. Cellular polyphosphate content in growing cultures of *P. shermanii.*
Polyphosphates were determined in hot perchloric acid extract (1), alkaline-soluble fraction
(2) and salt-soluble fraction (3). From Konovalova and Vorobjeva (1972).

Cells of *P. shermanii* contain polyphosphate kinase and polyphosphate
glucokinase, a small amount of polyphosphate phosphoglycerate kinase and
a large amount of ATP-3-P-glycerate kinase (Wood et al., 1985; Robinson
and Wood, 1986). Adding inorganic phosphate or short-chain polyphosphates (poly-P) stimulates the rate of polyphosphate synthesis from ATP
approximately 10-fold. Polyphoshate can be used by propionibacteria to
phosphorylate glucose. In *P. shermanii* the activity of ATP-dependent
glucokinase is 2.3 ± 0.6 U, the activity of poly-P glucokinase is 7.1 ± 1.5 U
per g of cells. It has been suggested (Wood et al., 1987) that the poly-P
kinase reaction is the sole pathway of polyphosphate synthesis in *P.
shermanii.* Poly-P glucokinase has a much lower K_M for long-chain than for
short-chain polyphosphates, therefore, since long-chain polyphosphates are
preferably used, short-chain polyphosphates will accumulate, which is
observed experimentally. If long-chain polyphosphates are growth limiting,
then it becomes clear why a certain level was required for the initiation of
growth (see above). In our experiments (Gaitan and Vorobjeva, 1981) the
exponential growth was conditional upon reaching a certain ATP threshold,
so that cells began dividing only when supplied with energy.

The fate of polyphosphates was traced in cells of *P. shermanii* incubated
in a medium deprived of nitrogen sources; these cells modeled nitrogen-
starved immobilized cells. We observed that in nitrogen-starved cells
incubated with lactate there was an increase in the content of all three
fractions of polyphosphates during the first 3 days (Fig. 4.17). Afterwards
the most polymeric fraction (extracted with hot perchloric acid) remained at

the same level, while the content of the salt-soluble and alkali-soluble polyphosphates continued to increase. The ATP pool steadily decreased in nitrogen-starved cells (Table 4.10), while the substrate (lactate in this case) was fermented to the usual end products. Apparently, the ATP formed under conditions of the growth arrest was diverted to polyphosphate synthesis, thus explaining the increase in polyphosphate content observed.

A different pattern was observed when glucose was added to nitrogen-starved bacteria. Glucose was metabolized during the whole period of nitrogen starvation (48 h), but the levels of both ATP and all three fractions of polyphosphates were reduced (Tables 4.11, 4.12; Fig. 4.18). The differences in polyphosphate levels between the cultures metabolizing lactate and glucose can be explained by the utilization of polyphosphates (and ATP) by phosphorylating enzymes of the glucose culture (Wood et al., 1985).

Table 4.10. Growth yield and ATP content in *P. shermanii* metabolizing lactate under conditions of nitrogen starvation

Cultivation time, h	Biomass, mg/ml	Medium lactate, mg/ml	Cellular ATP, μmol/ g dry weight
0	0.18	10.4	1.84
24	0.35	–	1.26
48	0.37	3.3	1.11
72	0.32	1.8	0.97
96 + $(NH_4)_2SO_4$	0.42	0.14	4.73

From Gaitan et al. (1982).

Table 4.11. Growth yield and ATP content in *P. shermanii* metabolizing glucose under conditions of nitrogen starvation

Cultivation time, h	Biomass, mg/ml	Medium glucose, mg/ml	Cellular ATP, μmol/ g dry weight
0	0.26	20.0	2.40
24	0.34	16.0	1.55
48	0.38	–	0.89
72 + $(NH_4)_2SO_4$	0.49	14.0	2.02

From Gaitan et al. (1982).

Table 4.12. Acid-insoluble polyphosphates in *P. shermanii* under different conditions of cultivation

Cultivation conditions	Cultivation time, h					
	0	19	24	48	66	74
	μmol P_i per g dry weight					
Nitrogen starvation with glucose	269.1	nd*	90.4	69.9	nd	28.8
Nitrogen starvation with lactate	514.7	nd	614.7	759.6	nd	1078.7
Complete medium with glucose	147.8	327.9	nd	nd	169.9	147.8

*nd, not determined.   From Gaitan et al. (1982).

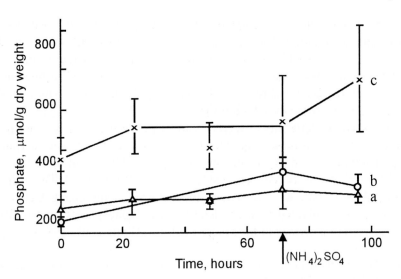

Figure 4.17. Polyphosphate content in nitrogen-starved cells of *P. shermanii* growing on lactate upon addition of ammonium sulfate. Acid-labile phosphate was assayed in alkaline-soluble fraction (a), salt-soluble fraction (b) and hot perchloric acid extract (c). Arrow indicates re-addition of ammonium sulfate. From Gaitan et al. (1982).

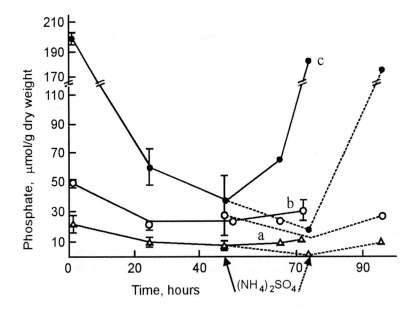

Figure 4.18. Polyphosphate content in nitrogen-starved cells of *P. shermanii* metabolizing glucose upon addition of ammonium sulfate. Symbols are same as in Fig. 4.17. Solid and broken arrows indicate re-addition of ammonium sulfate at 48 and 74 hours, respectively. From Gaitan et al. (1982).

It was shown that the formation of glucose phosphate in *P. arabinosum,* *P. freudenreichii* and *P. shermanii* occurs independently of the growth stage and is more effective in the presence of polyphosphates than ATP. Polyphosphates synthesis in these bacteria proceeded by the polyphosphate kinase pathway. The activity of phosphofructokinase in the three bacterial strains was 30 to 200 times higher in the presence of polyphosphates than in the presence of ATP. The activities of pyruvate kinase and pyruvate-phosphate dikinase were observed only with ATP. Carboxytransphosphorylases are specific with respect to PP_i and P_i, but not to polyphosphates or ATP. Glycerokinase is specific to ATP. No changes in enzyme specificity as a function of growth phase were observed.

When ammonium sulfate was added to nitrogen-starved cultures, then a decrease in the intracellular levels of salt- and alkali-soluble polyphosphates, and an increase of the highly polymeric fraction was observed (see Fig. 4.18). The salt-soluble polyphosphates could be utilized for the synthesis of nucleic acids and ATP, the pool of which was considerably increased. Addition of $(NH_4)_2SO_4$ to the nitrogen-starved glucose-culture resulted in increased intracellular levels of all three fractions of polyphosphates, with a subsequent resumption of growth. Thus, the fate of the salt- and alkali-soluble polyphosphates in *P. shermanii* depends on the nature of energy source, but in any case the initiation of growth is accompanied by the accumulation of high-molecular-weight polyphosphates.

4.4.4   Exopolysaccharides

The monosaccharides, most frequently identified in exopolysaccharides (EPS) isolated from *P. acidipropionici* growing in whey-based medium, are glucose, galactose, and mannose, with small amounts of fucose and rhamnose (Racine et al., 1991) The EPS produced by *P. freudenreichii* subsp. *shermanii* contained large amounts of methylpentoses along with lower amounts of glucose and galactose (Crow, 1988). Conditions favorable for EPS biosynthesis in propionibacteria (temperature and incubation time, medium composition with respect to carbon and nitrogen sources) are similar to those of lactic acid bacteria. An increase in viscosity over 100% due to exopolysaccharides was demonstrated in propionibacteria (Cerning, 1995) grown at 15°C rather than at 21°C (*P. zeae*) or 21°C instead of 32°C (*P. arabinosum*).

The optimal pH value for EPS synthesis by *P. acidipropionici* in whey-based medium was close to 6.0 (Racine et al., 1991). The molecular weight of propionibacterial EPS ranges from 200 to 5,000 (Cerning, 1995). The authors concluded that the EPS-producing activity is an unstable characteristic and may be linked with the presence of plasmids in propionibacteria.

4.4.5   Tetrapyrroles

Propionic acid bacteria can synthesize a number of tetrapyrrole compounds: corrinoids, heme, heme-containing enzymes, cytochromes, and linear tetrapyrroles. Biosynthesis of tetrapyrroles by all microorganisms starts with the formation of 5-aminolevulinic acid (ALA) (Corcoran and Shemin, 1957) and proceeds through the formation of porphobilinogen (PBG) and uroporphyrinogen (UPB) (Fig. 4.19A). Of the four possible isomers of uroporphyrinogenes, only the derivatives of UPB III are found in nature.

Modifications of UPB III lead to the formation of all hemes, chlorophylls, corrinoids and other tetrapyrrole pigments. Pathways leading to the synthesis of protoporphyrin IX and syrohydrochlorin (precorrin 2) diverge at the level of UPB III. Fe-containing complexes of protoporphyrin IX and syrohydrochlorin constitute the prosthetic group of heme proteins, including hemoglobin, myoglobin, leg-hemoglobin, peroxidase, catalase, chloroperoxidase, cytochrome P450, *a*- and *b*-type cytochromes. Mg-containing derivatives of protoporphyrin IX are represented by chlorophylls and bacteriochlorophylls. A methylated derivative of UPB III, chlorin, is used for the synthesis of syrohydrochlorin, from which other branches diverge: to syroheme (containing Fe), to corrinoids (containing Co), and to factor F_{430} (methanogenic factor, containing Ni).

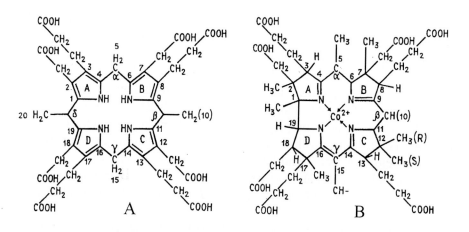

Figure 4.19. Chemical structures of uroporphyrinogen III (A) and cobyrinic acid (B).

4.5   Vitamin B$_{12}$ and Other Corrinoids

Vitamin B$_{12}$ was the first organometallic compound isolated from biological systems. Vitamin B$_{12}$ has the most complex structure of all the tetrapyrrole pigments, and even among non-polymeric compounds in general (Fig. 4.20A). The central atom of cobalt (Co^{3+}) is in the middle of a macrocycle, which differs from the other pigments by a direct linkage between ring A and ring D, and by a greater number of peripheral methyl groups. In cobalamins (Cbl) the lower ligand (Co-α) is represented by 5,6-dimethyl-benzimidazole (DMB) connected with ribose phosphate, which in turn is linked to the D-pyrrole ring via an aminoalcohol. Only cobalamins possess full biological activity.

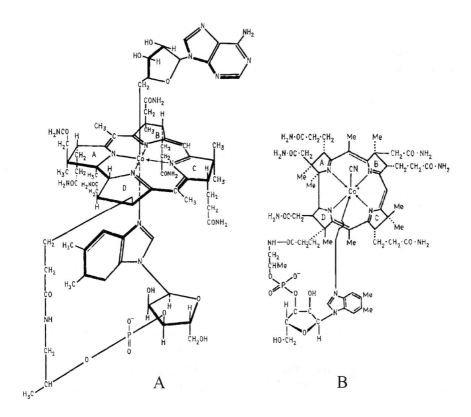

Figure 4.20. Structures of adenosylcobalamin (A) and vitamin B$_{12}$ (B).

In coenzymes the upper ligand (Co-β) is represented by 5′-deoxyadenosyl, the resulting adenosylcobalamin (AdoCbl) has a molecular mass of 1600 D (Fig. 4.20B). Actually, the upper ligand can vary widely with no effect on the association with proteins or on the biological activity in higher organisms. However, even small variations in the corrin ring or Co-α bond bring about serious changes in biological activity. Two phosphito-P-cobalamins, fluoride-methyl-phosphito-P-cobalamin and dimethyl-phosphito-P-cobalamin, isolated by partial chemical synthesis, were as active as CNCbl in experiments with *Ochromonas malhamensis* and *E. coli* 213. An Fe(III)-cobalamin and factor VA (with an altered corrin macrocycle), isolated from activated sewage (Kamikubo et al., 1982), display a very low activity. The microbiological activity correlated well with the ability to bind the mammalian intrinsic factor.

The corrin ring and the nucleotide moiety are synthesized only by bacteria and some algae, but the cobalt-binding β-ligands can be introduced by some enzymatic systems in animals and humans. The synthesis of precursors in adequate amounts is often impaired, that is why microorganisms, in addition to the biologically active forms, also contain incomplete forms of corrinoids, mainly those in which the nucleotide moiety is lacking: cobyric acid, cobinamide, cobinamide-P and cobinamide-GDP. Under conditions unfavorable for the synthesis of the nucleotide moiety incomplete forms may predominate, especially cobinamide. The content of complete forms increases in old (or aerated) cultures at the expense of incomplete forms.

Natural corrinoids can be divided into three groups according to the base they contain: (i) benzimidazole ring, (ii) purine base, and (iii) nitrogen-free compounds such as phenol and paracresol. Paracresol and phenol-containing cobamides have been isolated from the homoacetogenic bacterium *Sporomusa ovata* (Stupperich and Esinger, 1990). Purine (adenyl)-cobamide (pseudo-vitamin B_{12}) and 2-methyl-cobamide (factor A) are synthesized in small amounts by *P. shermanii* and *P. freudenreichii*. Cobamides without a lower base (factor B) are formed under conditions unfavorable for the synthesis of benzimidazole.

P. arabinosum and the bacteria that live in the gut of humans and animals synthesize mostly the purine-containing cobamides. If DMB is added to the feed, then the synthesis of cobalamins in the stomach of animals increases (Rickard et al., 1975). The finding of incomplete corrinoids in various bacteria and algae (Neujahr and Frires, 1966) shows that their biosynthesis may proceed with difficulty. Cobinamide, pseudo-vitamin B_{12}, factor A and factor III can act as growth factors for microorganisms; if they contain a nucleotide, they are also active in enzymatic reactions in animals.

4.5.1   Biogenesis

5-ALA, the common precursor in the biosynthesis of heme and vitamin B_{12}, is synthesized from L-glutamate in archaebacteria, anaerobic bacteria and facultative anaerobe *P. shermanii* (Bykhovsky et al., 1997), whereas in aerobic and aerotolerant microorganisms studied so far it is formed from succinyl-CoA and glycine (Shemin pathway) (Vogt and Renz, 1988). In the corrin biosynthesis two ALA molecules are condensed in the ALA dehydratase reaction, producing PBG. Then UPG III is formed by the condensation of four PBG molecules. At the next step the four acetyl residues of UPG III are decarboxylated, leading to coproporphyrin; its further transformation to protoporphyrin IX and then to heme and heme enzymes is described in several reviews (Bykhovsky and Saitzeva, 1983, 1989; Schneider, 1987a).

The *hemL* gene coding for the ALA synthesis from glutamate has been cloned (Murakani et al., 1993). The *hemD* gene encoding ALA dehydratase and the *cobA* gene encoding uroporphyrinogen III methyltransferase were isolated from *P. freudenreichii*. All three genes have been expressed in *E. coli* (Murooka et al., 1995). Corrinoids, the methylated derivatives of UPG III, are synthesized in considerable amounts by propionibacteria, and the first corrin structure is represented by cobyrinic acid (cf. Fig. 4.19B). To get a biologically active corrinoid from UPB III, it takes about 30 enzymatic transformations. These transformations include the following steps (Schneider, 1987a):

1. Removal of the CH_2-group attached to C_{20} carbon and formation of a direct bridge between rings A and D.
2. Insertion of cobalt.
3. Decarboxylation of acetyl at C_{12}.
4. Methylations at C_2, C_7, C_{20}, C_{17}, C_{12}, C_1 and C_{15} (Battersby, 1994). S-adenosylmethionine serves as donor of methyl groups. Methylations at C_{12} and C_{17} are accompanied by reduction of pyrrole rings C and D.
5. Ligation of 5'-deoxyadenosine with cobalt.

Methylated derivatives of UPG III were isolated from the culture liquid of resting cells of *P. shermanii* incubated in a cobalt-free medium (Bykhovsky et al., 1975a, 1976). The authors coined the name "corriphyrins" for these methylated tetrapyrrole pigments, meaning "between corrins and porphyrins". One of the isolated compounds carries two methyl groups (corriphyrin-2), while the other (corriphyrin-3) has an additional methyl group at C_{20}, which is subsequently removed with the formation of corrin structure. A compound carrying one methyl group was isolated from the suspension of *P. shermanii* incubated in a Cu^{2+}-containing medium (Müller et al., 1990). Tetrapyrrole pigments with one, two and three methyl groups are referred to as factor I, factor II and factor III, and their reduced forms are called precorrin-1, precorrin-2 and precorrin-3, respectively (Fig. 4.21).

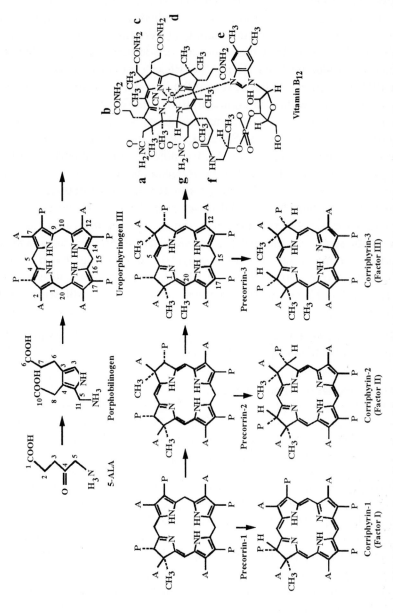

Figure 4.21. The first intermediates located between uroporphyrinogen III and the corrin ring of vitamin B₁₂. A and P represent acetyl and propionyl residues, respectively.

Several studies (Balachandran et al., 1994; Blanche et al., 1995; Scott, 1994) dealt with the next step of the corrinoid formation in propionic acid bacteria—insertion of cobalt. The authors came to the conclusion that the corrinoid biosynthesis in *P. shermanii* is specific to type II, or anaerobic, biosynthesis in which the insertion of cobalt occurs at an early stage of corrinoid synthesis; and this insertion is obligatory for the formation of cobyrinic acid. In type I (aerobic) biosynthesis the insertion of cobalt occurs at a late stage of corrinoid formation (Blanche et al., 1995). The mechanism of cobalt insertion in *P. freudenreichii* is not yet clear. In contrast with *E. coli* methyltransferase (encoded by *cysG* gene), the *P. freudenreichii* enzyme does not contain the additional peptide moiety with oxidase–ferrochelatase activity responsible for the synthesis of the cobalt-containing early intermediate (Sattler et al., 1995). The authors suggested that the cobalt insertion can occur spontaneously, since porphyrinoids are known as metal chelators. But they do not rule out the existence of a separate gene encoding an enzyme with properties similar to those encoded by the *cysG* gene.

Divalent metal ions were found (Schneider et al., 1995) to be competitive inhibitors of the cobalt binding by propionibacterial cells. Among the metals studied, Ni^{2+}, Mn^{2+} and Mg^{2+} were the strongest, while Pb^{2+} and Ca^{2+} the weakest inhibitors. KCl at concentrations above 0.2 M also significantly inhibited the binding of cobalt. In bacteria exposed to thermal shock at 60°C the cobalt-binding capacity increased several-fold.

Recently, additional intermediates were isolated from *P. shermanii* (Scott et al., 1996, 1998): Co-precorrin-3 and a compound called factor-4, which represents a further modification of Co-precorrin-3.

The pathway from cobyrinic acid to vitamin B_{12} includes the amidation of acetyl and propionyl groups (Bernhauer et al., 1968; Rapp, 1968; Friedman and Cagen, 1970). The amidation at ring C is an obligatory condition for the formation of coenzyme forms of vitamin B_{12}. A complete or at least partial amidation is also necessary for the incorporation of nucleotide-containing ligands in the B_{12} molecule (Friedman, 1975). Amino groups for the amidation can be derived from ammonia present in the growth medium; in the absence of NH_4^+ salts only aspartic acid, cysteine, asparagine, and glutamine supported a high production of vitamin B_{12} (Bykhovsky and Zaitseva, 1977). Other amino acids were either completely ineffective or showed a weak effect. Glutamine is the best source of amino groups, because glutamine in the cell can be directly utilized for the amidation of cobyrinic acid. The amide group of glutamine was utilized to a fourfold greater extent than the α-amino group. Propionibacteria have been shown to possess a glutamine-dependent synthetase that catalyzes the amidation of cobyrinic acid (Bykhovsky et al., 1982; Yeliseev et al., 1986a, b; 1988a, b). In anaerobic *Eubacterium limosum* (another genus of the family

Propionibacteriaceae) the amide group of glutamine is used not only for the amidation of the corrin ring, but also as a precursor for the N3 atom of DMB (Vogt and Renz, 1988).

The completely amidated cobyrinic acid without isopropanol residues is called cobyric acid. Cobyric acid containing isopropanol is named cobinic acid, and its amide, cobinamide. L-threonine is the precursor for 1-amino-2-propanol (Ford and Friedman, 1976). The complete corrinoids contain DMB as the lower (Co-α) nucleotide ligand. In anaerobic producers of corrinoids the synthesis of DMB differs from its aerobic synthesis. Aerobic and aerotolerant microorganisms form this base from FMN. Propionibacteria can use different forms of flavins: riboflavin, FMN and FAD as precursors of the vitamin B_{12} nucleotide ligand (Jaszewski et al., 1995).

In the cell, vitamin B_{12} functions in two coenzyme forms, 5'-deoxy-adenosylcobalamin (AdoCbl) and methylcobalamin (MeCbl, or CH_3Cbl), in which 5'-deoxyadenosine or methyl groups replace the upper cobalt ligand, represented by CN-group in vitamin B_{12} (cyanocobalamin) and by hydroxyl in other vitamin forms (hydroxocobalamins). Incorporation of the adenosine ligand in B_{12} molecule requires a reduced state of the cobalt atom (Co^{1+}). Adenosylating systems display broad specificity to corrinoid substrates and can adenosylate, in addition to cobyrinic acid and cobalamin, all cobalt-containing precursors of vitamin B_{12} and a large number of B_{12}-derivatives, including those that contain rhodium instead of cobalt in the corrin ring. A low specificity is also shown with respect to nucleoside donors (Corcoran and Shemin, 1957; Bray and Shemin, 1958). Adenosylating enzymes are present not only in bacteria, but also in animals. In congenital defects of the adenosylating enzyme in humans vitamin B_{12} fails to function as coenzyme, which causes serious disease.

Most of the corrinoids in propionic acid bacteria are represented by AdoCbl; MeCbl is usually present in low amounts. The principal form is pseudo-vitamin B_{12}, containing adenine nucleotide as the lower ligand. MeCbl is formed by the reaction of S-adenosylmethionine with the cobalamin that contains reduced Co(I).

4.5.2   Regulation

Early stages of the biosynthesis of tetrapyrrole compounds are inhibited by hemin and by the complete form of vitamin B_{12} (Lascels and Hatch, 1969). Vitamin B_{12} represses specifically its own synthesis (Bykhovsky et al., 1968, 1975a). The inhibiting effect of oxygen on vitamin B_{12} synthesis was attributed to the repression of ALA dehydratase and ALA synthase activities (Menon and Shemin, 1967), but now it is believed (Oh-hama et al., 1993) that in propionic acid bacteria ALA is formed in the C_5 pathway. Bykhovsky et al. (1974) showed that aeration specifically inhibits vitamin B_{12} biosynthesis, but not porphyrin synthesis. Aeration has a strong negative effect on cells grown under anaerobic conditions. If cultures were subjected to aeration after a preliminary growth for 2 to 3 days under anaerobic conditrions, ALA synthetase and ALA dehydratase were not inhibited, but the specific inhibition of vitamin B_{12} biosynthesis persisted. It was suggested (Bykhovsky, 1979) that this specific effect of oxygen is due to the inhibition of uroporphyrinogen III methylation.

Illumination of suspensions of *P. shermanii* by light (2000-2500 lux) for 48-72 h sharply decreases the production of vitamin B_{12} and its corrinoid precursors, cobyrinic acid and its amides, and is accompanied by a considerable increase in porphyrin synthesis (mainly coproporphyrins III) (Yeliseev and Bykhovsky, 1990). The amidation of cobyrinic acids is an important regulatory step of the vitamin B_{12} biosynthesis (Yeliseev et al., 1988). When methylation is impaired, polycarboxylic corrinoids accumulate, inhibiting steps that precede the synthesis of corrinoids, including the methylation of UPB III (Bykhovsky and Zaitseva, 1989).

4.5.2.1 Biosynthesis. General observations

The general pathway of biosynthesis has been subdivided into three principal sections on the basis of studying growth requirements of a series of mutants (Jeter et al., 1987) as shown in Table 4.13. The first section of the pathway (Cob I) is defined as the synthesis of cobinamide, containing the corrin structure, Co^{3+}, and the side chain (1-amino-2-propanol) to which DMB is added. Most enzymes catalyzing the synthesis of cobalamin coenzymes (totaling about 30) are evidently involved in this section of the pathway. This section does not function in the presence of oxygen, which is explained by the sensitivity of cobinamide precursors to oxidation (Battersby and Relter, 1984; Scott, 1998) or by the inhibiting effect of O_2. Many bacteria that synthesize vitamin B_{12} under aerobic conditions have some means of protection of these intermediates (Kusel et al., 1984).

Table 4.13. Phenotypes of *cobI*, *cobII* and *cobIII* mutants of *Salmonella typhimurium*

Genotype	Anaerobic growth in minimal medium plus*:					
	None	Methionine	B_{12}	Cobinamide	DMB	DMB and cobinamide
*metE***	+	+	+	+	+	+
metE cobI	–	+	+	+	+	+
metE cobII	–	+	+	–	+	+
metE cobIII	–	+	+	–	–	–

* +, growth, or –, no growth in minimal medium plus the indicated nutrient(s).
** MetE is a B_{12}-independent enzyme catalyzing the end stage of methionine synthesis.
Reproduced from Jeter et al. (1987); with permission.

The second section of the pathway (Cob II) includes the synthesis of DMB. The third section (Cob III) comprises the addition of DMB to cobinamide, forming cobalamin. Also the fourth section may be defined, comprising the synthesis of AdoCbl from exogenous vitamin. The Cob I, Cob II and Cob III mutants of *Salmonella typhimurium* have defects in a small region near the *his* operon, mapping at 42 min counterclockwise on the chromosome. Under anaerobic conditions in the presence of glycerol and fumarate the transcription of Cob-genes is 215 times higher than under aerobic conditions (Jeter et al., 1987).

The results of a comparative study (Pędziwilk, 1962) of the production of corrinoids by propionibacteria grown for 20 days at 30°C under anaerobic conditions are shown in Table 4.14. The most active producers are *P. shermanii* and *P. freudenreichii*, which are used for industrial production. Most of the corrinoids synthesized are represented not by cobalamins, but by factors A, B, and pseudovitamin B_{12}, which can be transformed into cobalamins in the presence of DMB by all strains, except *P. petersonii*. Certain strains of *P. freudenreichii* (ATCC 6207), *P. shermanii* (ATCC 8262, ATCC 9614-9617) and *P. technicum* (ATCC 14073-14074) are able to synthesize DMB under aerobic conditions (Perlman, 1978).

Table 4.14. Production of corrinoids by different species of propionic acid bacteria

Organism	Corrinoids, µg/ml	Analogues, µg/ml			
		CN-Cbl	Pseudo-B_{12}	Factor A	Factor B
P. shermanii	7.0	8.7	2.7	1.7	86.9
P. freudenreichii	5.3	7.6	1.9	0.0	91.1
P. petersonii	8.7	8.0	1.8	0.0	90.2
P. thoenii	4.7	1.9	86.1	6.5	5.4
P. rubrum	3.2	4.2	76.6	0.0	19.2
P. jensenii	2.7	0.0	43.4	19.6	37.0
P. sanguineum	5.6	0.0	61.4	17.9	20.7
P. raffinosaceum	1.2	2.8	78.5	0.0	18.7
P. arabinosum	4.7	3.7	85.1	4.2	7.0
P. pituitosum	2.5	6.9	62.0	0.0	31.1
P. wentii	1.4	12.9	34.7	9.9	42.5
P. pentosaceum	2.0	0.0	76.2	23.8	0.0
P. technicum	1.7	7.0	28.5	5.3	59.2
P. zeae	3.0	1.6	55.3	17.3	25.9

Reproduced from Pędziwilk (1962), with permission.

Strains producing elevated amounts of vitamin B_{12} (superproducers) have been isolated by chemical mutagenesis and serial selection (Vorobjeva et al., 1973; Gruzina et al., 1973; Pędziwilk et al., 1983). Strain M-82 synthesizes more than 50 mg/l of vitamin B_{12} (Gruzina et al., 1973) and is used to produce the vitamin in Russia. A Japanese patent describes a strain synthesizing 65 mg/l of vitamin B_{12} (Patent 87920 to Nippon Oil Co., Ltd., Eur., 1983). In Hungary, by selecting bacteria able to tolerate high concentrations of cobalt salts, a strain of *P. shermanii* 104V producing 28 mg/l of vitamin B_{12} was obtained (Patent 1032333, Brit., 1965).

A Japanese group (Maruhashi et al., 1996) has isolated the gene encoding uroporphyrinogen III methyltransferase from *P. shermanii*. The gene was expressed in a mutant strain of *P. shermanii*, with the resulting transgenic strain producing 115 mg/l of vitamin B_{12} as compared with 88 mg/l by the parental strain.

Growth and vitamin B_{12} production. Vitamin B_{12} may be regarded as one of the primary metabolites of propionic acid bacteria. It is required for growth and its biosynthesis occurs with a short lag in parallel with growth (Vorobjeva, 1976). Most prokaryotes, while being the only producers of corrinoids in nature, synthesize them in small amounts. An exception to this are certain wild-type strains of propionibacteria, which probably are natural mutants with an impaired regulation of corrinoid synthesis, or with a high physiological demand for corrinoids. We hold to the latter view.

As noted above, there is a lag in vitamin B_{12} production at the beginning of logarithmic phase (Fig. 4.22). Obviously, in the first 10 to 14 h the growth occurs at the expense of the endogenous vitamin stores. Thus, in the first hours of growth some conditions are lacking or some factors are present that prevent the *de novo* vitamin biosynthesis. In addition to the integrity of cell structures, the vitamin synthesis requires a certain level of such metabolites as methionine, NAD, ATP, GTP, flavins. Young developing cultures contain more of these compounds than the old ones, and it is young cultures, actively conducting anabolic processes, that utilize these compounds with higher rates. Thus, at an early stage of development there is competition between vitamin B_{12} biosynthesis and the other anabolic processes for the common precursors, since ATP, NAD, FAD are incorporated into the vitamin B_{12} molecule as structural units.

The addition of ATP (to a concentration of $3 \cdot 10^{-4}$ M) to the medium was shown (Kanopkaite and Gibavichyute, 1965) to increase vitamin B_{12} biosynthesis by 32%, and the addition of RNA (0.005 µg/ml) by 65%. These results show that the vitamin biosynthesis is limited by the availability of these compounds. One can suggest that by lowering the bacterial growth rate, while keeping the same level of energetic processes, a greater fraction of precursors can be diverted to the synthesis of vitamin B_{12}. The same result may be obtained by altering the coupling between catabolic and anabolic processes. Catabolism proceeds with a normal rate independent from the capacity of a microorganism to use energy for biosynthesis. Such factors as changes in temperature, medium composition, hydrogen ion concentration

(Ibragimova et al., 1971), excess of substrate (Ibragimova and Saccharova, 1972) can impair the coupling between energetic and constructive processes, so that a greater fraction of metabolites such as ATP, NAD, FAD can be utilized for vitamin B_{12} biosynthesis.

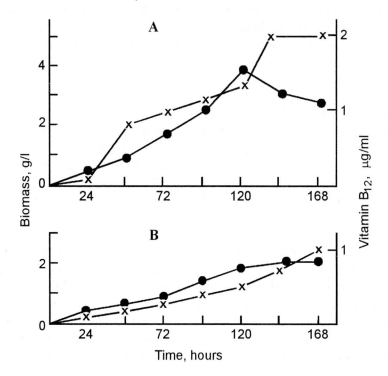

Figure 4.22. Effects of pre-cultivation of *Mycobacterium luteum* on subsequent growth and vitamin B_{12} production by *P. shermanii*. A: with pre-cultivation; B: without pre-cultivation. (●) - biomass; (✗) - vitamin B_{12}. From Vorobjeva and Baranova (1966).

We can present many facts in support of this suggestion. For instance, we showed (Konovalova and Vorobjeva, 1970) that lowering the initial growth rate with antibiotics, irrespective of their chemical nature and mechanism of action, resulted in an increased vitamin B_{12} content of the cells.

Eh and pH. An important condition for the synthesis and activity of enzymes involved in vitamin B_{12} biosynthesis is maintaining low Eh values. Vitamin B_{12} synthesis *de novo* begins at lower Eh values than cell division (Vorobjeva, 1976).

The effect of pH is a composite of separate effects on synthesis, on enzyme activities, and on vitamin B_{12} storage in the cell (below pH 5.5 the vitamin is washed out of the cell). In general, the process of biosynthesis is more sensitive to changes in pH than bacterial growth. This fact has a special significance for industrial applications, since it causes the problem of maintaining a constant pH during production, which is usually solved by automatic titration of the acids formed with an alkaline solution. This method has a drawback in that the alkaline solution is not sterile, which may lead to contamination and necessitate an increase in the volume of medium.

A nearly constant pH can be ensured by growing bacteria in lactate-based media, although the yield of biomass on lactate is usually lower than on glucose as carbon and energy source. We prepared a medium containing a buffer mixture based on citric acid; the pH changed insignificantly over 48 h of growth (from 7.3 to 6.9 and from 6.9 to 5.2 in the case of citrate-phosphate and citrate-alkaline buffers, respectively) (Vorobjeva, 1972). Maintaining a favorable pH is most important in the first days of bacterial growth. A time-course study showed that the addition of buffer mixtures containing salts of citric acid induced a small growth delay during the first 24 h. In the next 24 h the cells completely adapted to citrate, so that in 48 h the yield of biomass in the citrate-phosphate buffer was higher than in the control medium.

Temperature. In long-term cultures the temperature optimum for growth was observed at 18°C (Zodrow et al., 1967), at this temperature the largest accumulation of corrinoids was also found. But the maximal growth and vitamin accumulation were reached in 14 days, which is not economically sound for industrial production. We found (Vorobjeva and Kosireva, 1967) that at 48°C the growth is completely inhibited; at 37°C the growth rate is higher than at 30° or 22°C, but the maximal amount of vitamin B_{12} was found in cells grown at 28-30°C.

Effect of acids. Propionibacterial growth is limited by the formation of metabolic products: propionic acid at a concentration of 43 mM blocks the growth completely; acetic acid is less toxic. A favorable effect on the growth and vitamin B_{12} production is obtained by a partial replacement of culture liquid with a fresh portion of the medium. In this manner the yield of corrinoids was increased to 30 mg/l (Czarnoska-Roczniakowa, 1966). In the light of these data continuos fermentation may be considered, especially for *P. freudenreichii*, *P. shermanii* and *P. technicum* (Neronova and Ierusalimsky, 1959).

A system of two vessels has also been applied. In the first vessel anaerobic conditions were maintained, and sterile medium was introduced at a rate adequate for constant levels of cells and vitamin B_{12}; anaerobic conditions were maintained for 100 h. In the second vessel, connected with the first, aerobic conditions were created. This regime corresponds to two phases of the process of vitamin B_{12} formation. Under anaerobic conditions, favorable for growth, biomass containing incomplete factors accumulated; in the aerobic phase these incomplete factors were transformed into the complete corrinoids. In continuous culture the production of vitamin B_{12} was 213 $\mu g \cdot l^{-1} \cdot h^{-1}$, while in batch culture 141 $\mu g \cdot l^{-1} \cdot h^{-1}$ (Mervin and Smith, 1964). But the authors considered that continuous cultivation is reasonable only in cases when bacteria grow quickly and the product is formed in large amounts over relatively short times.

The growth, acid production and lactate utilization by propionibacteria (especially *P. freudenreichii* and *P. shermanii*) are inhibited by long-chain fatty acids (10-100 mg/l each): lauric ($C_{12:0}$), myristic ($C_{14:0}$), oleic ($C_{18:1}$) and linoleic ($C_{18:2}$) acids (Boyaval et al., 1995), present in milk lipids. However, the inhibitory effects of these acids were only observed in lactate-yeast extract medium (YEL), but not in milk or curd. The work was undertaken in connection with the often poor eyes formation in Swiss-type cheese.

Production in mixed cultures. In order to remove metabolic end-products from the culture liquid of propionic acid bacteria we suggested (Vorobjeva and Baranova, 1966) to cultivate some rapidly growing mycobacteria in spent medium. (Mycobacteria are persistent commensals of propionic acid bacteria in cheeses.) It was found that *Mycobacterium luteum* utilized completely propionic (2 g/l) and acetic (3 g/l) acids in 48-72 h and accumulated up to 15 g/l of dry biomass of a good amino acid composition (Samoilov et al., 1968). The biomass and culture liquid of mycobacteria contain vitamins of the B-group. If glucose was added to the culture liquid subsequent to the growth of mycobacteria, then a good growth medium was obtained for the vitamin production by propionic acid bacteria. Thus, a biological removal of toxic products of metabolism made it possible to recycle the medium three times for propionic acid bacteria by adding a fresh carbon source.

For the production of vitamin B_{12} we suggested a combined cultivation of propionic acid bacteria and mycobacteria. To create conditions favorable for both cultures, they were separated by a semipermeable membrane. Mycobacteria were grown in a tube made of cellulose film, inserted into a flask with growing propionic acid bacteria (Fig. 4.23A). Low-molecular-weight substances passed freely through the membrane, but proteins and bacterial cells could not. 48-h later sterile air was passed through the tube with mycobacterial culture. Under the conditions of side-by-side cultures the yield of vitamin B_{12} was twofold higher than in monoculture, but growth yield was 30-40% lower than in batch monoculture of *P. shermanii*, which may be due to a negative effect of oxygen. Stimulation by mycobacteria was observed mostly when propionic acid bacteria were grown in synthetic medium, in which $(NH_4)_2SO_4$ was the sole source of nitrogen. In these conditions the cell productivity in vitamin B_{12} was twice as high as in the industrial medium with corn-steep liquor.

The beneficial effect of mycobacterial culture liquid was due to its content of amino acids, vitamins, peptides, and possibly of specific stimulators of vitamin B_{12} synthesis, since the vitamin content of the cells was increased to a greater extent than their biomass. Although amino acids exert a significant effect on the synthesis of vitamin B_{12}, the complete culture liquid of mycobacteria was more effective, which can be explained by the complex influence of many metabolites released by mycobacteria in the exponential phase of growth. Of special interest are polypeptides secreted by mycobacteria into the medium, which may have a specific stimulatory effect on vitamin B_{12} biosynthesis. Chromatographic separation of mycobacterial polypeptides is shown in Fig. 4.23B. Specific stimulatory effects were displayed by fractions 10, 11 and 12.

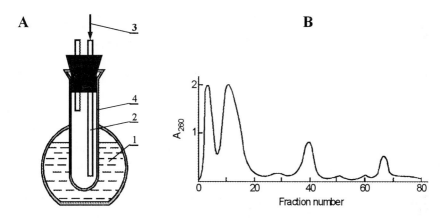

Figure 4.23. (A) The device for side-by-side culturing of propionic acid bacteria and mycobacteria. 1 - propionibacterial culture, 2 - mycobacterial culture, 3 - sterile air, 4 - semipermeable membrane. From Vorobjeva and Baranova (1966). (B) Separation of peptides from mycobacterial culture liquid by chromatography on DEAE-Sephadex. From Baranova and Vorobjeva (1971).

Trojanowska et al. (1995) studied the production of vitamin B_{12} in mixed cultures of propionic and lactic acid bacteria. Best results were observed when cultures of propionibacteria were grown for 48 h before being inoculated with lactic acid bacteria. Vitamin B_{12} production was greatest when *P. shermanii* was co-cultured with the following strains of lactic acid bacteria: *Lb. fermentum, Lb. arabinosus, Lb. helveticus* and *Lb. leichmannii*. In certain variants the vitamin production was more than 100% higher as compared with the respective controls, and the time of fermentation was significantly reduced. On the other hand, by applying a special technique for a continuous removal of propionic acid from the culture liquid of propionic acid bacteria, the yield of biomass was increased from 3.14 to 227 g/l, and of vitamin B_{12}, from 2.14 to 52 mg/l (Hatanaka et al., 1988).

In *P. shermanii* 94.6% of vitamin B_{12} was found in the cytoplasm in the form of Ado-, Me- and OH-cobalamins, and 5.4% in the fraction of membranes and ribosomes in the form of AdoCbl (Yeliseev et al., 1985). An AdoCbl-protein complex having a molecular mass of 10 kD was found in the cytoplasm of *P. shermanii*.

4.6   Porphyrins

Propionibacteria synthesize considerable quantities of compounds with the porphyrin structure, which accumulate both in the cell and in the medium (Müller et al., 1970). Porphyrins are represented mainly by coproporphyrins III, indicating high uroporphyrinogen III decarboxylase activity and relatively low activities of the enzymes catalyzing further steps of the coproporphyrin transformation into heme. Uroporphyrin III and corriphyrins are also found in growing cultures of *P. coccoides* (Bykhovsky et al., 1987).

Both *P. shermanii* and *P. technicum* accumulate significant amounts of porphyrins extracellularly in the form of free carboxylic acids and as metal complexes. They are also characterized by the high vitamin B_{12}-producing activity. In contrast, *P. rubrum*, *P. thoenii* and *P. jensenii* synthesize insignificant quantities of porphyrins and corrinoids. Thus, in propioni-bacteria the abilities to synthesize porphyrins and corrinoids are correlated. Strain M-82 (superproducer of vitamin B_{12}) of *P. shermanii* under optimal conditions synthesizes more than 2.0 mg/l of porphyrins with no precursor added, and about 3.5 mg/l in the presence of ALA. The maximum synthesis of porphyrins was observed in the medium containing 2% glucose, 0.5 mg/l $CoCl_2 \cdot 6H_2O$ and 20 mg/l ALA. In corn steep-glucose medium the bacteria synthesized twice as much porphyrins as in glucose-peptone medium (up to 10 mg/l under optimal conditions). Iron salts as well as aerobic conditions inhibited the formation of porphyrins by *P. shermanii* (Bykhovsky and Zaitseva, 1989), probably because of the inhibition of ALA dehydratase and ALA synthetase catalyzing the first steps of the common pathway of tetrapyrrole formation (Menon and Shemin, 1967).

Porphyrins are synthesized by growing, resting, and immobilized cells of propionic acid bacteria; this is essential for practical uses (see below). Suspensions of resting cells (*P. shermanii*, *P. technicum*, *P. coccoides*) accumulate coproporphyrin III and corriphyrins in the medium. The formation of porphyrins (mainly coproporphyrin III) from exogenous ALA was greatest in *P. shermanii* M-82, *P. technicum* and *P. coccoides*, whereas *P. rubrum* and *P. thoenii* virtually failed to utilize added ALA (Polulach et al., 1991). High accumulation of porphyrins correlated with high activities of

ALA dehydratase and porphobilinogenase complex. In resting cells of strain M-82 the synthesis of porphyrins was highest at a glucose concentration of 0.5% and at cell densities between 2.5 and 2.8 mg/ml (Polulach et al., 1991). These conditions promoted the greatest (up to 14.5%) transformation of ALA to tetrapyrroles.

At low cell densities most of the porphyrins was found outside the cell, but increasing the biomass to 2.8 mg/ml and more led to an increase in intracellular porphyrins from 40 to 2020 µg/l, with the total content in the medium increasing from 500 to 2940 µg/l. The accumulation of porphyrins was increased by increasing ALA content in the medium from 5.0 to 100.0 mg/l, but a linear correlation was observed only at concentrations of the precursor in the range of 5.0-20.0 mg/l. The amount of porphyrins and the extent of utilization of exogenous ALA were increased 2 and 3 times, respectively, when the cells were grown in media supplemented with cobalt as compared with media without cobalt.

Corriphyrins, reduced methylated derivatives of uroporphyrin III, are formed under strictly anaerobic conditions, in the complete absence of cobalt, and the presence of DL-methionine and reduced glutathione (Polulach et al., 1991). Interestingly, at longer terms of incubation and at elevated temperatures immobilized cells of *P. shermanii* synthesized coproporphyrin I (Polulach, 1987). In media containing hydrocarbons *P. shermanii* strain MU-512 can synthesize both intra-, and extracellular porphyrins (uro-, copro- and protoporphyrins) (Vygovskaya et al., 1990). It has been demonstrated (Poznanskaya and Yeliseev, 1984) that cobalt in the cobalt-porphyrin complex, isolated from the cytoplasm and membranes of *P. shermanii*, can accept electrons from an electrode. In general, the functions of porphyrins and porphyrin-protein complexes in propionibacteria are insufficiently studied. One can only suggest that they take part in electron transport, by analogy with the known functions of these compounds in *Desulfovibrio desulfuricans* (Poznanskaya and Yeliseev, 1984).

The cutaneous propionibacteria *P. acnes* and *P. granulosum* also form tetrapyrrole pigments, represented mainly by coproporphyrin III, with smaller quantities of uroporphyrin III and protoporphyrin IX (Lee et al., 1978). Anaerobically grown *P. granulosum* accumulated 3.2 mg/l of porphyrins in 80 h, whereas *P. acnes* produced only 0.46 mg/l. The production of porphyrins by cutaneous propionibacteria is greatly stimulated by ALA added to the medium. In this case the yield of porphyrins was up to 26.8 mg and 13.1 mg per g of dry biomass of *P. granulosum* and *P. acnes*, respectively. High yields of porphyrins produced by *P. granulosum* make it a prospective producer of porphyrins for industrial use (for applications of porphyrins, see Chapter 7).

Chapter 5

Significance of Vitamin B$_{12}$ to Propionibacteria

5.1   Vitamin B$_{12}$-dependent Reactions

As we know, corrinoids are a family of tetrapyrrole pigments that perform vital functions in living systems, being called "pigments of life" (Battersby, 1985); chlorophyll, hemoglobin, cytochromes and vitamin B$_{12}$ fall in this category. Metal ions in these pigments are complexed with organic ligands. In the coenzyme forms of vitamin B$_{12}$ (AdoCbl and CH$_3$Cbl) there is a direct bond between cobalt and carbon atoms. The two coenzymes of vitamin B$_{12}$, also referred to as cobamides, are the only organometallic compounds found in living organisms. AdoCbl is a unique biocatalyst. Organometallic compounds with Co-C bonds are usually very unstable; in contrast, AdoCbl is one of the most stable complexes of transient metals (Schrauzer, 1968).

Enzymatic cleavage of Co-C bonds leads to the formation of highly reactive substances, capable of abstracting hydrogen atoms from non-activated methyl groups. If these reactive intermediates are prevented from being stabilized, very fast side reactions will follow in solvents or gas mixtures, containing these highly reactive substances (Retey, 1998).

Corrinoids are distinguished from other related tetrapyrroles by their polyfunctionality. Most reactions catalyzed by cobamide coenzymes can be classified in two groups, depending on the type of cobamide involved, namely, AdoCbl or CH$_3$Cbl. Methylcobalamin-dependent reactions are characterized by heterolytic bond cleavage and represent the transfer of methyl groups. AdoCbl-dependent reactions are characterized by homolytic bond cleavage with protein-bound free-radical intermediates; in general, they constitute the transfer of an hydrogen atom to an adjacent carbon with a reciprocal replacement by another group.

174

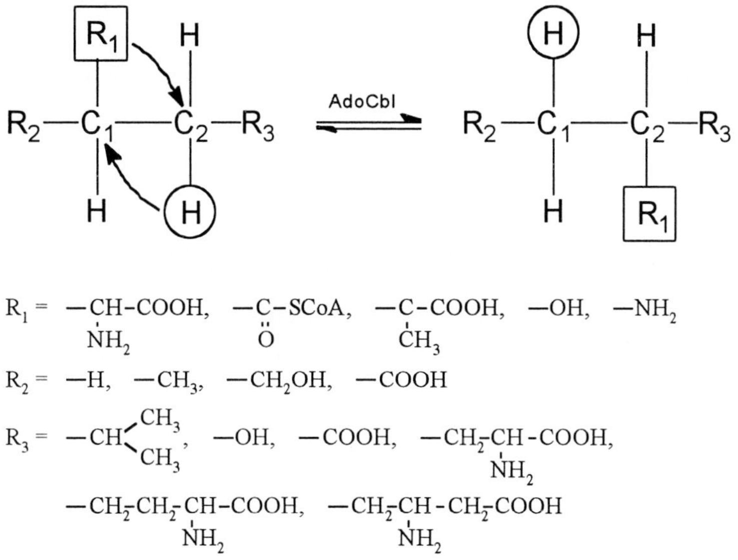

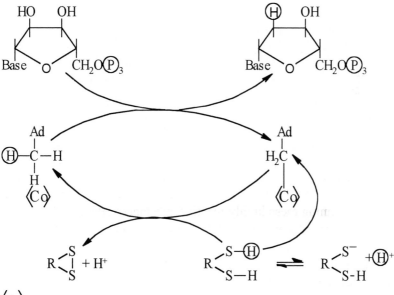

$R_1 = -CH-COOH, -C-SCoA, -C-COOH, -OH, -NH_2$
$\qquad\ \ \ NH_2 \qquad\quad\ \ddot{O} \qquad\quad\ \ CH_3$

$R_2 = -H, -CH_3, -CH_2OH, -COOH$

$R_3 = -CH\!\!<^{CH_3}_{CH_3}, -OH, -COOH, -CH_2-CH-COOH,$
$\qquad\qquad\qquad\qquad\qquad\qquad\qquad\qquad\ \ NH_2$

$\qquad -CH_2-CH_2-CH-COOH, -CH_2-CH-CH_2-COOH$
$\qquad\qquad\qquad\quad NH_2 \qquad\qquad\quad\ NH_2$

Figure 5.1. Isomerization reactions catalyzed by vitamin B_{12}-dependent enzymes.

$\langle Co \rangle$ = corrinoid-enzyme complex

Figure 5.2. Reaction mechanism of ribonucleotide triphosphate reductase.

Of AdoCbl-dependent reactions, the most important are: (*a*) isomerizations leading to a rearrangement of the amino acid skeleton (of glutamate, α-methyleneglutarate, D-α-lysine, L-β-lysine, ornithine) and of methyl-malonyl-CoA, (*b*) dehydration reactions, including dehydration of diols and glycerol, (*c*) deamination of aminoalcohols, and (*d*) reduction of ribonucle-otide triphosphate to 2-deoxyribonucleotide triphosphate. Reactions *a-c* can be schematically represented as shown in Fig. 5.1.

The reaction catalyzed by ribonucleotide triphosphate reductase (RNR) requires an exogenous hydrogen donor; various dithiols act in this capacity (Fig. 5.2). The reaction consists in a specific hydrogen atom exchange between ($5'$-5-H_2) of AdoCbl and H_2O via a thiol group, and hydroxyl group substitution on C-2 of the substrate by hydrogen from C-5 of AdoCbl with the retention of configuration. Interactions between AdoCbl/enzyme/dithiols result in the homolytic cleavage of the cobalt-carbon bond and the formation of Cob(II)alamin and deoxyadenosyl radical. Dithiols are involved in the cleavage of carbon-Co bonds. Catalysis of carbon-cobalt bond cleavage by the RNR of *Lb. leichmannii* was studied by Licht and Stubbe (1996) and Stubbe et al. (1998), who also presented a kinetic model.

AdoCbl-dependent reactions have been reviewed by Barker (1967), Hogenkamp (1968), Stadtman (1971), Schneider (1987b), and more recently by Golding and Buckel (1996) and Golding et al. (1998).

Methylcobalamine-dependent enzymes catalyze the transfer of CH_3-groups (reviewed by Mathew et al., 1990). CH_3Cbl-dependent reactions are carried out by the 'supernucleophile' cob(I)alamin, which abstracts methyl groups from substrates (e.g., N-methyltetrahydrofolate) of a relatively low reactivity (Poston and Stadtman, 1975; Drennan et al., 1998). These reactions are catalyzed by methionine, methane, and acetate synthases.

Methane synthases are present in methanogenic bacteria, where they catalyze the transformation of the following substances to methane:

$$CO_2$$
$$HCOOH$$
$$H_2CO$$
$$H_3COH$$ $$\longrightarrow CH_3-H_4\text{-folate} \longrightarrow CH_3Cbl \longrightarrow CH_4$$
$$CH_3COOH$$
$$CH_3COCOOH$$

Acetate synthases are found in *Clostridium thermoaceticum, Cl. formicum, Acetobacter woodii*. Active forms of the enzyme, containing CH_3Cbl, $AdoCH_3Cbl$, or ($5'$-methoxy-BZA)-CH_3Cbl, catalyze the reaction:

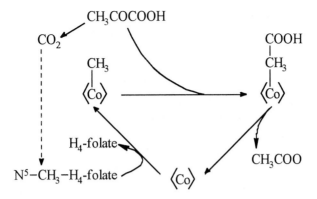

Methionine synthases are ubiquitous in both prokaryotes and eukaryotes; they catalyze the transfer of methyl groups from methyltetrahydrofolate to homocysteine, producing tetrahydrofolate and methionine:

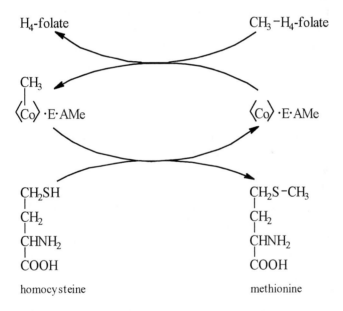

A number of new functions of cobamide coenzymes have been discovered (Frey et al., 1988), for example, cobamide-dependent reduction of epoxyquenosine to quenosine in tRNAs of *Escherichia coli* and *Salmonella typhimurium*. Quenosin and its derivatives can replace guanosine in the anticodon of tRNAs in eubacteria, and in cytoplasmic and mitochondrial tRNAs of lower and higher eukaryotes, except yeasts. The strong nucleophilic reactivity of Cbl(I) is used by certain dehalogenases, e.g., in the conversion of tetrachloroethene to 2-dichloroethene in *Dehalospirillum multivorans* (Golding and Buckel, 1996). More information about the new Cbl-dependent functions found in propionibacteria is given below.

5.1.1.1   B$_{12}$-dependent reactions in propionic acid bacteria

In propionic acid bacteria vitamin B$_{12}$ functions in the isomerization of succinyl-CoA to methylmalonyl-CoA, the key reaction of propionic acid fermentation. This reaction completes the fermentation and the formation of propionic acid as the end product. AdoCbl is also involved in DNA synthesis and in RNR activity; there is indirect evidence for the involvement of vitamin B$_{12}$ in methionine synthesis (Skupin et al., 1970), stabilization of thiol groups in enzymes (Vorobjeva and Iordan, 1976), methylation of cytosine residues in DNA (Iordan et al., 1979a) and in glutamate mutase activity (Toohey et al., 1961).

A number of wild strains of propionic acid bacteria, mostly those that prefer anaerobic way of life, are able to synthesize very high amounts of cobalamins. What are these strains, mutants with a perturbed regulation of B$_{12}$-synthesis, or organisms requiring high amounts of corrinoids for growth? The vitamin B$_{12}$-dependent reactions in propionic acid bacteria listed above provide evidence in favor of the latter. Additional support for this suggestion is given by the results of the experiments presented below.

5.2   Energy Metabolism and Vitamin B12

Fermentation is the main pathway of energy transduction in propionic acid bacteria; vitamin B$_{12}$-dependent methylmalonyl-CoA mutase catalyzes the interconversion of succinyl-CoA to methylmalonyl-CoA(b). The enzyme is insensitive to thiol reagents. Its molecular weight is 124,000. The enzyme exhibits high affinities to both the coenzyme and substrate: the K$_M$ values for AdoCbl, for succinyl-CoA and L-methylmalonyl-CoA are $3.5 \cdot 10^{-8}$ M, $3.45 \cdot 10^{-5}$ M and $8.0 \cdot 10^{-5}$ M, respectively. The equilibrium constant, K$_{eq}$ (with succinyl-CoA), is 23.5 (Schneider, 1987). The pH optimum is around 7.4 and the activity is only slightly lower at pH 6.0 and 8.0 (Kellermeyer et al., 1964).

The isomerization of succinyl-CoA to methylmalonyl-CoA is not directly linked to ATP generation. However, the B$_{12}$-dependent enzyme prepares the substrate for decarboxylation that leads to the synthesis of ATP and formation of CO$_2$ and propionate. Through the pivotal action of methylmalonyl-oxaloacetate transcarboxylase the propionic acid bacteria can gain 6 mol ATP from 1.5 mol fermented glucose. Thus, stimulation of methylmalonyl-CoA decarboxylation favors a general increase in energy output. If the decarboxylation cannot proceed, however, the bacteria carry out the PEP-carboxylation reaction, in which one molecule of the macroergic compound is consumed. This highlights the vital function of the vitamin B_{12}-dependent isomerization.

As shown previously (Vorobjeva and Iordan, 1976), the content of 2 μg vitamin B_{12} per g of dry biomass of *P. shermanii* represents a threshold for the isomerization to proceed. In these studies, a number of metabolic variables were investigated in vitamin B_{12}-replete cells (about 1000 μg vitamin B_{12}/g biomass), B_{12}-deficient cells (about 10 μg vitamin B_{12}/g biomass) and B_{12}-depleted cells (at most 2 μg vitamin B_{12}/g biomass). The cellular vitamin B_{12} content was varied by changing the concentration of a cobalt salt in the growth medium, or by using a mutant strain that produced only traces of corrinoids.

Normally, the strains of *P. pentosaceum*, *P. freudenreichii* and *P. shermanii* studied produced 10-11 mmoles of propionic acid per 100 ml, although the vitamin B_{12} levels in these strains varied greatly, being 56, 1650 and 1070 μg per g biomass, respectively. Inhibition of vitamin B_{12} biosynthesis resulted in a reduced propionic acid production, with acetic, or acetic and formic acids, becoming the main end products of fermentation. As a result, the ratio of propionic/acetic acid in the cells that produced insignificant quantities of vitamin B_{12} (growing in cobalt-free medium with methionine) was low, 1:4, as compared with the normal ratio of 2:1 (Vorobjeva, 1976).

Propionic acid bacteria grown under aerobic conditions contain considerably less corrinoids than do those grown under anaerobic conditions (Menon and Shemin, 1967). If AdoCbl was added to the cell-free extracts of propionic acid bacteria grown aerobically, the conversion of succinate to propionate was increased from 30 to 50% relative to the cell extract of the bacteria grown anaerobically (Menon, Shemin and 1967).

Under aerobic conditions vitamin B_{12}-depleted cells grew better than vitamin B_{12}-deficient or B_{12}-replete strains. The bacteria switch to aerobic mode of life by using the TCA cycle and respiratory chain. The suppression of fermentation in vitamin B_{12}-depleted cells may be due to a low activity (or absence) of methylmalonyl-CoA mutase, an SH-dependent fumarase (Ayres et al., 1962), or PEP-carboxytransphosphorylase. The latter enzyme requires and has a very high affinity to Co^{2+} (Davis et al., 1969), but cobalt was absent from the medium used to grow the B_{12}-depleted cells. The mutant of *P. shermanii* unable to synthesize coenzyme B_{12} failed to produce propionic acid, the main product of glucose fermentation being acetic acid (Mashur et al., 1971). These observations support the suggestion by V.N. Shaposhnikov that propionic acid is not an obligatory product of the propionic acid fermentation.

One of the factors regulating the synthesis of propionic acid is AdoCbl. As shown by Mashur et al. (1971), adding exogenous AdoCbl to the incubation mixture restored methylmalonyl-CoA mutase activity in the B_{12}^- mutant; normal B_{12}^+ cells were unaffected. The bacteria devoid of AdoCbl cannot convert succinyl-CoA to methylmalonyl-CoA. Therefore, transcarboxylation of pyruvate at the expense of methylmalonyl-CoA cannot proceed. The specific activity of CoA-transferase (EC 2.8.3.6) in such cells was very low, being 10 times lower than the corresponding activity in control cells. One can suggest that in this case oxaloacetate is formed by CO_2 fixation onto PEP. Comparative studies on the activity of PEP-carboxytransphosphorylase (EC 4.1.1.38) revealed that this activity was threefold higher in the B_{12}^- mutant than in the control B_{12}^+ strain. Also, the PEP-carboxytransphosphorylase activity was 2.5 times higher in *P. shermanii* grown in a medium devoid of cobalt (vitamin B_{12}-depleted cells) than in the presence of cobalt.

Furthermore, it was found that the growth of two B_{12}^- mutant strains is dependent on, or appreciably stimulated by, exogenous CO_2. The addition of $NaHCO_3$ had no significant effect on the growth of B_{12}^+ culture. Thus, under conditions of vitamin B_{12}-deficit, the PEP-transcarboxylation:

$$PEP + CO_2 + HPO_4^{2-} \longrightarrow oxaloacetate + PP_i$$

acquires special importance since it produces oxaloacetate, which is further utilized in the TCA cycle, leading to a more efficient energy transduction. In this context, the requirement for reduced sulfur compounds in vitamin B_{12}-deficient cultures (see below) has to be considered. In the presence of thiols, bicarbonate inhibits the reaction:

$$PEP + P_i \rightarrow pyruvate + PP_i$$

(Wood et al., 1969), so that more PEP is available for PEP-transcarboxylase. This reaction is irreversible under experimental conditions (Davies et al., 1969).

5.3   Anabolism and Vitamin B12

5.3.1   Growth of B_{12}-deficient bacteria

Vitamin B_{12}-deficient *P. shermanii* grows well under strictly anaerobic conditions. However, under microaerophilic conditions compounds containing reduced sulfur: methionine, cysteine, thiosulfate, reduced glutathione, or tryptone must be added to the medium (Fig. 5.3) (Vorobjeva and Iordan, 1976). These compounds are necessary to prevent the oxidation of thiol groups. A direct determination showed (Iordan et al., 1974) that the content of SH-groups in vitamin B_{12}-deficient cells is 1.5 times lower than in B_{12}^+ cells. The content of SH-groups was more significantly reduced in aging B_{12}-deficient cells than in aging B_{12}^+ cells. This indicates that vitamin B_{12} is directly involved in maintaining a constant level of SH-groups in the cell. We also found (Iordan et al., 1974) that the activities of some enzymes: 3-phosphoglycerol dehydrogenase, succinate dehydrogenase, isocitrate dehydrogenase and malate hydrogenase that have SH-groups in their allosteric and active centers were reduced in the absence of vitamin B_{12} in the cells. The results are consistent with the suggestion that the stabilization of SH-groups requires high levels of vitamin B_{12}. We believe that in this case corrinoids do not act catalytically but rather as chemical reactants.

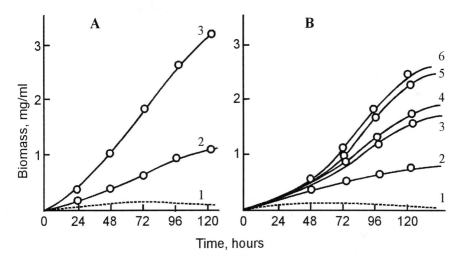

Figure 5.3. Growth of vitamin B_{12}-depleted and vitamin B_{12}-deficient cells of *P. shermanii* in different redox environments. A: wild-type cells deprived of cobalt (residual vitamin B_{12} about 10 µg/g) were grown in the absence of cobalt (1), under argon gas (2) or in the presence of $CoCl_2 \cdot 6H_2O$ at 3 mg/l (3). In 3, the final level of vitamin B_{12} was 1500 µg/g. B: vitamin B_{12}-deficient mutant cells were grown in the absence of cobalt (1), but in the presence of 0.03% each of the following: reduced glutathione (2), sodium thiosulfate (3), cysteine (4), methionine (5) or tryptone (6). From Iordan et al. (1984).

Survival rates of vitamin B_{12}-deficient cells irradiated with UV light were up to 10 times lower than of the B_{12}^+ cells (Vorobjeva and Iordan, 1976). It is known (Samoilova, 1967) that thiol groups can protect UV-irradiated cells. But the protective effect of vitamin B_{12}, as we found, was greater than that of cysteine. Accordingly, the survival rate of vitamin B_{12}-deficient cells, irradiated in the presence of AdoCbl, was higher than in the presence of cysteine but without AdoCbl, or in the absence of either. It follows that the protective effect of AdoCbl against UV-irradiation may be distinct from the effect of SH-groups, and may be due to the increased levels of nucleotides observed in the presence of corrinoids. As a consequence of vitamin B_{12}-deficiency, the protein and DNA synthesis might be reduced. These observations, together with those described earlier on the correlation between vitamin B_{12} and thiol groups in the cell (Vorobjeva, 1976) reinforce the suggestion of the stabilizing action of corrinoids on thiol-containing compounds.

The highest growth rates were observed (Iordan et al., 1984) when vitamin B_{12}-content in the cells was 800-1000 µg/g biomass, which corresponds to the concentration of $CoCl_2 \cdot 6H_2O$ in the medium of 1 mg/l (Fig. 5.4). Increasing or decreasing the content of corrinoids in the cell reduces the bacterial growth rate. The delay (in Co-deficient medium) or arrest (in the absence of vitamin B_{12}) of propionibacterial growth clearly indicate an important physiological role for vitamin B_{12}.

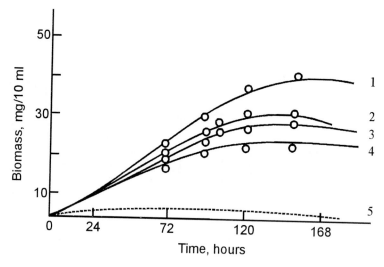

Figure 5.4. Effects of cobalt addition on growth of *P. shermanii* in minimal medium. Minimal medium was supplemented with $CoCl_2 \cdot 6H_2O$ at 1.0 mg/l (1), 3.0 mg/l (2), 5.0 mg/l (3), 0.5 mg/l (4) or was carefully depleted of cobalt (5). From Iordan et al. (1984).

On the other hand, the inhibitory effect of the excessive synthesis of vitamin B_{12} on the growth rate may be rationalized in terms of the competition for some common metabolites (Vorobjeva, 1976). A number of observations suggest the existence of a threshold level of vitamin B_{12} in the cell (1000 µg/g biomass), above which competition with other anabolic reactions for common intermediates occurs, since some of them, such as ATP, NAD, FAD, are sources of the structural units of the vitamin B_{12} molecule. Probably, this is the reason why factors that delay growth and decouple the anabolic and catabolic processes lead to an increased vitamin B_{12} synthesis (Konovalova and Vorobjeva, 1970; Ibragimova and Sakharova,

1974). On the other hand, it has been shown (Kanopkaite and Gibavichyute, 1965) that the addition of ATP to the medium increases vitamin B_{12} synthesis. Similarly, adding amino acids to minimal medium causes the growth rates of cultures producing excessive quantities of vitamin B_{12} to approach those of cultures synthesizing optimal amounts (Fig. 5.5) (Iordan et al., 1984).

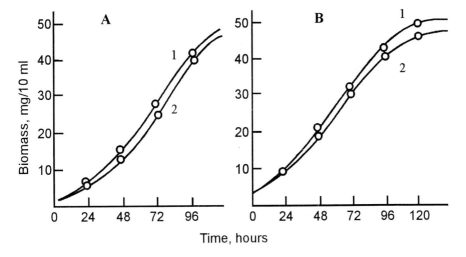

Figure 5.5. Effects of cobalt addition on growth of *P. shermanii* in minimal medium containing 0.03% tryptone. A: growth medium was supplemented with $CoCl_2 \cdot 6H_2O$ at 1.0 mg/l (1) or 0.1 mg/l (2). B: growth medium was supplemented with $CoCl_2 \cdot 6H_2O$ at 1.0 mg/l (1) or 3.0 mg/l (2). From Iordan et al. (1984).

5.3.2 Protein synthesis

The competition of vitamin B_{12} synthesis for amino acids has an effect on protein synthesis, as shown by the rate of incorporation of labeled leucine in cellular proteins (Fig. 5.6). In cells containing corrinoids at levels that are either higher or lower than the optimal (physiological) level, protein synthesis is reduced (Iordan et al., 1984). Apparently, vitamin B_{12} is indirectly involved in protein synthesis, since the protein content in vitamin B_{12}-deficient cells is 67% of the control cells. The decrease in protein content may be a consequence of the inhibition of synthesis and activity of a number of enzymes (first of all, SH-dependent enzymes) in vitamin B_{12}-deficient cells.

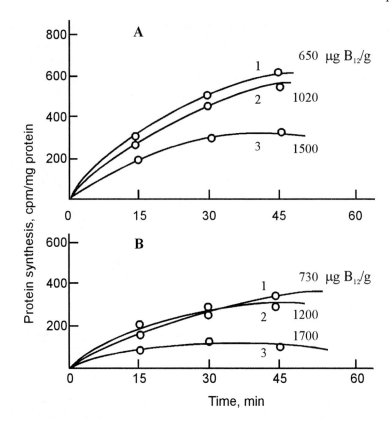

Figure 5.6. Protein synthesis *in vitro* by *P. shermanii* cells with different levels of vitamin B_{12}. Label incorporation was assayed in cells grown for 72 h (A) or 96 h (B). Cell vitamin B_{12} levels per g dry biomass were: in A, 650 μg/g (1), 1020 μg/g (2) and 1500 μg/g (3); in B, 730 μg/g (1), 1200 μg/g (2) and 1700 μg/g (3). From Iordan et al. (1984).

5.3.3   DNA synthesis

Vitamin B_{12}-deficient cells contained about 30-45% less DNA than cells with physiological levels of the vitamin (Vorobjeva and Iordan, 1976; Iordan, 1992). The DNA content in these cells increased by up to 80% when AdoCbl was added to cultures growing in cobalt-free medium (Iordan et al., 1983) (Table 5.1). Strains with a potential capacity for high corrinoid synthesis demonstrated a more significant stimulation by exogenous AdoCbl than strains with low synthetic capacity (Fig. 5.7). However, *P. acnes* represented an exception to this rule: it responded weakly to the addition of AdoCbl, despite having a high potential for vitamin B_{12} synthesis.

Table 5.1. Effects of exogenous AdoCbl on DNA content
of *P. shermanii* cells

Cellular B_{12}, µg/g dry weight	DNA**, mg/g dry weight
No AdoCbl	
1100	18.0
traces	9.5
+ AdoCbl* (µg/g)	
500	16.0
1000	17.0
2000	14.4

*AdoCbl was added on day 3 to cultures growing without
cobalt. **DNA was assayed in cells grown for 4 days.
From Iordan et al. (1983).

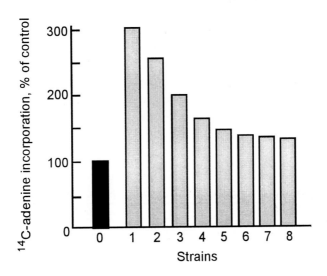

Figure 5.7. Adenosylcobalamin stimulation of DNA synthesis *in vitro* by different strains of
vitamin B_{12}-depleted cells. Results are expressed relative to the vitamin B_{12}-deficient mutant
of *P. shermanii*. Adenosylcobalamin was added to *P. freudenreichii* ssp. *globosum* (1), *P.
freudenreichii* ssp. *freudenreichii* (2), *P. freudenreichii* ssp. *shermanii* (3), *P. acidipropionici*
(4), *P. acnes* (5), *P. thoenii* (6), *P. jensenii* (7), *P. coccoides* (8). From Petukhova (1988).

The rate of incorporation of ^{3}H-adenine in DNA depends on vitamin B_{12}
content of the cells (Fig. 5.8, Table 5.2). Vitamin B_{12}-deficient cells show
lowest rates of the label incorporation into DNA (Iordan et al., 1983).

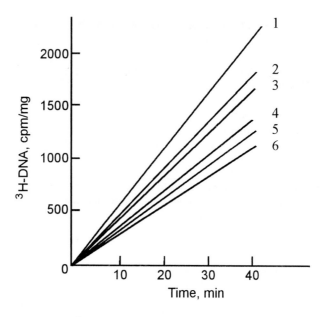

Figure 5.8. Incorporation of ^{3}H-adenine into DNA by *P. shermanii* cells with different levels of vitamin B$_{12}$. Cells grown for 96 h in different conditions contained the following levels of vitamin B$_{12}$: 1100 μg/g (1), 900 μg/g (2), 600 μg/g (3), 200 μg/g (4), 100 μg/g (5), 10 μg/g (6). From Iordan et al. (1983).

Table 5.2. DNA labeling of *P. shermanii* by a 40 min incubation with ^{3}H-adenine

Cellular B$_{12}$ content after 72 h of growth, μg/g	^{3}H-incorporation, cpm	
	per mg biomass	per μg DNA
1100	2110	116
600	1625	91
200	1340	79
100	1242	98
undetectable	1102	122

From Iordan et al. (1983).

It is known that in microorganisms ribonucleotide reductase (RNR) can use either AdoCbl, Fe or Mn as cofactors (Reichard, 1962, 1985; Thelander et al., 1983). The activity of the metal-dependent RNR is inhibited by hydroxyurea. The inhibition of RNR by hydroxyurea was demonstrated in *P. coccoides*, *P. jensenii* and *P. shermanii*. In other strains the inhibition has not been observed and this observation is an additional evidence for the presence of metal-independent, AdoCbl-dependent RNR in these strains.

The total DNA synthesis and the AdoCbl-dependent RNR activity was increased by 80% by adding exogenous DMB to the cells of *P. shermanii* in the exponential phase of growth (Iordan and Pryanishnikova, 1994). The stimulation of DNA synthesis was observed at DMB concentrations of 2 to 12 μg/ml with a maximum at 6 μg/ml. The addition of DMB at 20 μg/ml caused a decrease in the label incorporation in DNA. The authors conclude that the effect is mediated by the AdoCbl-dependent RNR, which normally is not saturated by the coenzyme. These observations show a possible way of controlling the DNA synthesis and its content in the cell by changing the amount of exogenous DMB.

The reduced DNA synthesis observed in vitamin B_{12}-deficient cells is not only due to the low RNR activity responsible for the synthesis of deoxyribonucleotide triphosphates. It is known (Andreeva, 1974) that CH_3Cbl is involved in the synthesis of the thymine intermediate of DNA. In vitamin B_{12}-deficient cells the DNA content increased upon adding thymine to the medium, which had no effect in the case of B_{12}^+-cells (Fig. 5.9, Table 5.3) (Iordan et al., 1979a). It follows, then, that the DNA of propionic acid bacteria is linked with B_{12}-coenzymes in two ways: through the synthesis of deoxyribosides and through the synthesis of thymine. In addition, we suggest that there is a third type of this connection, mediated by DNA methylation.

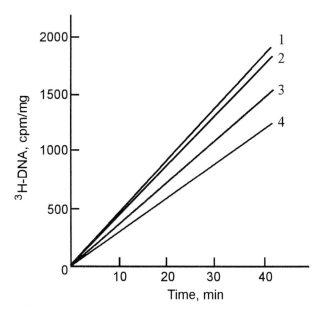

Figure 5.9. Effects of thymine addition on ^{3}H-adenine incorporation in DNA of *P. shermanii* with different levels of vitamin B_{12}. Cell vitamin B_{12} levels were: 900 μg/g (1, 2) or about 10 μg/g (3, 4). Thymine at 5 μg/ml was added (1, 3) or not added (2, 4) to the medium at 4 h before harvesting. Thymine was present in the assay mix at 1 μg/ml. From Iordan et al. (1983).

Table 5.3. Effects of exogenous thymine and thymidine on DNA content in B_{12}-deficient cells of *P. shermanii*

Thymine added*, µg/ml	DNA, mg/g dry weight	Thymidine added*, µg/ml	DNA, mg/g dry weight
None	9.68	None	9.68
1.0	9.42	1.0	9.60
10.0	12.02	10.0	9.10
20.0	9.50	20.0	9.30
B_{12}^{+}-cells	19.20		19.20

*Thymidine and thymine were added on day 3 of cultivation. DNA was assayed 24 h after the addition of thymine and thymidine.   From Iordan et al. (1983).

5.3.4   DNA methylation

We found (Iordan et al., 1979a; Antoshkina et al., 1981) that the DNA of vitamin B_{12}-deficient cells contains 0.28 mol% methylcytosine versus 0.59 mol% in the wild-type strain, indicating that methylation, thought to be required for DNA maturation, is significantly reduced. The DNA from vitamin B_{12}-deficient and control cells was identical in nucleotide composition. Incubation of the DNA isolated from B_{12}-deficient cells with CH_3Cbl in the presence of methylase from B_{12}^{+} cells resulted in an enrichment of CH_3 groups up to the level of B_{12}^{+} cells (Table 5.4). Obviously, the DNA of vitamin B_{12}-deficient cells is undermethylated. The addition of CH_3Cbl had no effect on the DNA methylation in control cells, indicating that all sites on the polynucleotide chain available to DNA-methylase have been saturated. Some incorporation of CH_3-groups from CH_3Cbl in the DNA of the deficient cells occurred in the absence of the enzymatic system, but the presence of methylase increased it considerably, suggesting an enzymatic nature of the process. In the absence of CH_3Cbl the methylation was incomplete. The methylation was blocked when an inhibitor of methylcobalamin, CF_2Cl-cobalamin, was added to the incubation mixture, thus pointing specifically to CH_3Cbl as the donor of methyl groups.

Table 5.4. DNA methylation *in vitro* in the presence of CH_3Cbl and B_{12}-derivatives

Addition		DNA methylation, mol % 5MC*	
Corrinoids	Other components	B_{12}-deficient cells	Control cells
None	None	0.28 ± 0.04	0.59 ± 0.05
CH_3Cbl	+ methylase	0.60 ± 0.05	0.62 ± 0.05
CH_3Cbl	None	0.35 ± 0.03	0.61 ± 0.06
None	+ methylase	0.29 ± 0.03	0.59 ± 0.04
AdoCbl	+ methylase	0.27 ± 0.05	0.58 ± 0.03
Factor B	+ methylase	0.26 ± 0.04	0.60 ± 0.06
CH_3Cbl	+ methylase + CF_2ClCbl	0.35 ± 0.05	0.61 ± 0.07
CH_3Cbl	+ methylase + S-adenosylhomocysteine	0.39 ± 0.06	0.57 ± 0.05

*5MC, 5-methylcytosine.   From Antoshkina et al. (1979).

The methylation also required a small amount of S-adenosylmethionine (AdoMet); its competitive inhibitor, adenosylhomocysteine, when added to the incubation mixture in the presence of CH_3Cbl, blocked the transfer of methyl groups to DNA. Furthermore, it appears that methylases of B_{12}^+ and B_{12}-deficient cells might use different donors of CH_3-groups. The methylase of vitamin B_{12}-deficient cells showed a higher affinity for AdoMet and catalyzed efficiently the methylation of both cytosine and adenine. The methylase of B_{12}^+ cells methylated adenine, but not cytosine with AdoMet. It is possible that AdoMet is used in the absence of CH_3Cbl as the natural donor of CH_3 groups for DNA methylation. In the presence of CH_3Cbl, additional methylation is specifically catalyzed by a vitamin B_{12}-dependent methyltransferase. It seems likely that either there are two separate methylases or one enzyme with two coenzyme sites, one binding AdoMet, and another CH_3Cbl.

Previously, AdoMet was considered as the sole and universal methyl donor for DNA methylation. Our suggestion of the involvement of CH_3Cbl in DNA methylation has been confirmed in *Micrococcus luteus* (Pfohl-Leszkowicz et al., 1991). The authors showed that vitamin B_{12}, CH_3Cbl and AdoCbl stimulated DNA methylation in cell-free extracts of *M. luteus* in the presence of rat spleen DNA-methylase and AdoMet at concentrations up to 1 µM; at higher concentrations, cobalamins acted as competitive inhibitors of the enzymatic methylation. In addition, the authors found that AdoMet did not inhibit the incorporation of CH_3-groups catalyzed by CH_3Cbl. This observation indicates that AdoMet and CH_3Cbl act at different sites on the enzyme. In fact, the nucleotide sequence of the cloned mouse DNA methylase has been determined, revealing two domains: one binding AdoMet and another, responsible for methyl transfer (Bestor et al., 1988).

In conclusion, the physiological level of corrinoids in *P. shermanii* is very high, equaling about 1000 µg/g biomass. This is an order of magnitude higher than in other good producers (Vorobjeva, 1976). B_{12}^+ cells have certain advantages over B_{12}-deficient cells in being more resistant to the toxic action of oxygen, to the lethal effects of UV-light, and to bacteriophages. They are characterized by higher specific rates of DNA and protein synthesis than B_{12}-deficient cells. But we should also recall that about 1.0 µg/l of $CoCl_2 \cdot 6H_2O$ was required to achieve the maximal level of vitamin B_{12}. In natural environments such amounts of cobalt salts are rarely found.

5.3.5   Ecological considerations

In the soil, cobalt is found at concentrations ranging from 1 to 40 ppm
(Mengel and Kirkby, 1980). Since plants require cobalt as a trace element,
they compete with bacteria for cobalt ions, particularly in soils poor in this
element. Meadow soils contain the highest amounts of cobalamins, up to
54.6 ng, while forest podzol soils, where fungi predominate, contain at most
0.05 ng per g dry soil. This corresponds to 2.3 ng and 0.08 ng cobalamins
per 10^6 bacterial cells, respectively. At the depth of more than 15 cm from
the surface, the content of corrinoids is still lower (Atlavinyte et al., 1982).
In natural waters (sea and fresh waters) the content of cobalamins is in the
range of 0.1 to 30 ng/l (Lochehead, 1958; Iwasaki et al., 1968), which is
about 1000 times lower than in soils. Due to the secretion by algae of
significant quantities of certain soluble proteins that bind vitamin B_{12} firmly
and irreversibly, cobalamins are virtually inaccessible to other organisms
(bacteria actively accumulate both cobalt and cobalamins).

Let us consider other natural habitats of propionic acid bacteria. In
grasses used as fodder for livestock the content of cobalt is often below a
certain limit (0.08 ppm), so that if cobaltous salts are not added to feeds,
animals will suffer from cobalamin deficiency. Animals are supplied with
corrinoids mainly through the biosynthetic activity of bacteria. If 1 mg of
cobalt a day is added to the feed, then the content of cobalamins in the dry
residue of lignin matter in rumen is 0.59 to 1.0 µg/g. When no cobalt is
added, the content of cobalamins is lowered by an order of magnitude, to
0.081-0.108 µg/g (Smith and Marston, 1970); correspondingly, the content
of vitamin B_{12} in all animal tissues and fluids is low. Therefore, the
concentration of vitamin B_{12} in meat, milk and other products obtained from
the animal will depend on the content of cobalt in the feed.

Other natural habitats of propionic acid bacteria are represented by
cheese and silage. If one recalls that there are at least 10^5 bacterial cells in 1
g of cheese, then it becomes clear that bacteria lead a cobalamin-deficient
existence. The same is true of bacteria that live in silage, where the
cobalamin content is about 0.1 to 2.0 µg per 100 g (Smith and Marston,
1970). In the rumen of ruminants, where cobalt is limited, the cobalamin
content is in the range of 0.14-0.41 ng/ml (Dryden et al., 1962). It is clear
that these bacteria live at a very low cobalamin level. Therefore, the
conclusion is obvious—most propionic acid bacteria lead a vitamin B_{12}-
deficient mode of life in nature, although they can readily attain high levels
of corrinoids under favorable conditions.

Assuming that the low-vitamin mode of life predominates in nature, it is
difficult to explain how the bacteria find reduced sulfur compounds in soil,
water and other habitats, or how they find strictly anaerobic conditions

required, as we showed above, for growth in these environments, depleted of cobalt. An alternative strategy of survival would necessarily imply the use of parallel systems, independent of vitamin B_{12}. It is known (Vorobjeva, 1976) that vitamin B_{12}-depleted propionic acid bacteria can transform their metabolism so as to reduce the concentration of inorganic phosphate. This is achieved by decreasing the utilization of phosphate from the medium and by increasing the amount of highly polymerized polyphosphates. As a consequence of this rearrangement, accompanied by a reduction in the activity of glycolytic enzymes (Labory, 1970), the pentose cycle is activated and NADPH and glutathione accumulate in the cell, facilitating the adaptation. Our results provide an additional, although indirect, indication of an important role of corrinoids in cell adaptation. We consider this role of corrinoids to be vital for the metabolism of propionic acid bacteria. Evidently, in the absence of vitamin B_{12} the whole metabolism is restructured to compensate for the loss of this function at the expense of some other systems. Meanwhile, the bacteria become auxotrophic in reduced sulfur compounds.

5.3.6 Back to DNA synthesis

Analysis of the results concerning the DNA synthesis in vitamin B_{12}-deficient cells allowed us to conclude that a vitamin B_{12}-independent system may operate in these cells. This system has been investigated in a series of studies (Iordan et al., 1986; Iordan, 1992; Iordan and Petukhova, 1995). After a number of passages in a medium carefully depleted of cobalt (which simulated natural habitats) the cells of *P. shermanii* contained less than 2 μg of cobalamins per g biomass. These cells, adapted to the vitamin B_{12}-free medium, are referred to as vitamin B_{12}-depleted cells. The synthesis and activities of ribonucleotide reductase (RNR) were compared in vitamin B_{12}-replete (B_{12}^+), B_{12}-deficient and B_{12}-depleted cells.

It was found that the DNA content in B_{12}^+ and B_{12}-depleted cells was approximately the same, but higher than in vitamin B_{12}-deficient cells. Hydroxyurea, a specific inhibitor of metal-dependent RNR enzymes, suppressed the RNR activity in vitamin B_{12}-depleted cells (Fig. 5.10). In addition, the RNR activity of vitamin B_{12}-depleted cells was inhibited by μM concentrations of AdoCbl in the incubation mix (Fig. 5.11). Furthermore, the RNR activity of vitamin B_{12}-depleted cells showed a requirement for manganese ions (Fig. 5.12) (Iordan and Petukhova, 1995); it probably used an endogenous H-donor since it did not require dithiothreitol, a regular H-donor in B_{12}^+-cells.

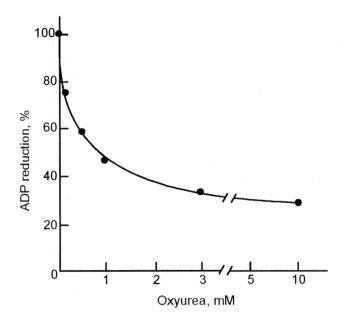

Figure 5.10. Effect of hydroxyurea on ribonucleotide reductase II activity in cell extracts of vitamin B_{12}-depleted cells of *P. freudenreichii* ssp. *shermanii* in the absence of hydrogen donors. From Iordan and Petukhova (1989).

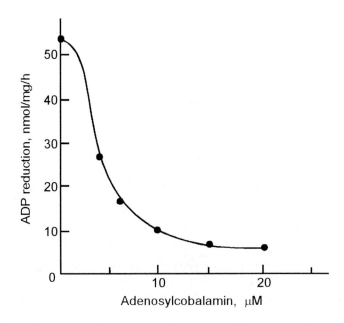

Figure 5.11. Effect of adenosylcobalamin on ribonucleotide reductase activity in cell extracts of vitamin B_{12}-depleted cells of *P. freudenreichii* ssp. *shermanii* in the absence of hydrogen donor. From Petukhova (1988).

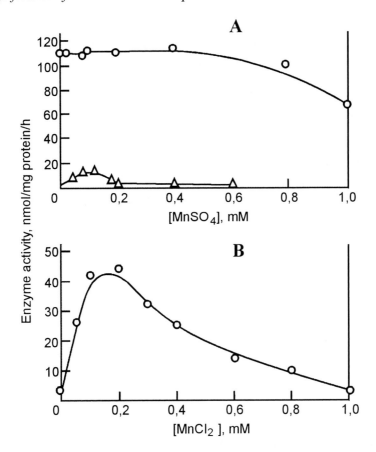

Figure 5.12. Influence of manganese ions on AdoCbl-independent ribonucleotide reductase activity in cell-free extracts of *P. freudenreichii.* A: before (○) and after (Δ) 8-hydroxyquinoline treatment of the cell extract with $MnSO_4 \cdot 5H_2O$. B: after EDTA-dialysis (○) due to $MnCl_2 \cdot 4H_2O$ addition. Magnesium ions were omitted. Incubation time was 40 min. Reproduced from Iordan and Petukhova (1995), with permission.

The AdoCbl-independent RNR had a pH optimum at pH 7.8-8.0, while the AdoCbl-dependent enzyme at pH 7.0. Under optimal conditions the RNR activity of B_{12}^+-cells was about 80 nmol ADP reduced per h per mg protein in the presence of 0.5 mM ADP and 30 mM DTT (reduced glutathione was an effective H-donor). Mg-ions practically were not required. In the same extracts the RNR activity with the pH optimum at 8.0 was also found (in the presence of 3 mM ADP, 30 mM DTT, 3 mM Mg-acetate), but it was twice as low (40-45 nmol ADP reduced per h per mg protein) as the activity of the AdoCbl-dependent RNR (pH optimum at 7.0).

In vitamin B_{12}-deficient cells the RNR activity at pH 8.0 was close to that of B_{12}^+-cells. The activity of AdoCbl-dependent RNR (pH 7.0) was undetectable in these cells. However, if AdoCbl was added to the incubation mix, this activity was recovered and reached the level of B_{12}^+-cells, indicating that the apoenzyme of AdoCbl-dependent RNR was synthesized in adequate amounts, although the vitamin B_{12} content of these cells was two orders of magnitude less than in B_{12}^+-cells. Exogenous AdoCbl, added to vitamin B_{12}-deficient cells, stimulated the DNA synthesis, and this effect was specific, since it was retained in the presence of inhibitors of protein and RNA synthesis (Iordan, 1992). However, in vitamin B_{12}-depleted cells (2 µg/g) the synthesis of the apoenzyme of AdoCbl-dependent RNR apparently was blocked.

Recently, it has been shown (Iordan and Petukhova, 1995) that the AdoCbl-independent RNR (metal-, probably manganese-dependent) is a molecular oxygen-consuming system (cf. Fig. 5.12), different from AdoCbl-dependent enzymes that function in the absence of air, but similar to mammalian RNRs that also require oxygen (Probst et al., 1989).

Two important conclusions may be drawn: *P. shermanii* exhibits both AdoCbl-dependent and AdoCbl-independent RNR activities. The latter system has a lower specific activity and is expressed in cells depleted of cobalamins (containing about 2 µg of vitamin B_{12} per g biomass). In cells containing vitamin B_{12} at 10 µg/g both systems may be functional, although the DNA synthesis is reduced by 50% as compared with the wild-type cells (B_{12} content about 1000 µg/g), which can be explained by a reduced activity of the B_{12}-dependent RNR and by a partial inhibition of the AdoCbl-independent enzyme by AdoCbl. Finally, in vitamin B_{12}-replete cells, the AdoCbl-dependent RNR enzyme predominates and its activity is stimulated by high levels of the coenzyme AdoCbl.

In this respect, the situation is similar to the regulation by cobalamins of AdoMet-dependent DNA-methylation, the expression of which is blocked under conditions favoring methylcobalamin-dependent DNA methylation system that uses CH_3Cbl as coenzyme and CH_3-donor (see above, section 5.3.4).

The observation that corrinoid-deficient cells of *P. shermanii* can grow and replicate normally pointed to the existence of alternative, corrinoid-independent, pathways of the above mentioned syntheses in propionibacteria. Indeed, as shown above, in corrinoid-deficient *P. shermanii* the AdoCbl-independent (probably metal-dependent) oxygen-requiring RNR enzyme functions, and DNA-methylation is catalyzed by the cobalamin-independent methylase which uses AdoMet as methyl donor. Conversely, in vitamin B_{12}-replete cells the AdoCbl-dependent RNR is more active than the metal-dependent RNR, which is inhibited by high concentrations of corrinoids. The discovery of two systems operating in DNA synthesis (ribose-to-deoxyribose conversion) is an example of the functioning in *P. shermanii* of alternative pathways for important biological processes.

Recently, it has been found that microorganisms, in contrast to macroorganisms, possess as much as four or five different types of ribonucleotide reductases (EC 1.17.4.) (Reichard, 1993; Auling and Follmann, 1994). Such a diversity of microbial RNRs might seem surprising, but it is readily explained by their polyphyletic origins and the importance of RNR enzymes, which catalyze the formation of deoxyribonucleotides and thus are vital for DNA synthesis.

To summarize, corrinoids in propionic acid bacteria are involved not only in fermentation, but also in such important anabolic processes as protein and DNA synthesis and DNA methylation. In this respect, corrinoids differ from other related tetrapyrrole compounds by their polyfunctionality. The involvement of corrinoid-dependent enzymes in different metabolic processes in propionibacteria explains the propensity of anaerobic strains of the classical propionic acid bacteria to synthesize large amounts of corrinoids under suitable conditions.

Chapter 6

Immobilized Cells

Immobilization is the method of cultivation of microorganisms that allows a repeated use of biocatalysts (be it enzyme or whole cells), creating prerequisites for the production of valuable products in an automated continuous mode. The most considerable problem in using biocatalysts is related to mass transfer. In aerobic systems, low solubility of oxygen in carriers, especially in some gels and polymers, can decrease the effectiveness of biocatalyst action. In this respect, propionic acid bacteria, which do not require aeration, show certain advantages over aerobic cultures. At present, about eight different processes that use immobilized enzymes and cells have found industrial applications. These are mainly one- or two-step processes used in the manufacture of foods and pharmaceutical preparations (Vorobjeva et al., 1978). An essential characteristic of a biocatalyst is productivity.

6.1   Physiology of Immobilized Propionibacteria

Several methods have been described for immobilization of the cells of *P. shermanii* (Scholl, 1976). Most often cells and enzymes are immobilized in polyacrylamide gel (PAAG) and collagen or are encapsulated in nylon capsules. In certain cases limited to one-step processes (production of aspartic acid and PBG) non-viable cells can be used. Non-viable cells usually have a productivity of 500 to 2000 moles of product per liter of the reactor volume for the period of two half-lives. For the stabilization and long life of a biocatalyst it is necessary to take into account physiological characteristics of immobilized cells.

196

Complex multi-step syntheses (of propionic acid, vitamin B_{12}, ribonucle-otides), however, are performed only by viable cells. Structural and functional integrity is an obligatory condition for this type of biocatalyst. Consequently, there is a problem of maintaining viability for a long time under conditions unfavorable for bacterial reproduction (in nitrogen-free medium). Such nitrogen-starved cells, as shown before (see Chapter 4), carry out endogenous metabolism maintaining a certain low level of ATP needed in particular for the turnover of proteins and some critical growth.

Propionic acid bacteria have been successfully used to obtain a number of practically valuable substances, products of synthetic and/or catabolic reactions. These products are volatile carboxylic acids (propionic and acetic), porphobilinogen, porphyrins, vitamin B_{12}, aspartic and malic acids, nucleotides and their derivatives. In general, these products can be produced both by free and by immobilized cells. The first process, in which immobilized cells of *P. shermanii* were used, was a multi-step synthesis of propionic acid (Vorobjeva, 1978).

6.1.1   Production of propionic acid

Having demonstrated that the immobilization of propionibacterial cells is possible in principle, we conducted a series of methodological and physiological investigations designed to optimize the process.

The cells were immobilized in PAAG by conventional procedures. Immobilized cells were incubated with periodic changes of the incubation mix, containing an energy source and $MgSO_4$, with neutralization of the acids produced. Before transferring the granules into a fresh solution they were repeatedly washed with saline. Immobilized cells can utilize glucose, lactate, lactose of cheese whey, or enzymatic hydrolysate of straw as fermentable substrates (Table 6.1) (Vorobjeva et al., 1984). As a result of fermentation the typical products: propionic and acetic acids were produced, with a small amount of pyruvic acid (Vorobjeva et al., 1977; Iordan et al., 1979b). The yield of organic acids was maximal (4.5 g/l) when lactate was fermented. The ratio of propionic to acetic acid was higher at 37°C than at 30°C.

Table 6.1. Substrates fermented to volatile acids by immobilized propionic acid bacteria

Substrate	Species
Sodium lactate, whey lactose, glucose	*P. technicum, P. shermanii, P. coccoides, P. arabinosum*
Enzymatic hydrolysate of straw	*P. technicum, P. pentosaceum*
Non-hydrolyzed starch	*P. technicum, P. arabinosum, P. coccoides*

From Vorobjeva et al. (1984).

The production of acids had a remarkable time course (Fig. 6.1). In contrast with a gradual and steady decrease in acid production by free cells, in immobilized cells the decrease was followed (after 240 h) by an increased acid production, and this pattern was repeated during 20-25 days of incubation. The curve relating cell viability and aspartase activity of the same biocatalyst (Fig. 6.2) was also waveform (Kalda and Vorobjeva, 1980, 1981). Such a character of acid production reflected cell viability and most likely was linked with a critical growth (rising acidity) at the expense of substances released from dying cells. Immobilized cells showed two peculiarities: (i) they retained viability much longer than free cells, and (ii) frequent changes of the incubation solution accelerated the reduction in cell viability, because more nutrient substances were washed out from the gel.

Immobilized cells of propionic acid bacteria have a number of advantages over immobilized enzymes: (i) fermentation can be performed without adding vitamins to the incubation medium; (ii) coenzymes necessary for fermentation are contained in the cells, though periodical treatment of the cells with a vitamin solution increases the acid production (Fig. 6.3). An especially sharp splash of the acid production is observed upon transferring the biocatalyst into a solution containing $(NH_4)_2SO_4$ or into rich corn steep-glucose medium (Fig. 6.4). A further periodic reactivation of the cells with these solutions resulted in a waveform character of the acid production, which thus could be maintained at 100% level for as long as 28 days.

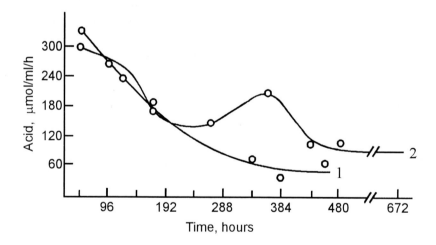

Figure 6.1. Production of acids by free (1) and immobilized cells (2) of *P. shermanii* during long-term incubation. From Ikonnikov (1985).

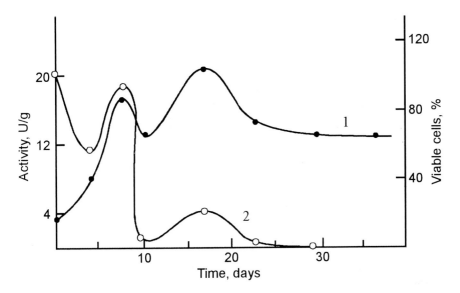

Figure 6.2. Activity and viability of *P. shermanii* cells immobilized in PAAG during incubation in the presence of 20 mM ammonium fumarate at 30°C. Curve 1 - activity; 2 - viability. From Kalda and Vorobjeva (1981).

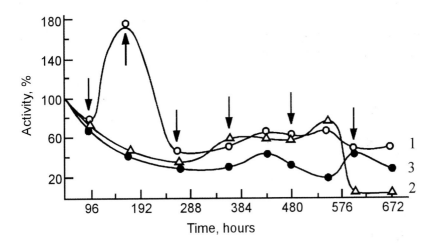

Figure 6.3. Biocatalyst reactivation by separate components of the nutrient medium. Curve 1 - 0.1% ammonium sulfate; 2 - vitamin mixture; 3 - control. Arrows show time of reactivation. From Iordan et al. (1979b).

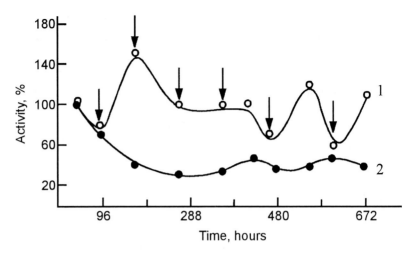

Figure 6.4. Biocatalyst reactivation by rich medium inducing cell division. Biocatalyst was treated with rich medium (1) or nutrient-free incubation solution (2). Arrows indicate time of treatment at 18 h. From Iordan et al. (1979b).

The observed time courses of changes in RNA-, DNA- and protein content of the immobilized cells (Figs. 6.5, 6.6) suggested the occurrence of the processes of degradation, resynthesis and renovation of cellular polymers (Ikonnikov et al., 1982). After 4 days of incubation the endogenous metabolism was accompanied by a reduction in cellular nucleic acids, which (especially RNA) could have been used for anabolic purposes.

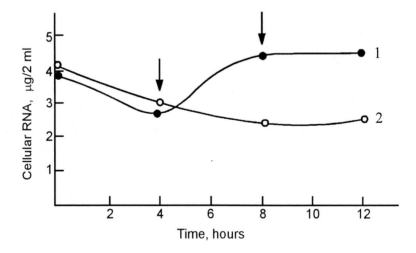

Figure 6.5. Time-dependent changes in RNA content of immobilized cells of *P. shermanii*. Biocatalyst was reactivated with rich medium (1) or remained without activation (2). Arrows indicate time of reactivation. From Ikonnikov et al. (1982).

After reactivation the levels of nucleic acids and proteins were considerably increased, suggesting that constructive processes were intensified. Immobilization increased thermal stability of *P. shermanii* cells, which retained their capacity for acid formation after heating at 70°C, while free cells completely lost this function upon heating at 62°C (Iordan et al., 1979b).

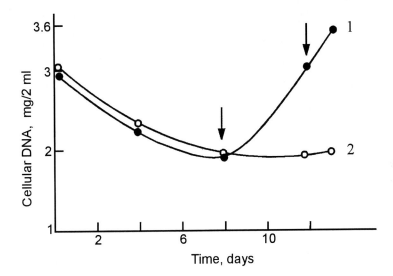

Figure 6.6. Time-dependent changes in DNA content of immobilized cells of *P. shermanii*. Biocatalyst was reactivated with rich medium (1) or remained without activation (2). Arrows show time of reactivation. From Ikonnikov et al. (1982).

Scanning and transmission electron microscopy showed that immobilized cells retained their normal appearance immediately after immobilization (Fig. 6.7A); 192 h later cell density in the gel was reduced, but still remained sufficiently high (Fig. 6.7B). The cells retained their structural integrity, although some injured cells were also present (Fig. 6.8). After 20 days of operation with reactivation some swollen cells with rarefied contents appeared in the population, but most cells showed normal morphology and cell structure.

How long in general can immobilized cells of propionibacteria work? We recorded that the cells still produced pH changes after 7 months of work (Vorobjeva et al., 1977), which was in accordance with the data (Meganathan and Ensign, 1976) showing a stable activity of glucose-metabolizing enzymes in microorganisms under conditions of long-term nitrogen- and carbon- starvation. After 9 days of functioning cells isolated from the biocatalyst were still capable of forming colonies typical for propionic acid bacteria under anaerobic conditions.

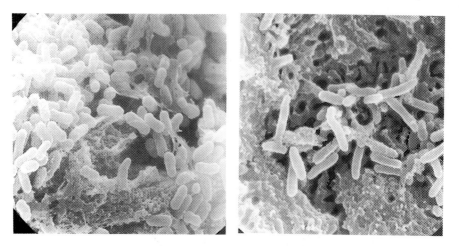

Figure 6.7. Scanning electron microscope view of *P. shermanii* cells immobilized in PAAG. (A) Immediately after immobilization. (B) After 192 h of functioning. × 6,612. From Ikonnikov (1985).

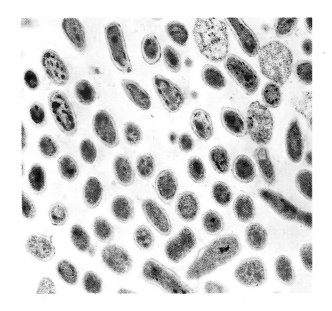

Figure 6.8. Morphology of *P. shermanii* cells immobilized and functional in PAAG for 192 h. Scanning electron microscope view. × 6,612. From Ikonnikov (1985).

Cells immobilized for 7 months differed in morphology: badly swollen cells could be found; cells with numerous indentations or broken at the ends; many empty cells and cells with damaged cell walls, apparently undergoing a deep lysis. However, intact cells that retained main structural features similar to the control cells were still found: one could distinguish the cell wall, cytoplasmic membrane and membrane structures. Degradation of ultrastructure in immobilized cells of *Bacillus megaterium* was apparent on day 5 after immobilization, and was accompanied by a reduction in 20-α-oxysteroid hydrogenase activity (Lusta et al., 1976). Evidently, the destruction of propionic acid bacteria proceeds more slowly and affects not all cells, inasmuch as they continued to produce volatile acids having been immobilized for such a long time.

These observations on the immobilized cells allowed us to develop an optimal regime (Table 6.2) for the operation of the biocatalyst column (16 × 500 mm): gel quantity, 40 g; flow rate, 0.25 h^{-1}; temperature, 43°C; glucose concentration, 0.8%; $MgSO_4$, 0.1% in 0.05 M K-Na-phosphate buffer, pH 7.0 (Iordan et al., 1978). The biocatalyst with immobilized propionibacterial cells can be recommended for continuous production of propionic and acetic acids, good preservatives for food industry and agriculture (see Chapter 7).

Table 6.2. Production of propionic, acetic and pyruvic acids by immobilized cells of *P. shermanii* under continuous incubation

Carbon source	Biomass immobilized in 10 ml, g	Volatile acids, μmol/100 ml			Pyruvic acid, μmol/100 ml
		total	propionic	acetic	
Glucose					
1%	1.75				17 ± 3
2%	1.00	1448 ± 25	872 ± 10	630 ± 9	
Na-lactate					
1%	1.75				38 ± 7
2%	1.75	5640 ± 21	2730 ± 12	2810 ± 16	
Dried whey					
10%	1.75	4480 ± 42	2958 ± 24	1450 ± 15	34 ± 4

*Granules with immobilized cells were preincubated with the respective carbon source for 18 h. Incubation conditions: t = 37°C; 40 g of gel per column; 0.1% $MgSO_4$ in 0.05M K-Na-phosphate buffer, pH 7.0. From Iordan et al. (1979b).

6.1.2 Release of nucleotides and their derivatives

As mentioned above, immobilized cells are studied mainly for practical reasons, since they show a number of economic advantages over the use of growing cells or cell suspensions. Production of organic acids is one of the prospective applications of immobilized cells. Another one is related to the release of nitrogenous bases and some nucleosides by immobilized cells. In nitrogen-starved immobilized cells the levels of all metabolites (first of all, nucleotides) are reduced (Leps and Ensign, 1979). It was shown (Ikonnikov et al., 1982) that immobilized cells of propionic acid bacteria, incubated periodically in nitrogen-free medium, released substances of protein and nucleic acid nature, whose quantity decreased with the time of incubation in the absence of glucose and upon heating at 70°C for 10 min. These observations showed that the release of proteins and nucleic acid derivatives is an energy-dependent process and not a simple degradation. The oscillatory character of the release of nucleic acids derivatives (Fig. 6.9) by both reactivated and non-reactivated cells suggested an active regulation of this process.

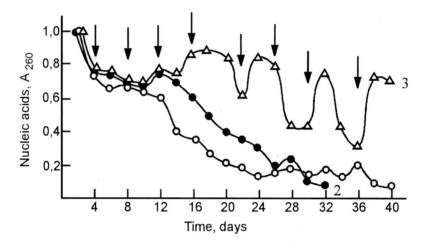

Figure 6.9. Release of nucleic acid derivatives by immobilized cells of *P. shermanii*. Untreated cells (1) or reactivated with rich medium (2). Arrows show time of reactivation. From Ikonnikov et al. (1982).

The main source of nucleic acid-derived substances appears to be ribosomal RNA (rRNA), since the rRNA content decreased faster than the total RNA content (Ikonnikov, 1985), and Mg^{2+} ions, known to have a stabilizing effect on ribosomes, prevented the release of nucleic acid derivatives. The main species released by the immobilized cells were nitrogenous bases: adenine, guanine, cytosine, uracil, as well as nucleosides: adenosine and guanosine (Fig. 6.10). The total yield of nucleic acid derivatives in the first 48 h reached 45 µg/ml, which is rather high. Reactivation of the biocatalyst, especially with rich medium, increased the release and led to its stabilization (cf. Fig. 6.9). Periodic reactivation allows maintaining 60-70% of the initial release level. Synthetic medium was much less effective.

As an alternative to immobilized cells, the use of resting cells of *P. shermanii*, made permeable to nucleotides, has been suggested (Iordan and Pryanishnikova, 1994). The cells were permeabilized by treatment with diethyl ether or Triton X-100. The best effect was obtained by using Triton X-100 in combination with subsequent freezing and thawing of the cells. The biocatalyst prepared in this way was permeable to both the substrate of

RNR and its reaction product, ADP and 2-deoxy-ADP, respectively. 2-deoxy-ADP accumulated in the incubation mixture. Moreover, inhibition of DNA synthesis in these cells by adding antibiotics, among which bleomycin (0.4-0.5 µg/ml for 2.5 h) was found to be the most effective, increased the RNR activity twofold as compared with the control; the yield was 9.8-10.5 nmoles of 2-deoxy-ADP per mg dry cells per h.

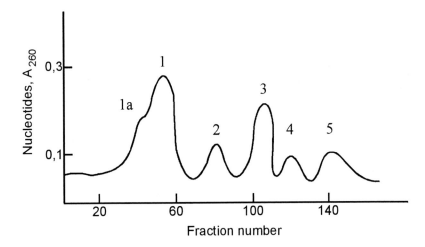

Figure 6.10. Elution of nucleotides, produced by immobilized cells of *P. shermanii*, from anion-exchange column. The column was packed with anionite AB-17 in COO⁻ form. 1 - AMP, 1a - CMP, 2 - GMP, 3 - GTP, 4 - ADP, 5 - various nucleosidephosphates. From Ikonnikov et al. (1982).

The nucleosides GMP, IMP and XMP are known to enhance food flavor. They display a synergistic action with sodium glutamate (Demain, 1968). Nucleotide derivatives are recommended for the treatment of thromboses. Inosine (under the name Riboxin) is a pharmaceutical agent used in the treatment of various heart and liver diseases in Russia. Ribosides and adenine support high levels of ATP in erythrocytes. Deoxyribonucleotides are used in chemical syntheses of some antitumor and antiviral preparations (Sidermene et al., 1985). In addition, 2-deoxyribonucleotides (-sides) serve as substrates for enzymatic and chemical syntheses of specific oligodeoxyribonucleotides, which at present are widely used as biosensors (Efremenko et al., 1990).

The main industrial method for the production of nucleotides is by enzymatic hydrolysis of yeast RNA to four nitrogenous bases with subsequent deamination of AMP to IMP, with CMP and UMP being waste products. Microbiological methods are being developed, based on the release of RNA derivatives, direct enzymatic production of nucleotides that are not RNA derivatives, and salvage synthesis (conversion of bases or nucleosides released by bacteria to nucleotides, for example, from hypoxanthene produced by a mutant strain of *Brevibacterium ammoniagenes*, IMP is produced). Usually mutants with a defective regulation of synthesis and permeability to nucleotides are used (Demain, 1968).

Further investigations of immobilized propionic acid bacteria may attract interest to these bacteria as possible producers of nucleosides and their derivatives for industrial use.

6.1.3   Production of aspartic and malic acids

Aspartic acid is produced on an industrial scale in Japan using non-viable immobilized cells of *E. coli* and ammonium fumarate as substrate of the one-step reaction:

$$\text{Fumarate} + NH_4^+ \xrightarrow{\text{aspartase}} \text{L-aspartate.}$$

The biocatalyst has a half-life of 120 days.

Immobilized cells of *E. coli* also produce malic acid in the following reaction:

$$\text{Fumaric acid} + H_2O \xrightarrow{\text{fumarase}} \text{malic acid.}$$

In general, endogenous metabolism of anaerobic bacteria was found to be more stable, when biocatalysts based on immobilized cells of *P. shermanii* and *E. coli* were compared with respect to the reactions shown above (Ikonnikov, 1985). *P. shermanii* had a higher aspartase activity than *P. pentosaceum*, *P. petersonii* and *P. technicum* (Kalda and Vorobjeva, 1981). After 3 days of incubation with continuous stirring at 37°C and pH 8.5, the extent of substrate conversion (ammonium fumarate) was 95-96% and 75-90% in the case of *E. coli* K-12 and *P. shermanii*, respectively. In addition to aspartic acid, the reaction mixtures of the two strains also contained malic acid. Heat treatment of the biomass of *P. shermanii* (50°C, 1.5 h, pH 5.0) resulted in a complete inactivation of fumarase, while the activity of aspartase was retained (Kalda and Vorobjeva, 1980, 1981). As a result of the elimination of fumarase activity, the yield of L-aspartic acid from ammonium fumarate was increased up to 96-98%; the incubation time was also shortened since no substrate was diverted to the side reaction forming malate.

Preparations of *P. shermanii* showed a number of advantages over *E. coli* as biocatalyst, being more stable (Kalda, 1984), not requiring aeration, and producing more biomass in a cheaper medium. Living cells of *P. shermanii* can also be used for malic acid production. The yield of malic acid in 5 days was up to 72% with respect to fumarate, when *P. shermanii* was incubated at 7 g biomass per one l of Na-fumarate solution at pH 8.5 (Kalda, 1984).

L-aspartic acid is used in pharmaceutical and food industries, and malic acid is used in food industry as a substitute for citric acid. The cells of *P. shermanii* immobilized in PAAG can be recommended for the production of L-aspartic acid with an expected cost reduction of 25%.

6.1.4 Production of porphobilinogen

Four methods are known for porphobilinogen (PBG) production: (i) isolation from human or animal urine; (ii) chemical synthesis from 5-aminolevulinic acid (ALA); (iii) enzymatic synthesis using ALA dehydratase (the enzyme is isolated from erythrocytes, *P. shermanii* or *Rhodopseudomonas sphaeroides*); and (iv) microbiological production using heat-treated cells of *P. shermanii*, incubated in a medium containing ALA.

Chemical synthesis of PBG, based on the formation of pyrroles, includes more than 10 steps and gives the product with a yield of 25%. In chemical synthesis from ALA, the yield of PBG is about 10%. Microbiological methods using ALA as a substrate have a good perspective, since the product yield can reach up to 54%. Propionibacteria present a special interest for PBG production, since they have a high natural capacity for the synthesis of tetrapyrrole compounds, for which PBG is a common precursor.

By incubating suspensions of *P. shermanii* (100 mg/ml in 0.05M tris-HCl buffer, pH 8.0) at 70°C, a 60-70% conversion of added ALA to PBG was observed (Scholl, 1976), but at this high temperature PBG is unstable and is converted to porphyrins. If the suspension is first incubated at 70°C for 30 min, then ALA is added and incubation continued at 30°C, the yield of crystalline PBG is increased from 21 to 54% (Scholl, 1976).

Therefore, we used thermally treated cells of *P. shermanii*, incorporating them in PAAG granules by a special method. The gel containing 1 g of *P. shermanii* cells (total weight, 10 g) was placed in 20 ml of an ALA-containing buffer solution. Immobilized cells were incubated with shaking for 30 h at 30°C. The cells were periodically reactivated by incubating them with shaking in a nutrient medium or in corn steep-glucose medium for 90 h. The yield of PBG was highest when the cells were reactivated in rich medium or in 12% corn meal. Maximal production of PBG was observed at 25 h when the pH was 8.7, and at 30-35 h when the pH was 9.2. After this period the cell productivity dropped.

It is noteworthy that with cells in suspension the pH-optimum was observed at pH 7.8 (buffered with Tris-HCl), but with immobilized cells it was higher and depended on the nature of the buffer used: pH 9.2 in the case of Bis-HCl and pH 11.0 in the case of NH_3/H_2O (Scholl, 1976). It shows once more (compare also the relation to temperature) that immobilization may change the properties of some enzymes.

6.1.5   Production of vitamin B_{12}

In 1982, a report appeared (Yongsmith et al., 1982) on vitamin B_{12} production by the cells of propionibacteria immobilized in urethane (prepolymer PU-9) and incubated periodically in a complete medium containing glucose, casamino acids, bactotryptone, yeast extract, vitamins, mineral salts, including $CoCl_2$, and the precursor DMB. After 18 days of incubation with changes of the medium every 3 days, 900 μg of vitamin B_{12} was obtained from 5 g of the cells of *P. arabinosum* AKU-1251; most of the vitamin was released into the medium. The vitamin was mainly represented by OHCbl. The authors showed that with immobilized cells it was possible to perform 5-6 consecutive production cycles while retaining the initial activity, and to obtain 180 μg of vitamin per 1 g of wet cells in 18 days. Afterwards the cell productivity was reduced by 50%.

It is remarkable that cells of the strain AKU-1251 secreted the vitamin synthesized, although the authors could not exclude that most of the vitamin released by the immobilized cells resulted from cell autolysis. The bacteria multiplied inside the gel, because the medium contained all the compounds required for growth.

The advantage of such a method of vitamin B_{12} production over the traditional methods is that the cells are used repeatedly; the vitamin is isolated from the culture liquid without separating the biomass, so that extraction and multi-stage purification of vitamin B_{12} is unnecessary.

Chapter 7

Economic and Medical Applications

Various products of propionibacterial metabolism have practical uses: propionic and acetic acids formed in fermentation, and other products formed in biosynthetic reactions, including their biomass. Depending on the manufacturing objectives, metabolically active or inactive biomass can be used.

Metabolically inactive biomass is used as a source of protein, which in propionic acid bacteria is rich in sulfur-containing amino acids, especially methionine, and also in threonine and lysine, as well as in group B vitamins (Stasinska, 1977). Propionic acid bacteria are approved for application in food industry and in animal husbandry. The biomass of *P. freudenreichii* has been recommended as an additive to fodder (Vuorinen and Mantere-Alhonen, 1982), to enrich the latter in trace elements, vitamins and proteins. The biomass of non-living (heat-treated) propionibacteria can serve as a source of vitamin B_{12}, which is resistant to heating. Cutaneous bacteria (*P. acnes* and *P. granulosum*) killed by heating are used in the production of immunostimulating preparations (see below), and *P. granulosum* as a source of porphyrins.

The second category of manufacturing is generally based on the production of metabolically active biomass. This includes the production of starter cultures for cheese making, for baking of bread, for silage, manufacturing of pharmaceutical and veterinary preparations, desaccharization of egg white. In other applications, metabolically active biomass can be used as a source for the isolation of superoxide dismutase (SOD) and catalase (Table 7.1).

Cheese making and vitamin B_{12} production are large-scale industries, operating in many countries. Propionibacteria are used to produce vitamin

B$_{12}$ in Russia, Great Britain, Hungary. Hard rennet cheeses (Swiss, Emmental, Soviet and others), in which propionic acid bacteria are involved, are produced almost everywhere.

Table 7.1. Current and prospective applications of propionic acid bacteria

Application	Current		Prospective	Recent improvement, reference
	World	Russia		
Cheese making	+	+		Multi-strain starter culture of propionic acid bacteria (Alekseeva et al., 1983)
Vitamin B$_{12}$	+	+		A mutant superproducer (Gruzina, 1974; Ganicheva, Vorobjeva, 1991). Rapid selection method for vitamin B$_{12}$ superproducers (Vorobjeva, 1976)
Preparation "Propiovit" for livestock	–	+		Sizova, Volkova, 1974; Mantere-Alhonen, 1995
Unicellular protein (biomass)	–	–	+	
Preparation for ensilage	+	+		Iljina, Besedina, 1966; Konoplev, Scherbakov, 1987
Leaven additive for baking	+	+		Bogatyreva et al., 1987
Desaccharization of egg white	–	–	+	Vorobjeva et al., 1979
Production of propionic acid	+	–		Boyaval, Corre, 1995
Production of superoxide dismutase	–	–	+	Kraeva, Vorobjeva, 1981

Applications of propionibacterial cultures as an additive to leavening for baking (USA, Russia) and for ensilage (Russia), production of propionic acid as a fungicide (Germany) are on a limited scale. There are good prospects for other manufactures based on propionibacteria, such as superoxide dismutase and catalase production, and desaccharization of egg white.

7.1   Cheese Making

Cheese making is the most ancient biotechnology capitalizing on biochemical activities of propionic acid bacteria, if we recall that the age of the first cheese may be as old as 9000 years ago, when sheep were domesticated in the Middle East. It has been suggested (Mair-Waldburg, 1974) that when bags made of animal stomachs were being used for the storage of milk, a lucky combination of rennin, some lactic acid bacteria incidentally present, and warm temperatures resulted in the first cheese-like product.

Nutritive value and taste of a good cheese were always highly appraised: "A dessert without cheese is like a girl without smile", said the celebrated French aphorist Anthelme Brillat-Savarin (1755-1826). In these words the pleasure of consuming this excellent product was expressed. The first investigations of propionic acid bacteria were conducted in relation to their role in cheese ripening. Excellent organoleptic properties and long storage times of hard rennet cheeses (produced with high-temperature second heating step) are due to the presence of propionibacteria.

A general rule concerning the use of propionibacteria in cheese ripening is as follows: both shortage and excess of propionibacteria are detrimental, but without these bacteria cheese of a proper quality cannot be made— "blind" cheese (Siewert, 1989) may result, i.e., cheese without eyes or with other defects. To ensure good conditions for propionic acid bacteria, it is recommended to use milk with a high protein content (3.3%); at urea concentrations below 4 mM it is necessary to ensure the prevalence of streptococci over lactic acid bacteria by inoculating a mixed thermophilic culture of these bacteria at a ratio of 10:1. There are other conditions as well, ignoring which may lead to the inhibition of propionibacterial activity in cheese.

Establishment of these conditions created a scientific basis for cheese making. Regardless of the type of cheese manufactured, the whole process generally consists of the following steps (Kosikowski, 1977):

   (1) setting milk (adding starter cultures and coagulants to prewarmed milk),
   (2) cutting the coagulum (curd),
   (3) shrinking the curd by cooking,
   (4) removing whey from the curd,
   (5) allowing curd particles to "knit,"
   (6) salting (this comes at different times in the procedure for different cheeses),
   (7) pressing, and
   (8) ripening of the finished cheese.

Ingredients used in the manufacture of cheese, besides milk, include selected bacteria, milk-clotting agents, and sodium chloride. Variations in these basic constituents, the use of additional ingredients, and variations in the physical conditions of the manufacturing process have given hundreds of varieties of cheese. A general scheme for the manufacture of a Swiss-type cheese (Soviet) with the high-temperature second heating step is shown in the following chart:

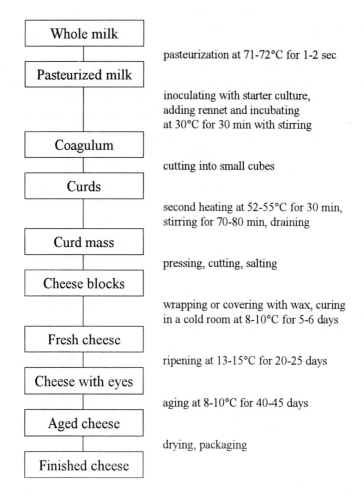

Ripening of cheese is a complex biochemical process, requiring the presence of rennin, milk enzymes, lactic and propionic acid bacteria. Enzymatic breakdown of proteins, lipids, nucleic acids occurs, forming the flavor, taste and consistency of the cheese. Rennin, a proteolytic enzyme, is contained in the preparation called rennet, which is obtained from the rumen of calves by extraction with salt brine. Salting inhibits the growth of undesirable cheese microflora, especially lactobacilli. High-temperature second heating (in the case of Swiss-type cheese) eliminates most of the mesophilic lactobacilli, and stimulates the growth of thermophilic lactic streptococci and propionic acid bacteria. The lactate formed at the preceding stages is actively fermented by propionibacteria. Physiological properties of propionibacteria, such as their relative heat resistance, tolerance to NaCl concentrations up to 4.5%, growth inhibition at 9°C, ability to ferment lactate, are advantageous for the manufacturing process of cheese making.

However, propionic acid bacteria play only a minor role in the ripening of cheeses processed at lower temperatures (second heating at 40-42°C), such as Cheddar or Gouda.

The taste and flavor of cheeses depend on microorganisms playing a major role in ripening. From the high-quality Soviet cheese and from cheese milk 25 strains of propionic acid bacteria were isolated (Alekseeva et al., 1973a), of which 17 strains represented *P. freudenreichii*, and the others were *P. acidipropionici* and *P. jensenii*; thus the principal strain of Soviet cheese is *P. freudenreichii*. In addition, propionic acid cocci were also isolated at an early stage of cheese ripening (see Section 1.2.2).

The main role of propionic acid bacteria in cheese ripening consists in the utilization of lactate produced by lactic acid bacteria as an end product of lactose fermentation. Lactate is then transformed into propionic and acetic acids and CO_2. The volatile acids provide a specific sharp taste and help preserve a milk protein, casein. Hydrolysis of lipids with the formation of fatty acids is essential for the taste qualities of cheese. The release of proline and other amino acids and such volatile compounds as acetoin, diacetyl, dimethylsulfide, acetaldehyde is important for the formation of cheese aroma. Carbon dioxide released in the processes of propionic acid fermentation and decarboxylation of amino acids (mainly) forms eyes, or holes. Propionic acid bacteria also produce vitamins, first of all, vitamin B_{12}. At the same time, an important condition is to keep propionibacteria from growing and producing CO_2 at low temperatures, since this would cause cracks and fissures in cheese.

A special medium has been suggested for the identification and enumeration of propionic acid bacteria in cheeses (Drinan and Cogan, 1992), having the following composition (%): sodium lactate, 1; tryptone, 1; yeast extract, 1; KH_2PO_4, 0.5; agar, 1; pH 7.0. The antibiotic cloxacycline (0.4 µg/ml) is added to inhibit the growth of mesophilic streptococci, a principal component of the natural microflora of cheeses with low-temperature second heating. Some bacteria, unrelated to the starter culture, such as mesophilic lactobacilli, enterococci, *Cl. tyrobutyricum*, also can grow in this medium, but their colonies are readily distinguished from the colonies of propionibacteria by size, color and the absence of catalase.

The classical manufacturing process of the Swiss-type cheese did not rely on special introduction of propionibacteria (as the starter), but sufficient numbers of propionic acid bacteria were present naturally in raw milk and rennet extract. At present, pasteurized milk is used in cheese making, and during heating at 71°C for 15 s most of the propionibacteria are killed (Alekseeva et al., 1983). A standard requirement for the content of propionibacteria is $2 \cdot 10^3$-$4 \cdot 10^3$ per g Soviet cheese, that is why it is necessary to add propionic acid bacteria with high acid-, gas- and lipolytic activities, resistant to various inhibitors (including foreign microflora), and compatible with lactic acid bacteria as a component of the starter cultures.

Interactions between propionic and lactic acid bacteria have a strong effect on cheese formation and ripening. As reported by Alekseeva et al. (1983), nine out of 22 strains of lactic acid bacteria were antagonistic to propionibacteria, with the greatest inhibition displayed by *Str. lactis* and *Str. diacetilactis*. On the other hand, *Str. cremoris*, *Str. thermophilus* and *Lb. helveticus* were found to be compatible with *P. freudenreichii* and *P. shermanii*. Piveteau et al. (1995b) observed that of the 14 strains of lactobacteria and 4 strains of propionibacteria tested the greatest stimulatory effect (assessed by growth rate and final biomass) was achieved by growing sequentially *Lb. helveticus* RR and *P. freudenreichii* KM. *Lb. helveticus* RR increased the levels of amino acids and peptides that were utilized during the subsequent growth of *P. freudenreichii* KM.

Certain natural inhabitants of cheese may exert a stimulatory action on biochemical activities of propionic acid bacteria. In the presence of cheese micrococci the CO_2 release by propionic acid bacteria increased by 20% (Ritter et al., 1967). *Micrococcus caseolyticus* inhibits the growth of *E. coli* during cheese making and ripening (Laipanov, 1989) and thus creates favorable conditions for propionic acid bacteria.

Propionibacterial phospholipases have an important role in the ripening and production of high quality cheeses. Propionic acid bacteria contain both intra- (type A) and extracellular (type C) phospholipases (Umansky and Melnikova, 1986). In most strains both lipases are present, but it is the extracellular phospholipase activity that is important for cheese making. Phospholipase C of propionic acid bacteria is specific to milk phospholipids: phosphosphingosides (sphingomyelin), phosphatidylcholine and phosphatidylethanolamine. By the action of phospholipase C a considerable hydrolysis of phospholipid components of cheese occurs (eight such components were found) without the formation of lysophospholipids, which are found in rancid samples of cheese in addition to phosphatidic acids (Umansky and Melnikova, 1986). The greatest number of strains with high phospholipase activity was found in *P. globosum* (Melnikova, 1987). The cheese, manufactured from the starter culture of a strain with a high level of phospholipase C, showed the best indexes of taste, consistency and texture; that strain of *P. globosum* was recommended to include in the starter culture for Soviet cheese.

A further improvement in the process of cheese making is related to the creation of a dry multi-strain starter of propionic acid bacteria (Alekseeva et al., 1983), consisting of three strains of *P. freudenreichii*. That starter culture was superior to the monoculture with regard to gas and acid production and could be inoculated directly in milk without preactivation. The dry starter culture is more convenient for transportation than the liquid one, used before. The multi-strain dry starter culture is being used in Russia in the manufacture of Soviet cheese.

7.2   Vitamin B12 Production

At present, only microbiological methods are used for the production of vitamin B_{12}. Chemical synthesis is not a practical alternative. Two types of vitamin B_{12} preparations are manufactured: pharmaceutical forms and crude concentrates for animal feed.

Pharmaceutical preparations are manufactured by using mutant strains of *P. shermanii* and *P. freudenreichii*, active producers of vitamin B_{12}. In the USA for the same purpose a mutant strain of *Pseudomonas denitrificans* is used. Vitamin B_{12} concentrate for feeds is prepared on the basis of the biomass of methanogenic bacteria.

Biosynthesis of vitamin B_{12} by propionic acid bacteria, as we have seen (Section 4.4.5), almost parallels growth under anaerobic conditions. Vitamin B_{12} accumulates in the cell (the above-mentioned case of vitamin B_{12} secretion apparently is peculiar to a single mutant strain) mainly as coenzyme forms of incomplete corrinoids. These features are taken into account in the industrial production of vitamin B_{12}. The culture is grown anaerobically, in a medium containing glucose, corn-steep liquor, ammonium sulfate, and a cobaltous salt, since Co is part of the vitamin B_{12} molecule. Acids formed by the culture are continuously neutralized with a solution of NaOH or $(NH_4)_2OH$.

To produce the complete, clinically active forms of corrinoids, the medium is supplemented, after 72 h of cultivation, with the vitamin B_{12} precursor 5,6-dimethylbenzimidazole (DMB). The cultivation is finished in 96-110 h. The precursor can also be added after the cultivation is terminated, and even to the suspension of non-growing cells, but not at the beginning of cultivation, for Cbl will repress its own synthesis. When the cultivation is finished, vitamin B_{12} is isolated from the separated biomass by extraction with acidified (to pH 4.5-5.0) hot water containing a stabilizer ($NaNO_2$ or KCN). If AdoCbl is being produced, no stabilizer is added and the extraction is carried out in a dark room or under red light.

The aqueous extract containing vitamin B_{12} is cooled, the pH is adjusted to 6.8-7.0, and soluble proteins are precipitated by adding $Al_2(SO_4)_3$ and $FeCl_3$. Then the solution is filtered and vitamin B_{12} is purified by ion-exchange chromatography. For this purpose, the solution is acidified to pH 2.5-2.7 and passed through a series of binding columns packed with a cation-exchange resin SG-1 (Russia). The vitamin bound to the column is eluted with an ammonium solution. The eluate is diluted tenfold with water, and activated charcoal and NaCN are added at a ratio of 0.9% NaCN relative to the charcoal. This mass is stirred for two hours, acidified with hydrochloric acid and, after filtration and washing, vitamin B_{12} is desorbed from the charcoal with isopropyl alcohol. Then the isopropyl alcohol is evaporated and vitamin B_{12} is subjected to an additional cycle of purification by the resorcin method and recrystallization.

This method of vitamin B_{12} isolation is accepted in Russia. Pharmaceutical preparations of vitamin B_{12} in Russia are produced at the Sintez plant in Kurgan, using a mutant strain of *P. freudenreichii*, which produces about 30 mg/l of vitamin B_{12}.

Apparently, the potential of propionic acid bacteria might exceed this level many times, if one takes into account the reported ability of a mutant strain of *P. shermanii* to synthesize vitamin B_{12} up to 216 mg/l (French patent No. 2209842, 1984). The strain was isolated by a combined treatment with mustard gas, antibiotics, and Mn salts that increase the permeability of the cell membrane for various compounds. Milk whey has been suggested as an inexpensive medium for the bacterial vitamin B_{12} production (Marwaha et al., 1983). Propionibacteria utilized lactose from the whey and produced more than 5 mg/l of vitamin B_{12}.

The world production of vitamin B_{12} was estimated at 12,000 kg in 1980, of which 54% is used for human consumption (pharmaceuticals and food additives) and 41% as a growth factor for animal feeds. The major share of this production is manufactured by three companies: Merck & Co., Glaxo Labs, and Rhone-Poulenck. Rhone-Poulenck produces more than 50% of the whole volume of the world market.

7.2.1   Applications of vitamin B_{12}

Medicine. In humans and animals AdoCbl is the coenzyme of methylmalonyl CoA-mutase that catalyzes the isomerization of methyl-malonyl-CoA into succinyl-CoA (in propionibacteria this reaction runs in the opposite direction). The reaction is linked with the catabolism of amino acids and lipids. If the activity of methylmalonyl CoA-mutase is blocked, the catabolism of some amino acids, fatty acids and thymine is inhibited (Fig. 7.1). Intracellular levels of methylmalonyl-CoA and propionyl-CoA are increased and may affect fatty acids synthesis. In some cases, an increase in the content of C_{15} and C_{17} odd-numbered fatty acids and branched-chain fatty acids in glycolipids of the nervous system is observed (Kishimoto et al., 1973).

Another important reaction of cobamide-dependent enzymes in mammals is the synthesis of methionine from homocysteine, in which the second coenzyme form, CH_3Cbl, is involved. A reduction or lack of CH_3Cbl-

dependent methionine synthetase activity cause a folate metabolism disorder, in which the transmembrane transport of N^5-methyltetrahydrofolate is impaired and folate accumulates as N^5-methyltetrahydrofolate (the methylfolate trap), resulting in the inhibition of thymidylate synthetase activity and reduced DNA synthesis (Fig. 7.2).

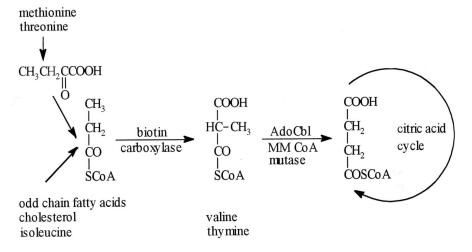

Figure 7.1. Dependence of catabolism of some amino acids, fatty acids and thymine upon methylmalonyl-CoA mutase. Reprinted with permission from A. Stroinski, Medical aspects of vitamin B_{12}, pp. 335-370 in: Z. Schneider and A. Stroinski (eds.) Comprehensive B_{12}: Chemistry. Biochemistry. Nutrition. Ecology. Medicine. 1987 © Walter de Gruyter.

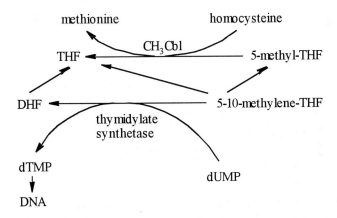

Figure 7.2. The methylfolate trap. THF, tetrahydrofolate; DHF, dihydrofolate.

AdoCbl and CH_3Cbl are metabolically interconvertible and are formed in the body from non-coenzyme forms, which must be supplied in a readily available form (Fig. 7.3). In humans, hydroxocobalamin (OHCbl), sulfitocobalamin (SO$_3$Cbl) and cyanocobalamin (CNCbl) have also been found. Vitamin B$_{12}$ is resistant to boiling in pure solutions, but is very labile in the presence of proteins, especially those containing thiol residues. AdoCbl and CH$_3$Cbl are the most stable; OHCbl is less stable, while CNCbl, being most resistant to heating, is absent in raw foods. It may be found in tobacco smokers, during phagocytosis, in patients on special diets and CNCbl injections.

Human body contains 3 to 5 g of cobalamins, mainly in the liver. The content of CH$_3$Cbl decreases with age, which dictates the necessity of increasing the vitamin content in the diet. Healthy persons receive vitamin B$_{12}$ with food (liver, meat). Vitamin B$_{12}$ is not synthesized by green plants, that is why it is not found in bread, oatmeal, rice, fruits, and vegetables. Algae and other components of the plankton contain significant amounts of cobalamins (Schneider, 1987a). Humans are unable to utilize corrinoids formed by the intestinal microflora.

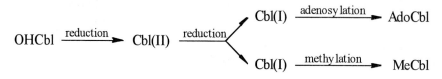

Figure 7.3. Scheme of the enzymatic synthesis of cobamides in the cell. AdoCbl and MeCbl are formed from the common precursor OHCbl via two reductive steps catalyzed by separate enzymes. Metabolic disorders due to enzyme defects may occur at the two reductions and additions of adenosyl or methyl residues. High urinary levels of methylmalonic acid (methylmalonic aciduria) indicate an impaired synthesis of AdoCbl, whereas high levels of urinary homocysteine (homocystinuria) indicate an impaired MeCbl synthesis. In patients with high urinary levels of both methylmalonic acid and homocysteine a defective reduction of cobalamin is likely (Rosenberg, 1983).

The absorption and transport of vitamin B$_{12}$ in humans and animals are mediated exclusively by proteins. The absorption of vitamin B$_{12}$ by bacteria is protein-mediated as well. Three groups of proteins can be distinguished on the basis of their source, biological function, relation to specific membrane acceptors, immunological differences and molecular weight. The three groups of protein carriers are intrinsic factors (IF), transcobalamins (TC), and cobalophilins (CP), all of which have been isolated and characterized (Stroinski, 1987).

IF and CP are glycoproteins. IF promotes the absorption of vitamin B$_{12}$ in human and animal digestive systems. It is capable of binding the tiny amounts of vitamin B$_{12}$ (3-5 mg per day in humans), present in the digested meal, in a complex resistant to digestion and to absorption by intestinal microorganisms. Cbl from the complex is transported in an energy dependent process across the intestinal epithelium. Once inside the ileal cell, it becomes trapped in lysosomes and subsequently binds to transcobalamin (TC). TC-Cbl complex is transported into the bloodstream. Appreciable amounts of Cbl in the digestive system remain bound to cobalophilin (CP), which has a much higher affinity to B_{12} than IF, particularly at acidic and neutral pHs. CP does not mediate the absorption of cobalamin. The binding of Cbl to IF is possible after its release from the CP-Cbl complex by digestion with pancreatic enzymes (Parmentier et al., 1979). Thus, under physiological conditions the greater part of Cbl is not bound to IF until in the intestine (Marcoullis et al., 1978).

Many disorders are caused not by the vitamin B_{12} deficit, but by deficiencies of the protein carriers. Various causes (not genetic) of the vitamin B_{12} deficiency in human body are listed in Table 7.2, which demonstrates that the causes are multiple. Vitamin B_{12} administration either intramuscularly or *per os* results in their elimination.

Table 7.2. Mechanisms of cobalamin metabolism disorders: Cobalamin malabsorption

Phase of absorption	Nature of abnormal events
A. Intragastric events	1. Nutritional cobalamin deficiency
	2. Defective cobalamin release from food
	3. Intrinsic factor (IF) deficiency
	– IF secretion defect
	– secretion of an abnormal IF
	– autoimmune reactions
	– surgical resection
B. Small bowel events	1. Competing parasites: bacterial overgrowth, tapeworm
	2. Pancreatic failure
C. Intestinal absorption	1. Ileal resection or ileal disease
	2. Lack of divalent cations
	3. Unfunctional transport system through membrane
	4. Drug interactions
	5. Deficiency of intracellular binders

Reproduced with permission from A. Stroinski, Medical aspects of vitamin B_{12}, pp. 335-370 in: Z. Schneider and A. Stroinski (eds.) Comprehensive B_{12}: Chemistry. Biochemistry. Nutrition. Ecology. Medicine. 1987 © Walter de Gruyter.

Several diseases have been described that arise as a consequence of the absence or deficit of vitamin B_{12} in the human body. Among such diseases is megaloblastic anemia, which is characterized by the following clinical signs: abnormal morphology of bone marrow cells, megaloblastic changes of red blood cell, granulocyte and platelet precursors (Fig. 7.4). The disease remained fatal until the 1920s, when two American physicians, G.R. Minot and W.P. Murphy, achieved positive results by treating pernicious anemia with liver diet. The antipernicious liver factor appeared to be vitamin B_{12}. Vitamin B_{12} was isolated in a pure crystalline state in 1948 by L. Smith. The discovery and application of the antipernicious factor was one of the most fascinating events in the history of medicine. Minot, Murphy and another physician, G.H. Whipple, received the Nobel Prize in 1934 for their life-saving contributions to medicine.

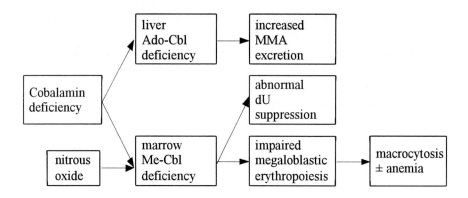

Figure 7.4. Consequences of AdoCbl and MeCbl deficiencies in man. MMA, methylmalonic acid. Reprinted with permission from A. Stroinski, Medical aspects of vitamin B_{12}, pp. 335-370 in: Z. Schneider and A. Stroinski (eds.) Comprehensive B_{12}: Chemistry. Biochemistry. Nutrition. Ecology. Medicine. 1987 © Walter de Gruyter.

Vitamin B_{12} is used in microgram quantities in therapeutics to treat pernicious anemia, malnutrition and peripheral neurologic disorders. Since vitamin B_{12} is not toxic, it is widely prescribed as an adjunct in the treatment of many types of chronic diseases, such as arthritis and psoriasis. It is also used to alleviate fatigue and other related symptoms. Taking into account the urgent need for drugs aimed at human acquired immunodeficiency syndromes (AIDS), derivatives of vitamin B_{12} were synthesized (Belozerov and Yurkevich, 1991) that contained azidothymidine, a known therapeutic agent against AIDS. Other Cbl-derivatives contained different cytostatic groups in their molecules. These derivatives were delivered to their target cells via the transport systems specific for cobalamins, whereupon the cytostatics were released by intracellular hydrolytic enzymes.

Livestock supplementation. Cobalt-deficient fodder such as grasses (contain less than 0.08 ppm) may cause serious losses of livestock and lead to vitamin B_{12} deficiency in humans consuming animal products. Farm animals utilize relatively small amounts of vitamin B_{12}, ranging from 0.05 to 0.5 mg per kg of their weight, but its content in animal feed is also very low. An effective utilization of the vitamin is due to the high affinity binding proteins. Low yields of cobalamins in the rumen of animals result from a relatively short period of bacterial transit and anaerobic conditions, unfavorable for the synthesis of the cobalamin precursor DMB. When DMB was added to sheep's feed, bacterial vitamin B_{12} production increased at the expense of the other forms of corrinoids, as did the absorption of vitamin B_{12} by the animals. Therefore, adding DMB to animal feeds is the preferred way of preventing vitamin B_{12} deficiency (Smith and Marston, 1970). Supplementing cobalamins is considered to be impractical, since most of the added cobalamins are transformed into the inactive forms of corrinoids by rumen microorganisms, even before the vitamin can be absorbed by the animal (Smith and Marston, 1970). Still, in cases when fodder consists exclusively of plant materials, vitamin B_{12} preparations may be added.

If propionic acid bacteria were used only for the production of vitamin B_{12}, they could be regarded as the most useful bacteria, but their application is much wider indeed.

7.3 Porphyrins

Some strains of propionic acid bacteria synthesize large quantities of porphyrins, the yield of which increases significantly when the precursor 5-aminolevulinic acid (ALA, see section 4.4.5) is added. In modern times, porphyrins and their metal complexes are widely used in practice as (i) dyes and pigments, including food dyes, (ii) catalysts for redox reactions, (iii) catalysts for radical-chain polymerization of metacrylates, (iv) catalysts for the oxidation of alkanes and mercaptans in oil and oil products, (v) semiconductors and photosemiconductors, (vi) chelators for the enrichment of radioactive isotopes (Bykhovsky and Zaitseva, 1989). Porphyrins may find application in medicine. They are found in the white matter of the brain and spinal cord (Kubatiev, 1973). Since in this tissue the cytochrome concentration is very low, it is suggested that porphyrins may act as functional substitutes for cytochrome oxidase there. They can be used as diagnostic and medical preparations. *P. granulosum* is the best candidate for the industrial production of porphyrins.

7.4 Probiotics. Animal Feed

Probiotics are health-promoting bacteria found in the gastrointestinal tract of humans and animals. Useful bacteria, inhabiting the gastrointestinal tract, play a significant role in the splitting of nutrients through their enzymatic activities; they synthesize amino acids, vitamins, and promote the assimilation of nutrients. A comprehensive overview of probiotics has recently appeared (Holzapfel et al., 1998).

Vitamin B_{12} might be called the animal-protein factor (Bukin et al., 1971). Adding vitamin B_{12} to plant feeds increases the assimilation of nutrients and promotes a significant increase in the productivity of livestock and poultry. Application of antibiotics and other medical preparations in modern industrial husbandry leads to the suppression not only of pathogenic bacteria, but also of useful and symbiotic ones; the normal intestinal microflora is perturbed and disbacterioses frequently arise.

The emphasis in the investigation of probiotics has been primarily on lactic acid and bifidobacteria; propionic acid bacteria were less prominent, although they are natural components of the gastric (rumen) microflora of ruminants. The usefulness of propionic acid bacteria for the host is beyond doubt: (i) they can reduce the excessive acidity of overfermented silage that causes ketoacidosis in farm animals; (ii) their fermentation products, propionic and acetic acids, can be reutilized; (iii) in the rumen, propionic acid is the main precursor for gluconeogenesis, on which ruminants that absorb only a small amount (10-15%) of the total glucose in the digestive tract, rely as the main source of glucose from C_3-compounds (Kurilov and Sevastjanova, 1978). Propionic acid deficit leads to the accumulation of ketone bodies in the blood, causing a disease called ketosis. It has been calculated that the oxidation of organic acids absorbed by the animal gives 23.5 kcal·h^{-1}. 40% of this amount of energy comes from the oxidation of acetic, 24% from propionic and only 10% from lactic acid (Kurilov and Sevastjanova, 1978). Lipolytic and proteolytic activities of some strains enhance the digestion of feed; and propionibacteria produce large amounts of vitamins B_{12} and B_2.

Recently, it has been shown (Roland et al., 1998) that dairy propionibacteria, especially *P. acidipropionici* (Pa1 and Pa3) and *P. freudenreichii* (Pf1 through Pf7), can produce and liberate NO by reducing nitrate or nitrite. Nitric oxide is accumulated in the culture tubes without being further reduced by the bacteria. The authors have selected NO-producing strains (Propiovitum®) particularly resistant *in vitro* to the digestive stresses (acidity and bile). In humans, nitric oxide is produced from arginine and has fundamental roles in many vital functions, such as neurotransmission, vasodilation and intestinal motility.

Propionic acid bacteria are resistant to penicillin, chlortetracycline, chloromycetin, erythromycin, gramicidin S, polymyxin and streptomycin; this makes possible a combined application of the indicated antibiotics and propionic acid bacteria for the treatment of some animal diseases.

Mineral composition of propionibacterial cells have been studied in *P. freudenreichii* (Vuorinen and Mantere-Alhonen, 1982). They found that the concentrations (in mg/kg) of Mn (267), Zn (159), Cu (102) and Fe (535) were higher than in the lactic acid and bifidobacterial strains tested.

Mantere-Alhonen (1982) has for the first time shown the probiotic and growth-promoting effects of pure propionibacteria on young piglets. The most effective probiotic was *P. freudenreichii* ssp. *shermanii*. The dose of propionibacteria fed to the test animal was $2 \cdot 10^9$ cfu/g. The weekly weight gain was higher by 9.2-14.5% and the feed consumption of the piglets was lower by 7.2-46.1% than in the control group. In addition, the piglets treated with propionibacteria had a lower incidence of diarrhea. In older pigs the effect was even more pronounced (Mantere-Alhonen and Myllymaki, 1985).

Thus, there is clear evidence that propionibacteria have probiotic effects. The probiotic effects are based on the production of propionic acid, bacteriocins, vitamin B_{12}, stimulation of growth of other beneficial bacteria, and the ability to survive gastric digestion. Recently it was shown (Roland et al., 1997) that the addition of propionibacteria led to a significantly higher maximal population of *Bifidobacterium bifidum*, and the cell-free supernatant of propionibacterial cultures greatly stimulated the growth rate and final cell mass of *B. bifidum*, as well as its fermentation activity. Bifidobacteria can hardly survive in the digestive tract, so that the use of 'prebiotics' (*P. freudenreichii*, Propiovit S1.26 and S1.41) is considered by the authors (Roland et al., 1997) to be the best way to increase the population of intestinal bifidobacteria.

In a clinical study the same authors showed that bifidobacteria are significantly increased by supplementing propionibacteria for 14 days (treatment period). The propionibacteria survived in the digestive tract of 61% of healthy human volunteers. After the treatment period the propionibacterial counts returned to the basal level below 5 log cfu/g. Propionibacteria were able to affect the composition of the intestinal microflora specifically by modifying the number of anaerobes and coliforms. Propionibacteria induce an increase in azoreductase, nitroreductase and especially β-glucuronidase activity (Perez-Chaia et al., 1998). No decrease in the concentration of *P. freudenreichii* was observed during an *in vitro* digestion test and this is an important characteristic for a probiotic microorganism.

The effect of *P. acidipropionici* on serum lipids was evaluated by feeding mice with 5 different diets, supplemented or not with milk and propioni-bacteria (Perez-Chaia et al., 1995). The results show that this strain tends to reverse the hyperlipidemic effect of a diet with high lipid content. An increase in both the phagocytic activity of peritoneal macrophages and in the phagocytic function of the reticuloendothelial system was observed on day 7 of feeding mice with milk containing viable cultures of propionibacteria. The data show that some strains of propionibacteria can grow in the gut and exert beneficial effects on the host. The useful properties of propionic acid bacteria listed above and the complete absence of toxicity allowed recommending them as preventive veterinary preparations.

In practice, the following methods can be employed to enrich farm animals' feeds with propionibacteria and products of their metabolism: (i) adding the preparation PABB, which represents a mixed culture of propionibacteria and acidophilic lactobacilli; (ii) the preparation Propiovit, containing living cells of *P. acnes*; (iii) feeding silage prepared with a propionibacterial preparation or with a binary concentrate of propioni-bacteria and amylolytic lactic acid streptococci; (iv) feeding a protein concentrate enriched with propionibacteria.

PABB and Propiovit. The preparation PABB contains cells of *P. shermanii* isolated from cheese and *Lb. acidophilus*. It is prepared in blood-whey broth and helps in the treatment of hypovitaminosis, alimentary anemia and gastro-intestinal disorders in young animals and poultry (Sergeeva et al., 1959).

The initial preparation was produced in the liquid form and had a short period of storage (2-3 months at 2-10°C). An additional shortcoming was the complex composition of the growth medium containing fresh blood. These shortcomings were eliminated by drying the preparation and using skim milk as the nutrient medium (Lagoda and Bannikova, 1970).

Another preparation, Propiovit, is based on the cells of *P. acnes* isolated from cow's rumen, and this is an advantage over PABB. The preparation contains rumen bacteria grown in the medium containing glucose-protein concentrate (industrial waste), enzymatic hydrolysate of yeasts, ammonium sulfate, and cobaltous salt (Volkova, 1980). One g of dry preparation contains up to $5 \cdot 10^9$ living cells, B-group vitamins (μg/g): B_{12}, 400-500; pyridoxine, 33-51; nicotinic acid, up to 350; pantothenic acid, up to 330; riboflavin, 60-140; folic acid, 3.2; water content less than 5-6%.

Temperature optimum for the growth of *P. acnes* is 37-38°C; the bacteria adapt well in the rumen of ruminants. A positive property of *P. acnes* in preparing the preparation is its resistance to high temperatures that permits to dry the cells without loss of biochemical activity. Propiovit provides animals with vitamin B_{12}, compensates for the deficit of vitamin B_1, B_2, B_3 and B_6, helps to overcome gastrointestinal disorders, reduces by half the losses from disease and deaths, increases weight gains and immune status. The preparation exhibits long-term effects and remains active for at least 12 months in the dry form.

Cerna et al. (1991) developed a preparation called Proma, composed of *Lb. plantarum*, *Enterococcus faecium*, *Lactococcus lactis* and *P. freudenreichii*. The concentration used was $2 \cdot 10^8$ cfu/g. The calves receiving Proma increased daily weight gains and decreased food intake in comparison with the control group of calves. The positive effect of Proma continued even after the withdrawal of the preparation.

Ensilage. Only a few bacteria isolated from different types of silage are able to ferment lactate (Woolford, 1973): *P. freudenreichii*, *P. jensenii*,

Micrococcus lactilyticus and some clostridia. The propionibacteria and *M. lactilyticus* fermented lactate better than clostridia and, in contrast with the latter, their fermentation activity was not inhibited by high concentrations of sugars. All the bacteria indicated above are anaerobic, and it is under anaerobic conditions that important biochemical processes occur in silage.

Under conditions modeling the ensiling process, apparently, propionibacteria fermented not only lactate, but also sugars at low pH values (below 4.0); in this case the pH was not increased significantly and the secondary growth of clostridia was absent (Woolford, 1975). No propionibacteria were isolated from fresh grasses, and from silages they were isolated in very low numbers; therefore, it was concluded (Woolford, 1973) that their role in natural ensilage is not significant. However, adding propionibacteria to the ensiling mass, especially to that (corn) containing high amounts of sugar, improved the quality of silage, which had a lower acidity, was enriched in vitamins B_2 and B_{12}, in propionic acid (Iljina and Besedina, 1966; Konoplev and Scherbakov, 1970), and was not moldy. As a result of feeding chicken with this silage for three months, the number of eggs laid and survival of hatchlings were increased; the content of β-carotene in blood was increased, and ammonia was reduced (Domracheva et al., 1970).

One gram of the bacterial concentrate 'Kazakhsil' contains 10^9 viable cells of propionic acid bacteria. It is recommended to add 1.5 g of the preparation per 1000 kg of the ensiling mass. A particularly high effect was achieved by combining three bacterial concentrates: PAB (propionic acid bacteria), AMS (*Streptococcus lactis* diastaticus) and PLB (pentose-fermenting lactic acid bacteria). These bacterial starters for ensilage are shown in Table 7.3. They were all developed at the Institute of Microbiology and Virology in Kazakhstan. Formerly, chemical preservatives were mostly used, consisting of 1 to 3 organic acids (Table 7.4). These acids are also produced by propionibacteria, although formic acid that predominates in chemical preservatives is very low in microbiological preparations.

Dawson et al. (1993) showed that propionic acid bacteria, inoculated into ensiled high-moisture corn grain, produced significant amounts of propionic acid and reduced yeast and mold counts. In this type of silage (dry matter more than 70%) the decline in pH was slow and propionibacteria could survive more easily. Weinberg et al. (1995a, b) tested *P. shermanii* in different silages. The bacteria produced propionic acid only in wheat silages, in which the pH decline was delayed and in that case aerobic stability was improved. In all other silages the pH decline was rapid and propionibacteria could not proliferate, and no improvement in aerobic stability was detected.

Table 7.3. Starter cultures for ensilage developed at the Institute of Microbiology and Virology, Kazakhstan Academy of Sciences

Name	Strain(s)	Plant material

AMS	*Streptococcus lactis* diastaticus (dried)	Hard to ferment (beans, cereals, grass mixes, sugarcane)
PAB	*Propionibacterium shermanii*	Sugar-rich, easily fermented (corn, sunflower)
PMB	*Lactobacterium pentoaceticus*	Straw and hard-stem grass
Mixed starters		
APP (AMS, PAB, PMB)	*S. lactis* diastaticus, *P. shermanii, L. pentoaceticus*	Corn straw
Silamp (AMS + PMB)	*S. lactis* diastaticus, *L. pentoaceticus*	Annual and perennial grasses, straw, beans

Table 7.4. Chemical preservatives for ensilage

Name	Component	%
VIK-1	Formic acid	27
	Acetic acid	27
	Propionic acid	26
	H_2O	20
ATV-2	Formic acid	80
	Phosphoric acid	2
	H_2O	18
VIK-11	Formic acid	80
	Acetic acid	9
	Propionic acid	11

Protein concentrates. This type of product has been suggested by the Latvian Institute of Microbiology (Ramniece et al., 1981). It is prepared from the biomass of *P. shermanii* grown in the brown sap of different plants: mixed grasses, corn, sugar-beet leaves, and alfalfa. Brown sap is obtained after the separation of proteins from the green sap of plants by thermal coagulation. It is used for several purposes: growing yeasts, as fertilizer in the fields, as a drink for animals (Ramniece et al., 1981). The authors showed that propionic acid bacteria can grow in the brown sap reaching $5 \cdot 10^9$ cells/ml in alfalfa sap (which is the best for growth). Acetic (0.1-0.3%) and propionic (0.3-0.4%) acids were accumulated in the sap. Thus, this type of product represents the sap of green plants enriched in volatile acids and proteins.

As an additive to animal feed a dried culture liquid of *P. acnes* was proposed (Sobczak and Komorowska, 1984), which was grown in nonsterile conditions in a medium with molasses, whey or a mixture of both. A strain of *P. acnes* isolated from the rumen produced more biomass (~4 g/l), volatile fatty acids (up to 2%) and vitamin B_{12} than the collection strains of *P. shermanii*.

7.5   Production of Propionic Acid

Propionic acid is used in industry and as an antifungal agent. An estimated world production in 1992 was about 100,000 metric tons (Boyaval and Corre, 1995). Propionic acid is used in the manufacture of some plastics, herbicides, fruit flavorings, perfume bases, and to improve processability and flame resistance of butyl rubber. Sodium propionate is used in veterinary medicine and in food industry as grain and silage preservative; propionic acid salts are added to suppress the growth of mold in bread, cakes, on the surface of cheese, meat, fruit, vegetables and tobacco. A mixture of propionic, lactic and acetic acids is recommended for the preservation of foods due to the synergistic effect of these acids in inhibiting the growth of *Listeria monocytogenes* in food.

Various pests sweep away about 15% of the world harvest during storage. When humidity exceeds 14%, the stored grain warms up with subsequent growth of mold. In moist environments seeds germinate, initiating respiration and stimulating the activity of microorganisms. Such methods of grain storage as drying, keeping at low temperature or in airtight conditions are costly or difficult to realize. But one additional method exists, which is applied in some countries and has been tested in Russia. This method implies the treatment of grain by sprinkling with a weak solution (0,5-1%) of propionic acid.

Propionic acid prevents the germination of grain and, being an antifungal agent, kills microorganisms such as molds. As shown above, the nutritive value of the feed treated with propionic acid is improved, and the incidence of animal diseases such as mycosis and mycotoxicosis is reduced. The preparation Propcorn based on propionic acid is applied in Great Britain for corn storage and in the USA for mixed fodder storage. Corn treated with propionic acid can be stored without losses in a shed in moist weather for 12 months.

In a typical case of propionic acid fermentation 2 moles of propionic acid and 1 mole of acetic acid are formed from 1.5 moles of glucose:

$$1.5 \text{ glucose} + 6P_i + 6ADP \rightarrow 6ATP + 2 \text{ propionic acid} + \text{acetic acid} + CO_2 + H_2O$$

When lactate is fermented, the reaction is as follows:

$$3 \text{ lactic acid} \rightarrow 2 \text{ propionic acid} + \text{acetic acid} + CO_2 + H_2O$$

A large share of propionic acid production is supplied by petrochemical methods. Nevertheless, microbiological production can co-exist and in some cases be preferred over chemical production, provided that a highly produc-tive strain and cheap substrates are available.

If lactose-fermenting *P. shermanii* is used, cheese whey can be used as a substrate. From the whey containing 12% of dry substances, only 50% of lactose is utilized by *P. shermanii*, producing 1.6-2.2% solution of propionic acid (Bodie et al., 1987). In the combined batch culture, composed of *P. shermanii* and *Lactobacillus casei*, lactose is completely utilized and a 3% solution of propionic acid is produced in 52 h. When the medium is partially replaced during cultivation, a 4.5% solution of propionic acid is obtained; by raising the concentration of dry substances in the whey up to 18% the production is increased up to 6.5% in mixed culture. For industrial production it is advisable to add a reducing agent to the medium (see above, Chapter 3), then the ratio of propionic to acetic acid will be increased (Emde and Schink, 1990).

Among propionibacteria there are strains capable of fermenting polymeric carbon substrates and pentoses, the main constituents of plants. Babuchowski and Hammond (1987) compared five strains of three species – *P. acidipropionici, P. thoenii, P. jensenii* – for their ability to convert maltose, starch and sodium lactate to propionic acid. The highest ratio of propionic to acetic acid was found during lactate fermentation, which correlates with the evidence, discussed before, for the stimulating effect of *Lb. casei*, producing lactate, on the production of propionate by *P. shermanii*. At the same time, the possibility was demonstrated (Babuchowski and Hammond, 1987) of obtaining propionic acid by using all five strains and hydrolyzed starch as a substrate in one-step process.

The filtrate of the culture liquid of propionibacteria used for vitamin B_{12} production can serve as a source of propionate and acetate. After the separation of biomass an anion exchange resin is placed in the fermenter to bind the acid products, which then can be eluted from the resin (Vorobjeva et al., 1979b). As the example of food preservation shows, the combination of propionic and acetic acid is useful: the latter had been used long ago to preserve food and is still widely used. If the effect of the mixture is greater than that of propionate alone, separation of the acids would not be required, making it possible to obtain a useful preservative as a by-product of the industrial production of vitamin B_{12}.

An economically profitable method of propionic acid production is by fermentation of the hydrolysate of fruit grasses by *P. acidipropionici* (Clausen and Gaddy, 1984; Tyree et al., 1991). In continuous culture a mixture of glucose and xylose was fermented, and the utilization of xylose was not inhibited by glucose. 80% of the two substrates were metabolized in 75 h. It was estimated that 32,000 metric tons per year of acetic and propionic acids (a mixture, equivalent to the actual demand for these acids in the USA) could be produced by this method at a cost less than ¼ of the market price of these products. If necessary, pure propionic acid can be obtained from the mixture (1-5%) by extraction or evaporation and extraction, but in this case the cost will be significantly higher. This method, producing the mixture of acids from grass hydrolysates, is suitable for the preservation of corn and fodder. In addition, foodstuffs such as bread or butter can be preserved by wrapping in paper impregnated with propionic acid. This method may represent a viable alternative to petrochemical methods.

A mixture of sodium propionate and acetate can be obtained by using immobilized cells of *P. shermanii* and *P. technicum* under continuous cultivation. Starch, either hydrolyzed by bacterial amylases or not hydrolyzed, is used as a carbon source. The rates of acid production by *P. shermanii* using glucose and hydrolyzed starch are similar. Polyacrylamide gel granules are permeable to nonhydrolyzed starch, which is fermented by immobilized cells of *P. technicum* producing propionic and acetic acids (Vorobjeva et al., 1979b). The acids can be separated (if required) from the respective salts by ion-exchange chromatography on a column with Amberlite-IRC-50 (for details, see Chapter 6). The process is carried out without aeration, which is advantageous for its scaling-up.

The cells of *P. thoenii* P.20 were immobilized in calcium alginate beads (Rickert et al., 1998); the beads contained approximately $2 \cdot 10^{11}$ cells per g wet weight. Glucose and lactate as carbon sources were evaluated in 12-h fed-batch fermentations. Substrate utilization rates and acid yields were highest at 75 g/l initial glucose and 42 g/l initial lactate; 34 g/l and 22 g/l propionate were produced at these substrate concentrations, respectively. The total acid yield (54%) was higher in glucose-fermenting batches. A continuos extraction of acids from the broth in the process of extractive fermentation improves acid production (Gu et al., 1998). A liquid consisting of 40% (v/v) Alamine 304-1 (trilaurylamine) in Witcohol 85NF (oleyl alcohol) was used as an extractant in a hollow-fiber extractor (Gu et al., 1998). The acid yield was doubled and higher overall acid productivity was obtained in the extractive fermentation. The extractant was selective for propionic over acetic acid, thus propionic acid was partially purified.

Lewis and Yang (1992) studied the process of propionate production from lactose by *P. acidipropionici* ATCC 4875 immobilized on surgical cotton fiber placed in the bioreactor. The productivity was increased by almost 100% by the continuous selective extraction of propionic acid with a tertiary amine solution (40% alamine in 2-octanol). Propionic acid was isolated and the extractant regenerated by adding small amounts of 1N NaOH solution. A concentrated propionate salt was obtained as the final product.

An alternative method of propionic acid production from sugars by propionibacteria has been suggested (Tyree et al., 1991). In this process two cultures, *Lactobacillus xylosus* and *P. shermanii*, are used in a system of two consecutive reactor vessels. Lactate, formed by *L. xylosus* in a batch process with constant stirring, enters the second vessel with immobilized cells of *P. shermanii*, where propionate is synthesized. The system is characterized by a high total productivity for propionate.

There are two technical problems that limit industrial production of propionic acid by fermentation:

a) a relatively low production of propionic acid due to inhibition by the acid accumulated in the medium;
b) difficulties of isolating the acid from dilute solutions.

At the end of a typical fermentation the concentration of propionic and acetic acids is about 2% and 1% (w/v), respectively. Yields of the acids can be increased by adapting the bacteria to the high content of propionic acid in the medium. Using this method, a strain of *P. acidipropionici* was selected that significantly surpassed the parental strain. Product inhibition of the acids formation was relieved by carrying out a continuous cultivation, particularly of immobilized cells, since growing cells are susceptible to contamination by foreign microorganisms.

Boyaval and Corre (1995), by using a high-density bioreactor, have been able to increase the low productivity of batch processes ($0.03 \ g \cdot l^{-1} \cdot h^{-1}$) to between 2 and 14 $g \cdot l^{-1} \cdot h^{-1}$. A continuously stirred tank reactor was coupled with cell recycling by ultrafiltration. Use of glycerol as a principal carbon source enabled to convert it exclusively to propionic acid. Such bioreactors allow to concentrate large amounts of cells within the system during continuous fermentation and to reduce the inhibitory effect of acid accumulation in the medium. The concentration of propionic acid in the bioreactor reached 25 g/l and biomass reached 100 g/l (dry weight). The authors employed such membrane processes as electrodialysis and electrodialysis with bipolar membranes to improve recovery and purification from the fermentation media. High volumetric productivity and improved downstream processing performance, achieved by the authors (Boyaval et al., 1994; Boyaval and Corre, 1995), resulted in an economical method of biological production of propionic acid, suitable for industrial application.

Isolation of organic acids from dilute solutions presents certain problems. The traditional method of distillation requires a large expenditure of energy, which noticeably increases production costs. The acids can be extracted by adding organic solvents to the culture liquid, but this may have an inhibitory effect on cells and the acid producing activity. An alternative technology implies the use of membranes, having pores filled with an organic solvent that selectively absorbs organic acids. In the course of such cultivation the culture circulates at one side of the membrane and the acids are absorbed by the solution (for example, a solution of alkali) at the opposite side of the membrane. Organic acids contact first the solvent, then the absorbing solution, from which they are finally isolated as salts.

Application of propionic acid as an antifungal agent is also not without problems. Propionic acid is highly corrosive for the equipment that supplies it to the corn. Besides, the addition of propionic acid or its salts to the food as a preservative is against the modern trend towards 'all-natural' food. For corn storage, it is unnecessary to use pure solutions of propionic acid. In our investigations (Vorobjeva et al., 1984), treatment of seeds with the culture liquid (0.5% volatile acids) of two strains of propionibacteria resulted in an almost complete elimination of bacteria, yeasts and molds. The mixture of acids produced by propionic acid cocci showed a stronger fungicidal effect than the mixture formed by *P. shermanii*, apparently due to the higher content of propionic acid, since the ratio of propionic to acetic acid in cocci is higher than in *P. shermanii*.

It is possible to cultivate propionibacteria in natural media, for example, in milk or cheese whey, then to dry the culture liquid containing bacteria and use it as a food preservative. Microgard, a fermentation product of skim milk, is an example of such a preservative (Glatz, 1992). It is used to preserve goat cheese since it inhibits growth of gram-negative bacteria and some fungi, including yeasts. Microgard possesses a wider antimicrobial spectrum than propionic acid alone, since besides the organic acids it contains other substances, such as diacetyl and bacteriocin, which inhibit the growth of various bacteria and fungi.

7.6   Food Industry

7.6.1   Baking of bread

Propionibacteria, together with lactic acid bacteria and yeasts, are added to some leavens for dough in order to enrich bread in propionic acid, in addition to lactic and acetic acids produced in the process of fermentation. Such bread contains up to 0.28% of propionic acid and has a prolonged storage life due to the suppressive action of propionic acid on mold growth. This effect can only be achieved if the number of propionibacteria is at least $2.5 \cdot 10^8$ cells per g raw dough (Spicher, 1983).

As a leaven for bread a mixed culture of *Lactobacillus brevis* TKK 105-6-1 and *P. acidipropionici* TKK 114-3-1 was used after three days of combined cultivation at 30°C in the medium containing hydrolyzed wheat flour, yeast extract and mineral salts (Javanainen et al., 1987); the pH of the culture was adjusted to 5.0-5.8 by adding calcium carbonate (20 g/l). Almost all of the lactic acid formed by the lactic acid bacteria was converted to propionic and acetic acids. Bread made from such leaven contained 0.1% acetic, 0.2% lactic and 0.1% propionic acid relative to flour weight. This quantity of propionic acid provides an effective antifungal action without noticeable effects on the flavor and aroma of baked bread. But when another mixed culture was used, composed of *P. acidipropionici* and *Lb. plantarum* and containing more than 4% lactic acid, propionic acid was not produced.

By adding propionic acid bacteria to leavens bread can be enriched in vitamin B_{12}; this is especially important for vegetarians and persons suffering from various diseases arising from vitamin B_{12} deficiency (see above). Vitamin B_{12} content in bacterial cells is regulated by the availability of cobaltous salts in the medium (see Chapter 5). Hence, the vitamin B_{12} content of bread can be modified by using an appropriate leaven.

Adding *P. shermanii* VKM-103 to leavens can prevent a form of bread spoilage called "potato blight", which is caused by a bacillus, *B. subtilis* (Bogatireva et al., 1993). The leaven must contain 4-10% propionic bacteria relative to the meal weight in dough; in this case the bread is resistant to contamination when challenged with 10^{10} spores per g of meal for 6-7 days. Without propionibacteria, the "potato blight" develops within 17 h of bread incubation. The antiseptic action is displayed by propionic acid and its salts. Another form of bread spoilage, caused by the bacterium *Bacillus pumilus*, is also prevented by *P. freudenreichii* ssp. *shermanii* (Odame-Darkwah and Marshall, 1993). A combination of *Lb. rhamnosus* LC705 and *P. freudenreichii* ssp. *shermanii* JS was found (Mayra-Makinen and Soomalainen, 1995) to be the most effective against yeasts and molds, and this mixed protective culture was active in fermented milks, bakery products, silage and spent grains.

7.6.2   Fermented milks and raw food

To enrich yogurt and other cultured dairy foods in vitamin B_{12}, Karlin (1966) recommended adding propionibacteria, thus increasing the nutritive properties and medical value of these products. In Russia, a yogurt-like product was prepared (Romanskaya et al., 1985) from a starter composed of the acetobacterium *Acetobacter lovaniense*, glutinous strains of lactic streptococci, and *P. shermanii* in ratios of (2.5-3.5):(9-11):(3-4). Mixtures of milk and whey in ratios of 9:1 through 7:3 served as substrate. Dilution of milk with whey, while saving milk, at the same time reduced the content of dry matter, vitamins and proteins. These inevitable losses could be compensated for by using starter cultures of bacteria capable of actively synthesizing protein, vitamins, and extracellular polysaccharides that increase viscosity.

Some strains of acetic acid bacteria can serve as active producers of extracellular polysaccharides. Formation of polysaccharides is increased by the symbiosis with streptococci. *P. shermanii* and *A. lovaniense* could not grow in milk as monocultures, but grew well as a mixed culture, causing a complete curdling of milk in 32-36 hours enriched in vitamins B_{12} and other B-group vitamins. However, in a mixed culture with lactic acid streptococci the content of vitamins was lower, apparently due to their utilization by streptococci. Therefore, to obtain the required product, acetic and propionic acid bacteria were grown together in whey, and lactic and acetic acid bacteria were cultured together in milk with subsequent mixing of these co-cultures. This method of milk processing was recommended for the production of fermented milks enriched in vitamins.

The growth of another milk-fermenting bacterium, *Bifidobacterium adolescens*, was found to be stimulated by *P. freudenreichii* (Kaneko et al., 1994). In mixed cultures the yield of bifidobacteria was increased from $3.7 \cdot 10^6$ cfu/ml in monoculture to $2.2 \cdot 10^8$ cfu/ml, and that of *P. freudenreichii* from $8.7 \cdot 10^7$ cfu/ml to $4.7 \cdot 10^8$ cfu/ml, respectively.

It has been shown (Nabukhotnyi et al., 1983; Kornyeva, 1981; Mantere-Alkonen, 1995) that raw foods (any fresh uncooked foods) have therapeutic effects on allergic, rheumatic and infectious diseases. *P. acidipropionici*, *P. freudenreichii* ssp. *shermanii*, *P. jensenii* and *P. thoenii* were identified in 24 unprocessed foodstuffs by Mantere-Alkonen (1995). *P. acidipropionici* was found in fermented products, while unfermented foods contained *P. jensenii* and *P. thoenii*. *P. freudenreichii* was found only in pickled mushrooms. The presence of propionic acid bacteria in raw food could have a role in its beneficial effects. The occurrence of propionibacteria together with lactic acid bacteria in those foods having probiotic effects, as well as a mutual growth stimulation of propionibacteria and bifidobacteria, extend the usefulness of these bacteria.

7.6.3 Desaccharization of egg white

This problem arose in connection with the storage of dried egg white. Fresh egg white contains lysozyme that has bactericidal activity, but during storage the enzyme activity decreases and egg white becomes susceptible to many germs, especially putrefactive bacteria, causing its spoiling. As a result of the accumulation of degradation products and oxidative processes, the egg white becomes unsuitable for food. Purified enzymatic preparations, catalase and glucose oxidase, were used in the presence of hydrogen peroxide for egg preservation. This method is directed at eliminating carbohydrates from egg white, improving its storage properties. However, not all carbohydrates are removed by this method and, in addition, some proteins are denatured.

Besides, the use of the purified enzymes greatly increases the production costs.

As an alternative, another method was proposed that envisages the use of propionibacteria (Stoyanova et al., 1979). It is based on the ability of propionibacteria to grow in liquid eggs and to ferment all the carbohydrates of egg white in 24 h, leading to the accumulation of propionic (more than 3.5 mg/g) and acetic acids (Vorobjeva et al., 1979b). Adding ammonium sulfate and sodium phosphate to the egg white stimulated the growth of propionibacteria, accelerated carbohydrate utilization and altered the fermentation in such a way that the production of propionic acid was increased (Fig. 7.5) (Stoyanova et al., 1979). Nutritional value of the egg white was improved due to the bacterial secretion of free amino acids, especially of glutamic and aspartic acids, cysteine, alanine, methionine, and vitamins B_2 and B_{12}.

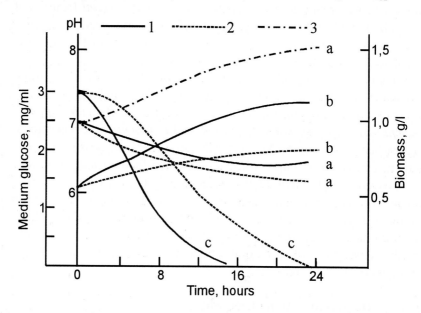

Figure 7.5. Growth parameters of *P. shermanii* growing in liquid egg white. Liquid egg white was inoculated with *P. shermanii* and incubated with (1) or without (2) $(NH_4)_2SO_4$ and K_2HPO_4. A control portion was left intact (3). The pH (a), biomass (b) and glucose concentration (c) were determined at indicated times. From Stoyanova et al. (1979).

Free amino acids prevent the formation of melanoidins, which may change the color, organoleptic and physical properties of egg white. *P. shermanii* did not affect the proteins of egg white; moreover, it was enriched in bacterial proteins, with the total amount of albumins increased by 5.5%. Due to the lipolytic activity of propionic acid bacteria, the fats were split that promote spoilage of dried eggs during storage. It was also found (Vorobjeva et al., 1979b) that the number of foreign bacteria was reduced almost threefold as the result of propionic acid fermentation. Bacterial spp. *Salmonella, Staphylococcus, Proteus* completely disappeared, and the titer of *E. coli* increased, in other words, the bacteriological status of dried egg white was improved. The fermented egg white acquired better physico-chemical properties for baking: the foam was lighter and more stable, its volume increased twofold. The volume of baked cakes depends directly on the foam stability. Cakes made with the egg white fermented by *P. shermanii* were characterized by a nice whiteness and aroma, good structure and porosity.

7.7 Bacteriocins

Antimicrobials continue to be one of the most important classes of food additives, and the food industry is especially interested in the application of naturally occurring and biologically derived preservatives. Among the metabolites of industrial importance produced by propionibacteria are proteins called bacteriocins (Paik and Glatz, 1995). Bacteriocins are defined as bactericidal proteins with a narrow spectrum of activity targeted against species related to the producer culture (Tagg et al., 1976). Many microorganisms associated with foods produce bacteriocins, which stimulates interest in the use of bacteriocins as natural food preservatives.

Numerous bacteriocins are produced by lactic acid bacteria (Klaenhammer, 1988), but only a few bacteriocins have been found in propionibacteria (Lyon and Glatz, 1991, 1993; Grinstead and Barefoot, 1992). Recently, 14 dairy propionibacteria were screened for catalase-insentitive, protease-sensitive inhibition of Gram-positive and Gram-negative bacteria (Ratnam et al., 1998). Bacteriocin production was identified in 57% of these cultures.

Bacteriocins are produced by *P. jensenii* 126 (jensenins) and *P. thoenii* 127 (propionicins PLG-1). Propionicin PLG-1 is a protein of molecular mass about 10 kDa, produced in late stationary phase. The activity of propionicin PLG-1 in culture broth was detectable only after about 10 days of incubation (Lyon and Glatz, 1993). Upon reaching a maximum at about 14 days of incubation, the bacteriocin activity dropped sharply, suggesting the presence of an inhibitor or the action of extracellular proteases. It has a wide spectrum of antimicrobial activity, inhibiting Gram-positive and some Gram-negative bacteria, some yeasts and molds.

It was shown (Paik and Glatz, 1997) that the synthesis of bacteriocin PLG-1 as well as culture growth and organic acids production were higher in two small-scale fed-batch fermentations (working volume, 1.5 l) than in batch culture. The authors suggested that propionicin PLG-1 production can be significantly increased by employing fed-batch rather than batch fermentation methods. Maximum viable cell concentrations were at least 10-

fold higher in the fed-batch fermentations than in batch culture. Maximum bacteriocin titers were 184 and 146 AU/ml (AU, activity unit) in the two fed-batch fermentations. Both fed-batch fermentations were performed with 1.2% sodium lactate, with lactate fed at 12-h intervals to give 0.3% sodium lactate in the fermenter. Upon reaching the maximum level, bacteriocin activity dropped sharply upon continued incubation. When the fermentation was scaled up to 14 1 (large-scale fermentation), the bacteriocin titer was lower (100 AU/ml) than in the small-scale fermentations.

Bacteriocin PLG-1, as shown by Gollop and Lindner (1998), does not act by forming pores in cell membranes, but inhibits metabolism, in contrast with most other antibacterial peptides. In general, bacteriocins form pores in the membrane, with the resulting efflux of ions and small metabolites leading to cell death. The presence of propionicin in skim milk, in which the strain *Propionibacterium* 127 had been grown, retarded the growth of certain psychrophilic spoilage and pathogenic organisms.

Propionicin PLG-1 activity was found (Hsien and Glatz, 1996) to increase by as much as 200% over the first 10 days of storage in liquid samples stored at 25°C or 4°C. Samples stored at 4°C retained high activity through 100 days of storage. The activity of samples stored at 25°C gradually decreased. It was shown (Glatz et al., 1995) that propionicin PLG-1 is composed of 99 amino acid residues. No homology of the propionicin sequence to other bacteriocins from lactic acid bacteria was found (Glatz et al., 1995).

Jensenin G is produced by *P. jensenii* 126, and is pH-tolerant (pH 2-11.0) and heat-stable, has a molecular weight of 4.5 kDa. It is active against propionibacteria, botulinal spores, lactobacilli, lactococci and *Streptococcus thermophilus* (Barefoot et al., 1995).

Jensenin P is a bacteriocin produced by *P. jensenii* B1264, which inhibits closely related propionibacteria and lactic acid bacteria. It was shown that jensenin P is a new anionic bacteriocin (Ratnam et al., 1998). Jensenin P is stable at 100°C for 45 min, at pH 2 to 10 for 225 min, and in 0.1 to 0.3M NaCl (pH 10) for 225 min (Barefoot et al., 1995). The resistance of jensenins to the extremes of pH and high temperature, their wide inhibitory spectrum allows to consider them as useful natural preservatives in thermal food processing. *P. jensenii* strain DF1 was found (Miesher et al., 1998) to markedly inhibit other *P. jensenii* strains and to suppress 15 out of 24 yeasts and 3 out of 4 molds tested. From the culture broth of *P. jensenii* DF1 an inhibitory protein substance, named SM1, was isolated and partially purified (Miesher et al., 1998). Crude propionicin SM1 was sensitive to proteolytic enzymes and stable to incubation at 30°C (more than 14 days), cold storage (more than 6 months at 4°C) and heat treatment (15 min, 100°C).

Concerning the cutaneous propionibacteria, the data on their bacteriocins are scarce. Two bacteriocin-like substances were partially purified from a strain of *P. acnes*: acnecin CN-8 and bacteriocin RTT108 (Fujimura and Nakamura, 1978). Acnecin was inhibitory to other strains of *P. acnes*, and bacteriocin RTT108 was active against both Gram-negative and Gram-positive anaerobes.

7.8   Superoxide Dismutase

Adverse effects of many types of irradiation and chemical mutagens on living organisms are associated with the formation of free radicals (Dubinin, 1976; Petkau, 1987; Poroshenko and Abilev, 1988). Lipids, proteins and nucleic acids are potential biological targets for radicals. Free radicals are often involved in the activation of many types of procarcinogens and promutagens, transforming them into the fully active carcinogens and mutagens which then react with DNA (Pryor, 1986). Peroxide radicals may cause DNA damage, and systems that produce superoxide radicals give rise to hydroxyl radicals (and dangerous singlet oxygen), all of which form radical sites on DNA. Radiation sickness, many forms of cancer and other serious diseases are connected, directly or indirectly, with the radical formation (Pryor, 1986).

Human serum and saliva contain superoxide dismutase (SOD), peroxidase and catalase—antioxidant enzymes that destroy H_2O_2 and O_2^- and represent a form of the antioxidant defense against mutagenic factors (Nishioka and Ninoshiba, 1986). Clinical investigations have shown that SOD administration has a significant beneficial effect in cardiac failure, in which the heart muscle is injured (Fass, 1987). SOD has prospects for application not only in medicine, but also in food industry, where, in combination with catalase and peroxidase, it may be used to prevent oxidation of lipids and other valuable components of food (Taylor and Richardson, 1979). The SOD isolated from certain marine bacteria could be used (Mickelson, 1977) to prevent autooxidation in several test systems.

SOD and other antioxidant enzymes are produced commercially on a limited scale, mainly for laboratory purposes, being isolated from red blood cells. Propionibacteria are a good source of SOD and their value in this respect rises due to the possibility of multi-purpose processing of the biomass. We have developed (Kraeva and Vorobjeva, 1981a, b) a simple method of isolation and purification of SOD to apparent homogeneity, which is shown in Table 7.5. Since SOD is a thermostable protein, heat treatment was used in the purification, thus significantly reducing the number of purification steps. In the course of purification no enzyme modification was observed. The SOD isolated from *P. globosum* is stable in 50% glycerol for 4 months at −13°C. Its properties are described above (Chapter 4).

Table 7.5. Isolation and purification of superoxide dismutase from *P. globosum*

Purification step	Total activity, units	Total protein, mg	Specific activity, U/mg	Yield, %
Cell-free extract	58560	2500	23	
Ammonium sulfate (50-80% saturation) precipitate	81820	216	379	100
Heat treatment (70°C, 5 min)	75750	180	416	92
Chromatography on DEAE-52	36826	36	1023	45
Preparative gel-electrophoresis and re-chromatography on DEAE-52	8213	7	1100	10

From Kraeva and Vorobjeva (1981a).

Since propionibacteria are approved for use in food industry, the following simple preparation can be recommended for this purpose. After the extraction of vitamin B_{12} with 20% ethanol, the biomass is disintegrated and cell debris removed by filtration; the filtrate contains catalase, peroxidase and SOD activities and can be used as an antioxidant preparation without further purification.

7.9   Propionibacteria in Experimental Medicine

7.9.1   Cutaneous propionibacteria in experimental and clinical oncology

First indications of a strong stimulation of the reticuloendothelial system (RES) by *Corynebacterium parvum* (subsequently reclassified as *P. acnes*, see Section 1.3) were reported in the early 1960s (Halpern et al., 1964, 1966). Over the years information was gathered concerning the immuno-stimulating properties of these bacteria: stimulation of antibodies production, cell-mediated stimulation of immunity, and antiviral and antibacterial activities. In several laboratories *C. parvum* was found to inhibit or significantly delay the growth of different (including malignant) tumors as well as tumor invasion by enhancing natural defenses of the organism. But even more striking was the fact that *C. parvum* could inhibit the spread of metastases. Immunotherapy of tumors using *C. parvum* was especially effective (Bomford and Olivotto, 1975) after a surgical removal of the primary source of tumor cells, and also after the chemotherapy of leukemia.

In experimental studies, cells of *C. parvum* (or other bacteria studied), killed by heating and formalin treatment and washed with saline, were administered to experimental animals. The animals were then inoculated with cultures of tumor cells and their responses were evaluated by comparison with control animals. These experiments had clinically important implications, since no system or organ toxicity was detected in different animals, even at doses significantly exceeding those used in the therapy of chronic diseases. Since dead bacteria were used, any danger of bacterial contamination to the host could be excluded. Subsequent experience with human patients confirmed the safety of *C. parvum* for clinical applications.

It is important to note that most of the results were obtained by treating animals with *C. parvum* before, simultaneously or just after the inoculation of tumor cells, although in one report (Fisher et al., 1975) antitumor effects of *C. parvum* were observed on well-implanted tumors. The antitumor effects were shown to be associated with the stimulation of cell-mediated immunity: macrophage numbers were increased, their chemotaxis was enhanced, as well as the secretion of lysosomal acid hydrolase (Wilkinson, 1975). Not only quantitative, but also qualitative changes of immune adjuvants were observed. Activated macrophages acquired new biological activities: they were shown to modulate the activity of T-cells and to increase the number of T-lymphocytes in spleen (Toujas et al., 1975). In addition, they stimulated the activity of B-cells and inhibited the proliferation of tumor cells by a direct action on the immune system: macrophages activate lymphocytes, which act as effector cells and kill target tumor cells by direct contact (Baum and Breeze, 1976).

Close membrane contacts between activated macrophages and tumor cells YC-8 were observed by electron microscopy (Puvion et al., 1975). Many tumor cells, fixed by macrophages, had cavities and probably were dead. Such a picture has never been observed with untreated macrophages. *C. parvum* was most effective against immunogenic tumors induced by chemical carcinogens.

C. parvum was found to stimulate the proliferation of spleen cells and the formation of bone marrow cell colonies *in vitro* (Dimitrov et al., 1975). The weight of spleen and the size of colonies were increased (Baum and Breeze, 1976). It is particularly important to restore the precursors of blood-forming cells in cases when they are reduced by chemo- and immunotherapy. Application of *C. parvum* permits to use higher doses of cytotoxic agents (Dimitrov et al., 1975) without the risk of depleting bone marrow cells. It was suggested (Baum and Breeze, 1976) that the stimulation of bone marrow cell proliferation is the key to the understanding of the unique antitumor properties of *C. parvum*. This effect on mouse bone marrow cells is not observed with bacteria that do not show antitumor activity.

By phenol extraction of the dead cells of *C. parvum* (*P. acnes*) an insoluble middle fraction was obtained (Saino et al., 1976) which, when injected intraperitoneally, prevented tumor formation by sarcoma-180 cells, showed an adjuvant effect on immunological responses and stimulated the host RES. Chemical analysis showed (Saino et al., 1976) that the active fraction contained mainly polysaccharides and mucopeptides, components of the cell wall (Azuma et al., 1975).

Promising results were obtained by treating human patients. In a group of 414 oncological patients that did not receive chemotherapy, intravenous injections of the biomass of *C. parvum* at 4 g per day led to an improvement of their condition and to a decrease in death rates (Israel, 1975). Beneficial effects of *C. parvum* were synergistic with BCG-vaccine, as well as with chemotherapeutic agents. *C. parvum* recruited more white blood cells, whereby the tolerability of chemotherapy was increased (Israel, 1975). Also, as mentioned above, the bacteria activate macrophages and lymphocytes and enhance the specific immunological responses against cancer. The results of these preliminary tests established the nonspecific stimulation of the host immune system by *C. parvum*, *C. granulosum* and *C. avidum* (Halpern, 1975). This discovery had a major impact on fundamental and applied immunology as well as oncology.

From the microbiological point of view, it is interesting that anaerobic coryneforms, when administered to animals, significantly (more than twofold) stimulated the phagocytic K-index, whereas classical, or dairy, propionibacteria were almost without effect (White, 1975). Enzymatic activities of macrophages were similarly affected: only the cutaneous propionibacteria stimulated the activities of acid phosphatase, β-D-galactosidase and phospholipase A. As an example, the data concerning acid phosphatase activity and chemotactic activity of macrophages are shown in Tables 7.6 and 7.7, respectively.

It has been suggested (Wilkinson, 1975) that the activating factor of cutaneous propionibacteria, at least in part, is represented by a peptide that has affinity to the membrane of mononuclear phagocytes, in which it activates such processes as exocytosis, endocytosis, motility. After a short period of activation an increase in protein synthesis is observed in macrophages. But the molecular basis of the immunostimulating properties of anaerobic coryneforms requires further studies.

Many mycobacteria, both saprophytic and pathogenic, and most nocardia also show some adjuvant-like effects on guinea-pigs, similar to cutaneous coryneforms, but strange as it may seem, aerobic corynebacteria do not share this property. Cutaneous propionibacteria differ from the other bacteria listed above by their mechanism of macrophage activation. They do not activate thymus lymphocytes, differing in this respect also from mycobacteria and nocardia. Hence it was suggested (White, 1975) that the biological effects of mycobacteria and cutaneous coryneforms are based on quite different types of biochemical action.

Table 7.6. Pulmonary acid phosphatase levels in chicken, 6 days after an intravenous injection of different heat-killed microorganisms

Organism	Serological group	Acid phosphatase, K.A. units/ml (mean)
Mycobacteria		
Mycobacterium avium		238
Anaerobic coryneforms		
P. avidum 4982	4	196
C. parvum 3085	1	166
C. avidum 0589	4	162
C. parvum 0208	1	125
C. liquefaciens 814	1	118
C. parvum B	0	96
P. granulosum 0507	3	69
Classical propionibacteria		
P. freudenreichii 10470		41.6
P. jensenii 5960		28.6
P. rubrum 8901		25.6
P. arabinosum 5958		19.6
Saline control		10.6

Reproduced from White (1975), with permission.

Table 7.7. Chemotactic activities of anaerobic corynebacteria and propionibacteria for guinea-pig peritoneal macrophages

Strain	Microns migrated toward bacterial suspension*	Strain	Microns migrated toward bacterial suspension*
Group I (*P. acnes* group)		Group III (*P. granulosum* group)	
C. parvum 0208	40	*C. parvum* 10387	40
C. parvum 3085	90	*C. parvum* C	60
C. parvum 1383	120	*P. granulosum* 0507	40
C. liquefaciens 814	30	Group IV (*P. avidum* group)	
P. acnes 737	50	*P. avidum* 0575	50
C. parvum A	50	*P. avidum* 4982	120
Group II		*P. avidum* 0589	90
C. diphtheroides 2764	10	Classical propionibacteria	
C. anaerobium 578	40	*P. freudenreichii*	30
C. parvum 10390	30	*P. rubrum*	10
C. parvum B	20	*P. arabinosum*	20
Negative control, Gey's solution	20	Positive control, casein, 5 mg/ml	140

* Results are shown for bacterial suspensions at $1500 \cdot 10^4$ cells/ml. Positive response is defined as migration distance of ≥ 60 μm in 130 min through 8 μm filter.
Adapted from White (1975), with permission.

An additional clinically important activity was discovered in cutaneous propionibacteria. Phospholipid substances isolated from the cells stimulated not only the digestive capacity of liver and spleen macrophages, but also their bactericidal effects against *Salmonella typhimurium* and *Listeria monocytogenes* (Fauve, 1975). When mice were treated with *C. parvum*, their resistance not only to tumors, but also to bacterial infections was increased (Adlam et al., 1972). These antitumor and bactericidal actions can not be explained solely by the stimulation of the RES.

Both *in vivo* and *in vitro* it was observed (Ceruti, 1975) that peritoneal exudates of mice treated with *C. parvum* contained a factor that inhibited vesicular stomatitis virus and encephalomyocarditis virus. The inhibitory factor was different from interferons by a number of characteristics. For example, interferon is thermolabile, but the inhibitor is heat-resistant.

The results of experimental and clinical trials of cutaneous propionibacteria are regularly discussed at international colloquia, one of which entitled "Bacteria and Cancer" was held in Bologna in 1982.

In clinical studies of Szmigielski (1982), *P. granulosum* KP-45 was shown to display immunomodulating, antineoplastic and antiviral effects, which were associated with the activation of monocyte-macrophage system, induction of interferon synthesis and/or activation of killer cells. Stimulation of interferon synthesis was also observed in human tissue cultures. The author suggested that these effects are mediated by the cell wall components, in particular, by peptidoglycans and teichoic acids.

In addition, a stimulation of hemopoietic systems and bone marrow cell proliferation was observed. When *P. granulosum* was administered to mice several days before the irradiation with a lethal dose of X-rays (850 rad), the life span of the animals was significantly prolonged. For the treatment of patients it was suggested to combine immunotherapy with intensive radio- and chemotherapy. Roszkowski et al., (1982) concluded that *P. granulosum* KP-45, having a relatively low toxicity, could be used as a common adjuvant in anticancer therapy.

C. parvum is unique in another aspect, connected with medicine, namely, in its ability to synthesize prostaglandin-like compounds (PLC).

Prostaglandin-like compounds. Lipids extracted from *P. acnes* (*C. parvum*) were found to contain vasoactive prostaglandin-like compounds (PLC), with aliphatic chains typical of the family of prostaglandins. These substances of *P. acnes* represent a new class of bacterial prostaglandins (Hellgren and Vincent, 1983). Biological activities of the PLC were compared with prostaglandins PGE2, PGF2 and PGL2, and arachidonic acid. The spectrum of biological actions of PLC was similar to that of PGL2, but at the same time PLC had a unique property: they inhibited platelet aggregation. *In vitro* the PLC of *P. acnes* induced changes in blood capillaries (in animals) (Gulbic et al., 1981) and displayed a potent chemotactic activity on human polymorphonuclear leukocytes (Belsheim et al., 1979).

7.9.2   Dairy propionibacteria and medicine

Some dairy propionibacteria also may find application in medicine, besides being a source of vitamin B_{12} for industry (see Section 7.2.1). *P. shermanii* is a useful model for studying the regulation of ribonucleotide reductase (RNR) activity, which has a direct role in the control of cell division. On the basis of RNR inhibitors tested in *P. shermanii*, antitumor preparations were developed (Elford et al., 1979a, b).

In medical practice a number of metabolites of propionibacteria may find application. As shown above, some strains (e.g., *P. granulosum*) synthesize large amounts of porphyrins, which can be applied in medicine as diagnostic and medical preparations. Immobilized cells of *P. shermanii* can be used as sources of the nucleotides AMP, GMP, GTP, ADP etc., and their derivatives (see Section 6.2). Nucleotide derivatives can be used in the prevention and treatment of human thrombotic diseases. Purine bases and ribosides are maintenance factors for erythrocytes, ensuring high ATP levels (Demain, 1968). Inosin is effective in treating different heart and liver diseases; it is produced in Russia under the trade name Riboxin (Chagin et al., 1981). IMP is used as a starting material for the synthesis of a number of nucleic acid derivatives, which are valuable as medical preparations of antimicrobial and anticarcinogenic actions.

Recently, it has been shown (Perez-Chaia et al., 1998) that a strain of *P. acidipropionici* stimulated immunological responses of the host like the cutaneous propionibacteria discussed above, and could be useful in the prevention of intestinal infections or tumor development. Immuno-modulatory effects of some classical propionibacteria, i.e., stimulation of the cell populations involved in nonspecific resistance, have also been reported (Roszkowski et al., 1990).

Conclusion

Propionic acid bacteria represent a distinct and ancient evolutionary branch ascending from an actinomycetous ancestor. It is a bridge that links the past and present. Three subgroups of propionic acid bacteria: classical, cutaneous, and *P. propionicum* comprise the genus *Propionibacterium*. The classical propionibacteria inhabit mainly milk and cheese (hence their second name—dairy propionic acid bacteria), the cutaneous species inhabit human skin and the rumen of ruminants, and *P. propionicum* dwells in the soil.

Application of molecular taxonomy methods has changed the quantitative and qualitative composition of the genus and made it possible to find close and distant relatives. High levels of DNA homology between representatives of some species resulted in the reduction of the number of classical species from 11 to 5: *P. freudenreichii, P. thoenii, P. jensenii, P. acidipropionici* and *P. cyclohexanicum*.

On the other hand, the boundaries of the genus were extended by including four species of anaerobic corynebacteria that inhabit human skin. Using the methods of traditional, numerical and molecular taxonomy, anaerobic corynebacteria were shown to be different from the other members of the genus *Corynebacterium* by a number of essential characteristics and to have genomic similarities with propionic acid bacteria. The anaerobic corynebacteria have been reclassified as *P. acnes, P. avidum, P. granulosum* and *P. lymphophilum* and are called cutaneous propionibacteria now.

Human skin is inhabited by another species of propionic bacteria, *P. innocuum*, with a phenotype typical for classical propionibacteria, but differing from the latter and from the cutaneous bacteria in several aspects (Pitcher and Collins, 1991). DNA hybridization of *P. innocuum* and the other species of the genus *Propionibacterium* gave negative results and could not find any close genetic homology at the species level, despite the similarity in nucleotide composition. Partial sequencing (95%) of the 16S rRNA gene confirmed its affinity with the genus *Propionibacterium*, placing *P. innocuum* on a distant line at the periphery of the unrooted phylogenetic tree. The shortest evolutionary distance is between *P. innocuum*, *P. thoenii* and *P. propionicum*.

Yokota et al. (1994), however, pointed out that the differences between *P. innocuum* and the other propionibacterial species are too profound to classify them in the same genus. They proposed to transfer *P. innocuum* to a new genus, *Propioniferax*, under the name *Propioniferax innocua* gen. nov. comb. nov. The closest phylogenetic neighbor of the new species is *Luteococcus japonicus*.

P. cyclohexanicum sp. nov. was isolated recently from spoiled orange juice (Kusano et al., 1997). The results of phylogenetic analysis of the 16S rRNA gene indicated the highest level of homology with the classical propionibacterium *P. freudenreichii*, to which it is also similar phenotypically. ω-Cyclohexyl undecanoic acid, an alicyclic fatty acid rarely found in bacteria, is the major cellular fatty acid in the new propionibacterium. *P. cyclohexanicum* produces large amounts of lactic acid, is acid-tolerant and heat-resistant, which differentiates the new species from the other propionibacteria.

Taking into consideration the transfer of *Arachnia propionica* (now called *P. propionicum*) from the group *Actinomyces* to the genus *Propionibacterium*, at present the genus includes 10 species: *P. freudenreichii*, *P. thoenii*, *P. jensenii*, *P. acidipropionici*, *P. cyclohexanicum*, *P. acnes*, *P. avidum*, *P. granulosum*, *P. lymphophilum* and *P. propionicum*.

Intergeneric relationships of propionibacteria and various actinomycetes were established by comparing 16S rRNA sequences (Charfreitag and Stackebrandt, 1989). The results showed that the genus *Propionibacterium* stands isolated among other major groups of the actinomycetes. Its phylogenetic neighbor is *Nocardioides*. The most obvious common phenotypic character of propionibacteria, *Nocardioides* and *Terrabacter* is the presence of the rare L-DAP-glycine peptidoglycan type, supporting the inclusion of these genera in a phylogenetic cluster.

The taxonomy of propionibacteria was given special consideration at the 1st International Symposium "Dairy propionibacteria" held in Rennes, France, in May 1995.

The genus *Propionibacterium* is not a uniform group. Undoubtedly, environmental influences made its mark on the properties of representatives of the subgroups. But even the classical propionibacteria, studied better than the other subgroups, do not represent a uniform group and differ in many respects, including the chemical composition of cell walls, fatty acids of membrane lipids, relation to oxygen, mechanism of sulfate utilization, storage conditions, and others. To say more, a 'mix' of functions is found even within a single species and many functions are duplicated by separate systems; e.g., energy can be generated by substrate level phosphorylation and by oxidative phosphorylation; DNA synthesis and methylation may proceed via vitamin B_{12}-dependent and by B_{12}-independent pathways. The microbes activate those enzymes that allow them to realize vital necessities with maximal effectiveness in each specific situation. It has been said that clever subjects are those, who are able to make the right decision at the right time in the right place.

In creating propionibacteria Nature made almost no mistakes. The 'mixing' of functions in propionibacteria determines their another peculiarity — high adaptability, which other dairy bacteria such as *Lactobacillus* lack. Plasmids discovered in propionibacteria, although insufficiently studied, might strengthen their adaptive possibilities. It is possible that the conservation of genes (many of them may be silent under ordinary conditions) made the metabolism of propionibacteria flexible and plastic, and slowed down the evolutionary rate of this form of life, compared with evolutionary rates of the other descendants of the actinomycetous line.

All representatives of the genus preserved a common pathway of energy generation under anaerobic conditions: the propionic acid fermentation, which is the most harmonious and energetically profitable way of ATP generation due to the involvement of AdoCbl in this pathway. It is the only bacterial fermentation capable of gaining about 6 moles of ATP from 1.5 moles of fermented glucose. Oxidative phosphorylation, discovered in propionibacteria, proceeds with a low efficiency, but contributes to the ATP synthesis also under anaerobic conditions: fumarate-linked respiration takes place in parallel with fermentation. The preference for anaerobic style of life is also explained by a strong repression of cytochrome synthesis in the presence of oxygen, including cytochrome *b* functional in the fermentation.

Biosynthetic capabilities of propionibacteria are well developed, although they vary in different strains. The finding of nitrogen fixation, of the ability to utilize paraffins (alkanes), of independent syntheses of many vitamins have widened our ideas of the life boundaries of propionic acid bacteria.

All the propionibacteria synthesize vitamin B_{12}, which has diverse functions in fermentation, protein and DNA synthesis, in regulation of DNA-synthesis and some other reactions. 'Anaerobic' strains synthesize greater amounts of vitamin B_{12} than 'aerobic' strains when grown under identical anaerobic conditions. This observation shows that the 'supersynthesis' of vitamin B_{12} by 'anaerobic' strains is genetically determined. In anaerobic forms (as shown experimentally) metabolism is tuned to the high level of vitamin B$_{12}$, which is down-regulated, however, when its level is too high in the cell.

Cutaneous propionibacteria can be considered as a biological defense for their human host and as a useful natural microflora in the rumen of ruminants. They stimulate the immune system in humans, have beneficial effects on farm animals and poultry, and therefore may find application as components of medical and preventive preparations. However, further investigations are needed before propionibacteria can be used in medicine.

Genetic investigations represent an important trend in studying propionibacteria. In Russia, they are concentrated at the Institute of Genetics of Industrial Microorganisms, where a complete genomic library of *P. shermanii* has been prepared and the genes thrB and recA of *P. shermanii* have been cloned and expressed in different strains of *E. coli* (Pankova and Abilev, 1993; Pankova et al., 1993a, b). Bacteriophages found in *P. freudenreichii* can be used as vectors for gene transfer (Borisova et al., 1973; Gautier et al., 1995). A total of 19 bacteriophages have already been isolated from *P. freudenreichii* (Cassin et al., 1995). However, no major genomic project on a scale, comparable to those carried out with *E. coli* and some other bacteria, has yet emerged.

A new perspective is opened by the discovery of antimutagenic and reactivative properties of propionibacteria. A protein substance (active factor, AF) responsible for the protection and reactivation of cells damaged by UV light has been identified in *P. shermanii* (Vorobjeva et al., 1995a, b, 1996b). The factor is active with both prokaryotic and eukaryotic cells, e.g., bacteria and yeasts. The activity seems to be specific to the damaging effects of the short-wave UV light (254 to 320 nm), similar to the AFs isolated from other propionibacterial species (Vorobjeva et al., 1996b). The factor from *P. shermanii* also has a protective activity against some environmental stress factors other than UV light: heating, exposure to ethanol and divalent cations (Cd, Zn, Cu, Co). These data are consistent with the suggestion that a single gene-mediated system, present in propionibacteria, can respond to a variety of different environmental stress factors (Van Bogelen et al., 1987; Piper, 1995).

Cellular responses to external stresses are often mediated by heat-shock proteins with chaperonin functions (Hartl, 1995; Mesger-Lallemand et al., 1995). It is possible that the AF of *P. shermanii* is related to or operates in conjunction with cellular heat-shock proteins. Further characterization of the individual 44-kDa polypeptide, isolated from the AF of *P. shermanii* in our laboratory, will shed light on its mechanism of action. The importance of these studies is underscored by the depletion of the ozone layer and industrial pollution of the environment, leading to an increased incidence of mutations and other adverse effects.

Propionibacteria are among the most valuable anaerobic species. They are active producers of vitamin B_{12} and porphyrins. But the production of large quantities of vitamin B_{12} is only one of the peculiarities of propionic acid bacteria that are used in practice. From ancient times these bacteria are used in cheese making and many defects of Swiss-type cheeses are caused by the absence or weak growth of propionic acid bacteria. Therefore, it is still important to develop multi-strain starter cultures of propionic acid bacteria, designed specifically for cheese making. Propionibacteria such as *P. acidipropionici* and *P. freudenreichii* can be added to leavenings for bread baking (Arnoux et al., 1995); propionic acid produced by these bacteria inhibits the growth of mold, improves aroma of the bread and keeps it fresh for a longer period.

New applications that emerge from the current research include the use of propionibacteria as probiotics and growth promoters (Mantere-Alhonen, 1995). These bacteria can easily adapt to live in the gut of humans and animals and are able to reduce genotoxic effects of a number of chemical substances and UV light. In addition, propionibacteria produce bacteriocins that suppress the growth of other microorganisms. These proteins are thermostable and resistant to the extremes of pH (Glatz et al., 1995; Paik and Glatz, 1997), so they can be recommended as antimicrobials for food preservation.

To sum up the long-term investigations of propionibacteria, it is necessary to emphasize once more the heterogeneity of the classical (dairy) species (except *P. freudenreichii*) revealed by modern methods of molecular identification (De Carvalho, 1995; Riedel and Britz, 1996; Riedel et al., 1998; Dasen et al., 1998). Since strain peculiarities are at the basis of all technologies using propionibacteria, it is important to realize their true status and biochemical potentials. In this complicated area, as well as in several others that have been reviewed in this book, still much remains to be clarified and better defined. Great interest in propionic acid bacteria that gathered scientists from different countries at the International Symposia in May 1995 (Rennes, France) and in June 1998 (Cork, Ireland) gives basis to conclude that many still unsolved questions will be answered, and that new secrets of the original form of life chosen by propionibacteria will be revealed.

References

Adlam C, Broughton E and Scott MT (1972) Enhanced resistance of mice to infection with bacteria following pre-treatment with *Corynebacterium parvum*. Nature New Biol 235: 219-224

Adler HI, Fisher WD, Hardigree AA and Stapleton GE (1966) Repair of radiation-induced damage to the cell division mechanism of *Escherichia coli*. J Bacteriol 91: 733-742

Alekperov UK (1984) Antimutagenesis. Nauka, Moscow.

Alekseeva MA, Anischenko IP, Schlegel AH, Ott EF and Vorobjeva LI (1983) Improving the criteria of selection of propionibacteria for cheesemaking. In: Shlegel AH (ed) Nauchno-Technichesky Progress, pp 117-129. Barnaul

Alekseeva MA, Klimovsky II and Anischenko IP (1973a) Species content of propionic acid bacteria in Soviet cheese. Moloch Promyshl 12: 12-13

Alekseeva MA, Vorobjeva LI, Chan TT, Baranova NA, Aleksandrushkina NI and Gavrina TP (1973b) Physiologo-biochemical properties of propionic acid cocci. Mikrobiologiya 42: 464-467

Allaker RP, Greenman J and Osborne RH (1986) Histamine production by *Propionibacterium acnes* in batch and continuous culture. Microbios 48: 165-172

Allen JM and Beard ME (1965) α-Hydroxy acid oxidase: localization in renal microbodies. Science 149: 1507-1509

Allen SHG and Linehan BA (1977) Presence of transcarboxylase in *Arachnia propionica*. Int J Syst Bacteriol 27: 291-292

Allen SHG, Kellermeyer RW, Stjernholm R, Jacobson B and Wood HG (1963) The isolation, purification and properties of methylmalonyl racemase. J Biol Chem 238: 1637-1642

Allen SHG, Kellermeyer RW, Stjernholm R and Wood HG (1964) Purification and properties of enzymes involved in the propionic acid fermentation. J Bacteriol 87: 171-187

Andreeva NA (1974) Enzymes of Folic Acid Metabolism. Nauka, Moscow

Antila M (1954) Über die Propionsäurebakterien im Emmentaler Käse. Meijerit Aikakausk 16: 1-132

Antila M (1956/1957a) Der Aminosäurebau durch Propionsäurebakterien. Meijerit Aikakausk 18/19: 1-6

Antila M (1956/1957b) Die Bildung von Azetoin und Diazetyl bei Propionsaurebakterien. Meijerit Aikakausk 18/19 : 7-14

Antila M and Hietaranta M (1953) Growth inhibition of propionic acid bacteria by propionate. Meijerit Aikakausk 15: 3-10

Antila V and Antila M (1968) The content of free amino acids in Finnish cheese. Milchwissenschaft 23: 597-602

Antoshkina NV, Vorobjeva LI and Iordan EP (1979) Methylcobalamin involvement in DNA-methylation in cell-free extracts of *Propionibacterium shermanii*. Mikrobiologiya 48: 217-220

Antoshkina NV, Vorobjeva LI and Buryanov YJ (1981) Methylcobalamin-dependent enzymatic DNA methylation in cell-free extracts of *Propionibacterium shermanii*. Mikrobiologiya 50: 462-467

Arima K and Oka T (1965) Cyanide resistance in *Achromobacter*. J Bacteriol 90: 734-743

Arkadjeva ZA, Olsinskaya NL and Alekseeva MA (1988) Storage of propionic acid bacteria. Prikl Biokhim Mikrobiol 24: 839-845

Arnoux P, Machado ES, Kirane D, Cochet N, Nonus M and Lebeault J-M (1995) Propionibacteria: bread making and cheese aroma production. Abstr 1st Int Symp Dairy Propionibacteria, Rennes, France, C21

Atlavinyte O, Galvelis A, Dacinlyte J and Luganskas A (1982) Effects of entobacterium *Bacillus thuringiensis* var. *gallerie* on earthworm activity. Pedobiologia 23: 372-379

Auling G and Follmann H (1994) Manganese-dependent ribonucleotide reduction and overproduction of nucleotides in coryneform bacteria. In: Siegel H and Siegel A (eds) Metal Ions in Biological Systems, vol 30, pp 131-161. Marcel Dekker, New York

Axelsson LT, Chung SH, Dobrogosz WJ and Lindgren SE (1989) Production of a broad spectrum antimicrobial substances by *Lactobacillus reuteri*. Microbiol Ecol Health Disease 2: 131-136

Ayres C, de Mello GC and Lara FJS (1962) Fumarate hydratase from *Propionibacterium pentosaceum*. Biochim Biophys Acta 62: 435-437

Azuma I, Sugimura K, Taniyama T, Aladin AA, Yamamura Y and Ribi E (1975) Chemical and immunological studies on the cell walls of *Propionibacterium acnes* strain C7 and *Corynebacterium parvum* ATCC11829. Jap J Microbiol 19: 265-275

Babuchowski A and Hammond EG (1987) Production of propionic acid by *Propionibacterium* spp. grown on various carbohydrate sources. Abstr Annu Meet Am Soc Microbiol Atlanta, 1-6 March, 1987, Washington, DC, p 265

Babuchowski A, Glatz BA and Hammond EG (1987) Effect of carbohydrate source on growth yield coefficient of selected strains of propionibacteria. Abstr Annu Meet Am Soc Microbiol Atlanta, 1-6 March, 1987, Washington, DC, p 265

Baehman LR and Glatz B (1986) Protoplast formation and regeneration in *Propionibacterium*. Abstr Annu Meet Am Soc Microbiol, Washington, DC

Baer A (1987) Identification and differentiation of propionibacteria by electrophoresis of their proteins. Milchwissenschaft 42: 431-433

Bala S and Grover IS (1989) Antimutagenicity of some citrus fruits in *Salmonella typhimurium*. Mutat Res 222: 141-148

Balachandran S, Vishwakarma RA, Monaghan SM, Prelle A, Stamford NPJ, Leeper FJ and Battersby AR (1994) Biosynthesis of porphyrins and related macrocycles. Part 42. Pulse labelling experiments concerning the timing of cobalt insertion during vitamin B_{12} biosynthesis. J Chem Soc Perkin Trans I: 487-491

Baranova NA and Gogotov IN (1974) Nitrogen fixation by propionic acid bacteria. Mikrobiologiya 43: 791-794

Baranova NA and Vorobjeva LI (1971) Influence of peptides on vitamin B_{12} biosynthesis in *Propionibacterium shermanii*. Nauchn Dokl Vys Shkoly Biol Nauki 3: 97-99

Barefoot SF, Garver KI, Ratnam PR, Grinstead DA and Prince LD (1995) Jenseniins G and P, heat-stable bacteriocins produced by *Propionibacterium jensenii*. Abstr 1st Int Symp Dairy Propionibacteria, Rennes, France, C13

Barker HA (1967) Biochemical functions of corrinoid compounds. Biochem J 105: 1-15

Barker HA and Lipmann F (1944) On lactic acid metabolism in propionic acid bacteria and the problem of oxido-reduction in the system of fatty-hydroxyketo acids. Arch Biochem 4: 361-370

Barker HA and Lipmann F (1949) The role of phosphate in the metabolism of *Propionibacterium pentosaceum*. J Biol Chem 179: 247-257

Barksdale L (1970) *Corynebacterium diphtheria* and its relatives. Bacteriol Rev 34: 378-422

Battersby AR (1985) Biosynthesis of the pigments of life. Proc R Soc Lond B 225: 1-26

Battersby AR (1994) How nature builds the pigments of life: the conquest of vitamin B_{12}. Science 264: 1551-1557

Battersby AR and Reiter LA (1984) Synthetic studies relevant to biosynthetic research on vitamin B_{12}. Part 3. An approach to isobacteriochlorins via nitrones. J Chem Soc Perkin Trans I: 2743-2749

Bauchamp CO and Fridovich I (1971) Superoxide dismutase: improved assay and an assay applicable to acrylamide gels. Anal Biochem 44: 276-287

Bauchop T and Elsden SR (1960) The growth of microorganisms in relation to their energy supply. J Gen Microbiol 23: 457-469

Baum M and Breese M (1976) Antitumour effect of *Corynebacterium parvum*. Possible mode of action. Br J Cancer 33: 468-473

Beerens H, Neut C and Romond C (1986) Important properties in the differentiation of Gram-positive non-sporing rods in the genera *Propionibacterium, Eubacterium, Actinomyces* and *Bifidobacterium*. In: Barnes EM and Mead GC (eds) Anaerobic Bacteria in Habitats Other than Man, pp 38-42. Blackwell Scientific, Oxford

Beevers H (1961) Respiratory Metabolism in Plants. Row, Peterson & Co, Evanston, Illinois/ White Plains, New York

Belozerov EV and Yurkevich AM (1991) Vitamin B_{12} derivatives and human immunodeficiency. Abstr 9th All-Union Symp "Screening of medicinal substances," Riga, p 11

Belsheim J, Dalen A, Gnarpe H, Hellegren L, Iverson OJ and Vincent J (1979) The effect of acidic polysaccharides and prostaglandine-like substances isolated from *Propionibacterium acnes* on granulocyte chemotaxis. Experientia 35: 1587-1589

Bernhauer K, Wagner F, Michna H, Rapp P and Vogelmann H (1968) Biogenesewege von der Cobyrinsäure zur Cobyrsäure und zum Cobinamide bei *Propionibacterium shermanii*. Hoppe-Seyler's Z Physiol Chem 349: 1297-1309

Berry DB (1982) Biology of Yeasts. Edward Arnold

Bestor T, Laudano A, Mattaliano R and Ingram V (1988) Cloning and sequencing of a cDNA encoding DNA methyltransferase of mouse cells. The carboxyl-terminal domain of the mammalian enzymes is related to bacterial restriction methyltransferases. J Mol Biol 203: 971-983

Blanca BR, Ester L-B-P and Horacio ST (1989) Oncolytic activity of *Clostridium* M-55 in tumors induced by SV-40 transformed cells. Rev Latinoam Microbiol 31: 159-165

Blanche F, Cameron B, Crouzet J, Debussche L, Thibaut D, Vuilhorgne M, Leeper FJ and Battersby AR (1995) Vitamin B_{12}: how the problem of its biosynthesis was solved. Angew Chem Int Ed Engl 34: 383-411

Bobik MA (1971) Discovery of a new enzyme of polyphosphate metabolism. PhD thesis, Moscow State University, Moscow

Bodie EA, Anderson TM, Goodman N and Schwartz RD (1987) Propionic acid fermentation of ultra-high-temperature sterilized whey using mono- and mixed-cultures. Appl Microbiol Biotechnol 25: 434-437

Bogatyreva TG, Iordan EP and Vorobjeva LI (1993) A means of preventing potato blight of bread. Patent No 1608849, Russian Federation

Bogdanoff I, Popkhristov P and Marinov L (1962) Anticancer effect of *Antibioticum bulgaricum* on sarcoma-180 and on solid form of Ehrlich carcinoma. Abstr 8th Int Cancer Congr, Moscow

Bomford R and Olivotto M (1975) Inhibition by *Corynebacterium parvum* of lung-nodule formation by intravenously injected fibrosarcoma cells. In: Halpern B (ed) *Corynebacterium parvum*. Applications in Experimental and Clinical Oncology, pp 268-275. Plenum Press, New York

Bonarceva GA, Krainova OA and Vorobjeva LI (1973a) Pathways of terminal oxidation in propionic acid bacteria. Mikrobiologiya 42: 583-588

Bonarceva GA, Taptykova SD, Vorobjeva LI, Krainova OA and Bruchatcheva NL (1973b) Aerobic metabolism of propionic acid bacteria. Mikrobiologiya 42: 765-771

Borisova TG, Baranova NA and Vorobjeva LI (1973) Phage of propionic acid bacteria. Prikl Biokhim Mikrobiol 9: 246-249

Borst-Pauwels GWFH (1981) Ion transport in yeast. Biochim Biophys Acta 650: 88-127

Boyaval P and Corre C (1995) Production of propionic acid. Lait 75: 453-461

Boyaval P, Corre C and Madec M-N (1994) Propionic acid production in a membrane bioreactor. Enzyme Microb Technol 16: 883-886

Boyaval P, Corre C, Dupuis C and Roussel E (1995) Effects of free fatty acids on propionic acid bacteria. Lait 75: 17-29

Bradley DE (1967) Ultrastructure of bacteriophages and bacteriocins. Bacteriol Rev 31: 230-314

Bray R and Shemin D (1958) The biosynthesis of the porphyrin-like moiety of vitamin B_{12}. Biochim Biophys Acta 30: 647-648

Breed RS, Murray EGD and Smith NR (1957) Genus I. *Propionibacterium* Orla-Jensen 1909. In: Breed RS, Murray EGD and Smith NR (eds). Bergey's Manual of Determinative Bacteriology, 7th ed, pp 569-576. Williams & Wilkins, Baltimore

Brennan P and Ballou CE (1968) Phosphatidylmyoinositol monomannoside in *Propionibacterium shermanii*. Biochem Biophys Res Commun 30: 69-75

Britz TJ and Riedel K-HJ (1995) Numerical evaluation of the species groupings among the classical propionibacteria. Lait 75: 309-314

Brock DW, George LK, Brown JM and Hicklin MD (1973) Actinomycosis caused by *Arachnia propionica*. Am J Clin Pathol 59: 66-77

Brovko LY, Ugarova NN, Vasiljeva TE and Dombrovsky VA (1978) Application of immobilized luciferase for quantitative determination of ATP and enzymes that synthesize and degrade ATP. Biokhimiya 43: 798-805

Bruchatcheva NL, Bonarceva GA and Vorobjeva LI (1975) Oxidative phosphorylation in propionic acid bacteria. Mikrobiologiya 44: 11-15

Bukin VN, Bykhovsky VY and Panzchava ES (1971) Vitamin B_{12} and its application in husbandry. Nauka, Moscow

Bykhovsky VY (1979) Biogenesis of tetrapyrrole compounds (porphyrins and corrinoids), and its regulation. In: Zagalak B and Friedrich W (eds) Vitamin B_{12}, pp 293-314. Walter de Gruyter, Berlin

Bykhovsky VY and Zaitseva NI (1977) Role of ammonia nitrogen in the biosynthesis of vitamin B_{12} by *Propionibacterium shermanii*. Prikl Biokhim Mikrobiol 13: 16-23

Bykhovsky VY and Zaitseva NI (1983) Microbiological synthesis of tetrapyrrole compounds. Prikl Biokhim Mikrobiol 19: 163-175

Bykhovsky VY and Zaitseva NI (1989) Microbiological syntheses of tetrapyrrole compounds. Itogi Nauki i Techniki Ser Biol Khimiya 32

Bykhovsky VY, Zaitseva NI and Bukin VN (1968) Some aspects of the regulation of vitamin B_{12} and porphyrin biosynthesis in *Propionibacterium shermanii*. Dokl Akad Nauk SSSR 185: 459-461

Bykhovsky VY, Zaitseva NI and Yavorskaya AN (1974) Regulation of biosynthesis of tetrapyrrole compounds in microorganisms. In: Chagovets RV (ed) Vitaminy, v 7, pp 120-125. Naukova Dumka, Kiev

Bykhovsky VY, Zaitseva NI and Bukin VN (1975a) Biosynthesis and some properties of a cyclic tetrapyrrole precursor of vitamin B_{12}. Dokl Akad Nauk SSSR 224: 1431-1434

Bykhovsky VY, Zaitseva NI, Gruzina VD, Malyarova ZA, Ponomareva GM and Yavorskaya AN (1975b) Influence of certain natural analogs and derivatives of vitamin B_{12} on its biosynthesis by *Propionibacterium shermanii*. Prikl Biokhim Mikrobiol 11: 179-184

Bykhovsky VY, Zaitseva NI, Umrichina AV and Yavorskaya AN (1976) Spectral and photochemical properties of a tetrapyrrole pigment (corriphyrin) synthesized by *Propionibacterium shermanii*. Prikl Biokhim Mikrobiol 12: 825-833

Bykhovsky VY, Zaitseva NI, Yeliseev AA, Pushkin AV, Evstigneeva ZG and Kretovich VL (1982) Glutamine-dependent amidation of the corrin ring of vitamin B_{12}. Dokl Akad Nauk SSSR 267: 250-254

Bykhovsky VY, Zaitseva NI and Polulach OV (1987) Biosynthesis of porphyrins by microorganisms. Prikl Biokhim Mikrobiol 23: 725-739

Bykhovsky VY, Demain AL and Zaitseva NI (1997) The crucial contribution of starved resting cells to the elucidation of the pathway of vitamin B_{12} biosynthesis. Crit Rev Biotechnol 17: 21-37

Cai MY, Lu DS, Wang DS, He ZZ and Wang LC (1989) A new sarcomostatic species of *Pseudomonas, Pseudomonas jinanensis* sp. nov. Acta Microbiol Sin 29: 155-160

Cancho FG, Vega MN, Diaz MF and Buzcu NJY (1970) Especies de *Propionibacterium* relacionades con la zapateria. Factores que influyen en su desarrollo. Microbiol Esp 23: 233-252

Cancho FG, Navarro LR and de la Borbolla y Alcala R (1980) La formacion de acido propionico durante la concervacion de las aceitunas verdes de mesa. III. Microorganismos responsables. Grases Aceites 31: 245-250

Canzi E, Del Puppo E, Brusa T, Galli A and Ferrari A (1993) Influence of oxygen on growth and fermentation activity of propionic acid bacteria. Ann Microbiol Enzimol 43: 147-157

Cassin D, Rouault A and Gautier M (1995) Isolation and characterization of bacteriophages infecting dairy propionibacteria. Abstr 1st Int Symp Dairy Propionibacteria, Rennes, France, A6

Castberg HB and Morris HA (1978) The pyruvate oxidizing system of *Propionibacterium freudenreichii* subsp. *shermanii*. Milchwissenschaft 33: 541-544

Cerdá-Olmedo E and Hanawalt PC (1967) Macromolecular action of nitrosoguanidine in *Escherichia coli*. Biochim Biophys Acta 142: 450-464

Cerna B, Cerny M, Betkova H, Patricny P, Soch M and Opatrna I (1991) Effect of the Proma probiotics on calves. Dairy Sci Abstr 55: 1735

Cerning J (1995) Production of exopolysaccharides by lactic acid bacteria and dairy propionibacteria. Lait 75: 463-472

Cerutti I (1975) Antiviral properties of *Corynebacterium parvum*. In: Halpern B (ed) *Corynebacterium parvum*. Applications in Experimental and Clinical Oncology, pp 84-90. Plenum Press, New York

Chagin BA, Kusmenok VA and Korolev NV (1981) Control of riboxin biosynthesis. Abstr. 3rd Conference "Theory and practice of the controlled cultivation of microorganisms", Naukova Dumka, Kiev, p. 24

Chaix P and Fromageot C (1939) Action du groupe -SH sur la fermentation et la respiration des bactéries propioniques en présence de glucose. Enzymologia 6: 33-45

Chaix-Audemand P (1940) Sur les relations existant entre respiration et fermentation chez *Propionibacterium pentozaceum*. PhD thesis, Faculty of Science, Lion

Chan TT (1973) Study of physiological and biochemical properties of representatives of different species of propionic acid bacteria and their mutant forms. PhD thesis, Moscow State University, Moscow

Charakhchyan IA and Vorobjeva LI (1984) Sulfate assimilation by propionic acid bacteria. Mikrobiologiya 53: 38-42

Charfreitag O and Stackebrandt E (1989) Inter- and intrageneric relationships of the genus *Propionibacterium* as determined by 16S rRNA sequences. J Gen Microbiol 135: 2065-2070

Charfreitag O, Collins MD and Stackebrandt E (1988) Reclassification of *Arachnia propionica* as *Propionibacterium propionicus* comb. nov. Int J Syst Bacteriol 38: 354-357

Clausen EC and Gaddy JL (1984) Organic acids from biomass by continuous fermentation. Chem Eng Prog 80: 59-63

Collins MD, Dorsch M and Stackebrandt E (1989) Phylogenetic studies on the genera *Pimelobacter* and *Nocardioides*: transfer of *Pimelobacter tumescens* to a new genus *Terrabacter*. Int J Syst Bacteriol 39: 1-6

Cooper TG, Tchen TT, Wood HG and Benedict CR (1968) The carboxylation of phosphoenolpyruvate and pyruvate. I. The active species of "CO_2" utilized by phosphoenolpyruvate carboxykinase, carboxytransphosphorylase, and pyruvic carboxylase. J Biol Chem 243: 3857-3863

Corcoran JV and Shemin D (1957) On the biosynthesis of the porphyrin-like moiety of vitamin B_{12}: the mode of utilization of δ-aminolevulinic acid. Biochim Biophys Acta 25: 661-662

Corre C, Quelen L, Dupuis C and Boyaval P (1995) Prolinase (EC 3.4.13.8) activity of *Propionibacterium freudenreichii* subsp. *shermanii*. Abstr 1st Int Symp Dairy Propionibacteria, Rennes, France, B2

Cox GB, Newton NA, Gibson F, Snoswell AM and Hamilton JA (1970) The function of ubiquinone in *Escherichia coli*. Biochem J 117: 551-562

Crow VL (1988) Polysaccharide production by propionibacteria during lactose fermentation. Appl Environ Microbiol 54: 1892-1895

Cummins CS (1985) Distribution of 2,3-diaminohexuronic acid in strains of *Propionibacterium* and other bacteria. Int J Syst Bacteriol 35: 411-416

Cummins CS and Hall P (1986) Acetate and pyruvate in cell wall polysaccharides of *Propionibacterium acnes*, *P. avidum* and *P. granulosum*. Curr Microbiol 14: 61-63

Cummins CS and Johnson JL (1974) *Corynebacterium parvum*: a synonym for *Propionibacterium acnes*. J Gen Microbiol 80: 433-442

Cummins CS and Johnson JL (1981) The genus *Propionibacterium*. In: Starr MP, Stolp H, Trüper HG, Balows A and Schlegel HG (eds) The Prokaryotes. A Handbook on Habitats, Isolation, and Identification of Bacteria, 1st ed, vol 1, pp 1894-1902. Springer-Verlag, Berlin

Cummins CS and Johnson JL (1986) Genus *Propionibacterium* Orla-Jensen 1909. In: Sneath PHA, Mair NS, Sharpe ME and Holt JG (eds) Bergey's Manual of Systematic Bacteriology, 1st ed, vol 2, pp 1346-1353. Williams & Wilkins, Baltimore

Cummins CS and Johnson JL (1992) The genus *Propionibacterium*. In: Balows A, Trüper HG, Dworkin M, Harder W and Schleifer K-H (eds) The Prokaryotes. A Handbook on the Biology of Bacteria: Ecophysiology, Isolation, Identification, Applications, 2nd ed, vol 2, pp 834-849. Springer-Verlag, New York

Cummins CS and Moss CW (1990) Fatty acid composition of *Propionibacterium propionicum (Arachnia propionica)*. Int J Syst Bacteriol 40: 307-308

Cummins CS and White RH (1983) Isolation, identification and synthesis of 2,3-diamino-2,3-dideoxyglucuronic acid: a component of *Propionibacterium acnes* cell wall polysaccharide. J Bacteriol 153: 1388-1393

Czarnoska-Roczniakowa B (1966) Semi-continuous corrinoid biosynthesis by *Propionibacterium petersonii*. Acta Microbiol Pol 15: 349-356

Dahl TA, Midden WR and Hartman PE (1988) Some prevalent biomolecules as defenses against singlet oxygen damage. Photochem Photobiol 47: 357-362

Dasen G, Smutny J, Teuber M and Meile L (1998) Classification and identification of propionibacteria based on ribosomal RNA genes and PCR. System Appl Microbiol 21: 251-259

Davis JJ and Wood HG (1966) Inorganic pyrophosphate and the formation of phosphoenolpyruvate. Fed Proc, Abstr 50th Annu Meet, Atlantic City, New Jersey, April 11-16, 1966, p. 500

Davis JJ, Willard JM and Wood HG (1969) Phosphoenolpyruvate carboxytransphosphorylase. III. Comparison of the fixation of carbon dioxide and the conversion of phosphoenolpyruvate and phosphate into pyruvate and pyrophosphate. Biochemistry 8: 3127-3136

Dawson TE, Rust SR and Yokoyama MT (1993) Manipulation of silage fermentation and aerobic stability by propionic acid producing bacteria. Silage production from seed to animal NRAES-67. Northeast Regional Agric Eng Service, pp 96-105. Syracuse

De Carvalho AF (1995) Molecular identification of dairy *Propionibacterium* spp. Abstr 1st Int Symp Dairy Propionibacteria, Rennes, France, C2

De Carvalho AF, Guezenec S, Gautier M and Grimont PAD (1995) Reclassification of *Propionibacterium rubrum* as *P. jensenii*. Res Microbiol 146: 51-58

De Vries W, van Wijck-Kapteijn WMC and Stouthamer AH (1972) Influence of oxygen on growth, cytochrome synthesis and fermentation pattern in propionic acid bacteria. J Gen Microbiol 71: 515-524

De Vries W, van Wijck-Kapteijn WMC and Stouthamer AH (1973) Generation of ATP during cytochrome-linked anaerobic electron transport in propionic acid bacteria. J Gen Microbiol 76: 31-41

De Vries W, Aleem MIH, Hemrika-Wagner A and Stouthamer AH (1977) The functioning of cytochrome *b* in the electron transport to fumarate in *Propionibacterium freudenreichii* and *P. pentosaceum*. Arch Microbiol 112: 271-276

Delwiche EA (1948) Mechanism of propionic acid formation by *Propionibacterium pentosaceum*. J Bacteriol 56: 811-820

Delwiche EA (1949) Vitamin requirements in the genus *Propionibacterium*. J Bacteriol 58: 395-398

Delwiche EA and Carson SF (1953) A citric acid cycle in *Propionibacterium pentosaceum*. J Bacteriol 65: 318-321

Demain AL (1968) Production of purine nucleotides by fermentation. Prog Indust Microbiol 8: 35-71

Demple B and Linn S (1982) 5,6-Saturated thymine lesions in DNA: production by ultraviolet light or hydrogen peroxide. Nucl Acids Res 10: 3781-3789

Dills SS, Apperson A, Schmidt MR and Saier MH (1980) Carbohydrate transport in bacteria. Microbiol Rev 44: 385-418

Dimitrov N, Andre S, Eliopoulos G and Halpern B (1975) Comparative studies of the effect of *Corynebacterium parvum* on bone-marrow cell colony formation *in vitro*. In: Halpern B (ed) *Corynebacterium parvum*. Applications in Experimental and Clinical Oncology, pp 173-180. Plenum Press, New York

Dimroth P (1988) Bakterielle Energieübertragung über einen Natrium-Cyclus. Forum Mikrobiol 5: 180-187

Doelle HW (1981) New developments in the elucidation of the mechanisms of the Pasteur and Crabtree effects in bacteria. Adv Biotechnol Proc 6th Int Ferment Symp, London (Canada), July 20-24, 1980, vol 1, pp 249-254

Domracheva GI, Kononov YV and Maidanyuk AE (1970) Influence of propionibacteria on the quality of silage, growth and development of young animals. Sci Proc Sibir Inst of Husbandry 15: 173-177

Dose K (1989) The bioenergetics of anaerobic bacteria: evolutionary concepts. Adv Space Res 9: 93-100

Douglas HC and Gunter SE (1946) The taxonomic position of *Corynebacterium acnes*. J Bacteriol 52: 15-23

Drennan CL, Dixon MM, Hoover DM, Jarret JT, Goulding CW, Matthews RG and Ludwig ML (1998) Cobalamin-dependent methionine synthase from *Escherichia coli*: structure and reactivity. In: Kräutler B, Arigoni D and Golding BT (eds) Vitamin B_{12} and B_{12}-proteins, pp 133-156. Wiley-VCH, Weinheim, Chichester, New York

Drinan FD and Cogan TM (1992) Detection of propionic acid bacteria in cheese. J Dairy Res 59: 65-69

Dryden LP, Hartman AM, Bryant MP, Robinson JM and Moore LA (1962) Production of vitamin B_{12} and vitamin B_{12} analogues by pure cultures of ruminal bacteria. Nature 195: 201-202

Dubinin NP (1976) General Genetics. Nauka, Moscow

Dupuis C, Corre C and Boyaval P (1993) Lipolytic and esterasic activities of *Propionibacterium freudenreichii* subsp. *freudenreichii*. Appl Environ Microbiol 59: 4004-4009

Dykstra GJ, Drerup DL, Branen AL and Keenan TW (1971) Formation of dimethyl sulfide by *Propionibacterium shermanii* ATCC 9617. J Dairy Sci 54: 168-172

Efremenko VI, Stolbin SV and Trekov LI (1990) Biosensors and some aspects of their uses. Prikl Biokhim Mikrobiol 26: 11-18

El Soda M (1995) The esterolytic and lipolytic activities of dairy propionibacteria. Abstr 1st Int Symp Dairy Propionibacteria, Rennes, France, C7

El-Hagarawy IS, Slatter WL, Halper WJ and Gould IA (1954) Factors affecting the organic acid production of the propionibacteria used in manufacture of Swiss cheese. J Dairy Sci 37: 638-645

El-Hagarawy IS, Slatter WL and Harper WJ (1957) Organic acid production by propionibacteria. I. Effect of strains, pH, carbon source, and intermediate fermentation products. J Dairy Sci 40: 579-587

Elford HL, Freese M, Passamani E and Morris HP (1979a) Ribonucleotide reductase and cell proliferation. I. Variations of ribonucleotide reductase activity with tumor growth rate in a series of rat hepatomas. J Biol Chem 245: 5228-5233

Elford HL, Wampler GL and van't Riet B (1979b) New ribonucleotide reductase inhibitors with antineoplastic activity. Cancer Res 39: 844-851

Emde R and Schink B (1990a) Oxidation of glycerol, lactate, and propionate by *Propionibacterium freudenreichii* in a poised-potential amperometric culture system. Arch Microbiol 153: 506-512

Emde R and Schink B (1990b) Enhanced propionate formation by *Propionibacterium freudenreichii* in a three-electrode amperometric culture system. Appl Environ Microbiol 56: 2771-2776

Evans HJ and Wood HG (1968) Phosphoenolpyruvate synthesis from pyruvate. Fed Proc, Abstr 52nd Annu Meet Atlantic City, New Jersey, April 15-20, 1968, 2089

Fass S (1987) Superoxide dismutase: human efficacy trials for heart attack indication. D-M-Enzyme Rep 6: 2

Fauve RM (1975) Stimulating effect of *Corynebacterium parvum* and *C. parvum* extract on the macrophage activities against *Salmonella typhimurium* and *Listeria monocytogenes*. In: Halpern B (ed) *Corynebacterium parvum*. Applications in Experimental and Clinical Oncology, pp 77-83. Plenum Press, New York

Ferguson DA and Cummins CS (1978) Nutritional requirements of anaerobic coryneforms. J Bacteriol 135: 858-867

Fernandez F and Collins MD (1987) Vitamin K composition of anaerobic gut bacteria. FEMS Microbiol Lett 41: 175-180

Field MF and Lichstein HC (1958a) Growth stimulating effect of autoclaved glucose media and its relationship to the CO_2 requirement of propionibacteria. J Bacteriol 76:485-490

Field MF and Lichstein HC (1958b) Influence of casein hydrolysates and amino acids on glucose fermentation by *Propionibacterium freudenreichii*. J Bacteriol 76:491-494

Fisher B, Wolmark N and Fisher ER (1975) Results of investigations with *Corynebacterium parvum* in an experimental animal system. In: Halpern B (ed) *Corynebacterium parvum*. Applications in Experimental and Clinical Oncology, pp 218-243. Plenum Press, New York

Flavin M, Ortiz PJ and Ochoa S (1955) Metabolism of propionic acid in animal tissues. Nature 176: 823-826

Ford SH and Friedmann HC (1976) Vitamin B_{12} biosynthesis: *in vitro* formation of cobinamide from cobyric acid and L-threonine. Arch Biochem Biophys 175:121-130

Forrest WW and Walker DJ (1964) Change in entropy during bacterial metabolism. Nature 201:49-52

Foschino R, Galli A, Ponticelli G and Volonterio G (1988) Propionic bacteria activity at different culture conditions. Ann Microbiol 38: 207-222

Fraikin GY, Vorobjeva LI, Khodjaev EY, Pinyaskina EV and Ponomareva GM (1995) Protective and reactivating effects of cell extract of a propionic acid bacterium on UV-inactivated *Candida guilliermondii* and *Escherichia coli*. Microbiology 64: 640-644

Francalanci F, Davis NK, Fuller JQ, Murfitt D and Leadlay PF (1986) The subunit structure of methylmalonyl-CoA mutase from *Propionibacterium shermanii*. Biochem J 236: 489-494

Freer J, Kim KS, Krauss MR, Beaman L and Barksdale L (1969) Ultrastructural changes in bacteria isolated from cases of leprosy. J Bacteriol 100: 1062-1075

Frey B, McCloskey J, Kersten W and Kersten H (1988) New function of vitamin B_{12}: cobamide-dependent reduction of epoxyquenosine to quenosine in tRNAs of *Escherichia coli* and *Salmonella typhimurium*. J Bacteriol 170: 2078-2082

Friedmann HC (1975) Biosynthesis of corrinoids. In: Babior BM (ed) Cobalamin: Biochemistry and Pathophysiology, pp 75-109. John Wiley & Sons, New York

Friedmann HC and Cagen LM (1970) Microbial biosynthesis of vitamin B_{12}-like compounds. Annu Rev Microbiol 24: 159-208

Frings W and Schlegel HG (1970) Über reaktionen der heterotrophen kohlendioxidfixierung. Biol Rundsch 8: 219-232

Fromageot C and Chaix P (1937) Respiration et fermentation chez *Propionibacterium pentosaceum*. Enzymologia 3: 288

Fujii K and Fukui S (1969) Relationship between vitamin B$_{12}$ content and ratio of monounsaturated fatty acids to methyl-branched fatty acids in *Corynebacterium simplex* cells grown on hydrocarbons. FEBS Lett 5: 343-346

Fujimura S and Nakamura T (1978) Purification and properties of a bacteriocin-like substance (acnecin) of oral *Propionibacterium acnes*. Antimicrob Agents Chemother 14: 893-898

Fukui GM (1952) Studies on growth and respiration mechanisms of the propionic acid bacteria. PhD thesis, Cornell University, Ithaca, New York

Gaitan VI and Vorobjeva LI (1981) ATP pool in *Propionibacterium shermanii* cells under different conditions of cultivation. Mikrobiologiya 50: 949-954

Gaitan VI, Vorobjeva LI and Kovrizhnykh VA (1982) The content of ATP in *Propionibacterium shermanii* cells under conditions of nitrogen deficiency. Mikrobiologiya 51: 747-750

Galesloot TE (1957) Involved van Nisine op die Bacterien Welka Betrokken Zijn of Kunnen Zijn bij bacteriologische Processen in Kaas en Smeltkaas. Netherlands Milk Dairy J 11: 58-73

Ganicheva TV and Vorobjeva LI (1991) *Propionibacterium shermanii* variability under the action of chemical mutagens. Microbiology 60: 101-106

Gautier M, Rouault A, Sommer P, Briandet R and Cassin D (1995) Bacteriophages infecting dairy propionibacteria. Lait 75: 427-434

Gautier M, Rouault A, Hervé C, Sommer P, Leret V, Jan G, Fraslin J-M, Prévot F and Coste A (1998) Bacteriophages of dairy propionibacteria. Abstr 2nd Int Symp Propionibacteria, Cork, Ireland

Gelman NS, Lukoyanova MA and Ostrovsky DN (1972) Bacterial Membranes and Respiratory Chains. Nauka, Moscow

Gershanovich VN (1965) Investigation of synthesis, regulation and enzyme activity of carbohydrate metabolism and biological oxidation in different cell systems. DSc thesis, Moscow State University, Moscow

Gerwin BI, Jacobson BE and Wood HG (1969) Transcarboxylase. VIII. Isolation and properties of a biotin-carboxyl carrier protein. Proc Natl Acad Sci USA 64: 1315-1322

Glatz BA (1992) The classical propionibacteria: their past, present, and future as industrial organisms. ASM News 58: 197-201

Glatz BA and Anderson KI (1988) Isolation and characterization of mutants of *Propionibacterium* strains. J Dairy Sci 71: 1769-1776

Glatz BA, Paik H-D and Hsieh H-Y (1995) Further characterization of propionicin PLG-1, a bacteriocin produced by *Propionibacterium thoenii* P127. Abstr 1st Int Symp Dairy Propionibacteria, Rennes, France, C9

Golding BT and Buckel W (1996) A mechanistic overview of B$_{12}$-dependent processes. Abstr Proc Eur Int Symp "Vitamin B$_{12}$ and B$_{12}$-Proteins", Vienna, Austria, L34

Golding BT, Anderson RJ, Ashwell S, Edwards CH, Garnett I, Kroll F and Buckel W (1998) A mechanistic overview of B$_{12}$ dependent processes. In: Kräutler B, Arigoni D and Golding BT (eds) Vitamin B$_{12}$ and B$_{12}$-proteins, pp 201-216. Wiley-VCH, Weinheim, Chichester, New York

Gollop N and Lindner P (1998) The propionicin PLG-1, an antibacterial peptide with an unique mode of action. Abstr 2nd Int Symp Propionibacteria, Cork, Ireland

Grüter A, Friederich U and Würgler FE (1990) Antimutagenic effects of mushrooms. Mutat Res 231: 243-249

Greenman J, Holland KT and Cunliffe WJ (1981) Effects of glucose concentration on biomass, maximum specific growth rate and extracellular enzyme production by three species of cutaneous propionibacteria grown in continuous culture. J Gen Microbiol 127: 371-376

Gregory EM and Fridovich I (1974) Visualization of catalase on acrylamide gels. Anal Biochem 58: 57-62

Grinstead DA and Barefoot SF (1992) Jenseniin G, a heat-stable bacteriocin produced by *Propionibacterium jensenii* P126. Appl Environ Microbiol 58: 215-220

Grusina VD, Erokhina LI and Ponomareva GM (1973) Effects of mutagenic factors on variability of *Propionibacterium shermanii*, producer of vitamin B_{12}. Genetika 9: 158-161

Grusina VD, Erokhina LI and Ponomareva GM (1974) Selection of propionic acid bacteria: role of morphological mutants in the selection of active variants. Genetika 10: 121-127

Gu Z, Rickert DA, Glatz BA and Glatz CE (1998) Feasibility of propionic acid production by extractive fermentation. Abstr 2nd Int Symp Propionibacteria, Cork, Ireland

Gulbic E, Galand N, Dumont JE and Schell-Frederick E (1981) Prostaglandin formation in bacteria: a reevaluation. Prostaglandins 21: 439-441

Haase CF, Beegan H and Allen SHG (1984) Propionyl-coenzyme A carboxylase of *Mycobacterium smegmatis*. An electron microscope study. Eur J Biochem 140: 147-151

Halankar PP and Blomquist GJ (1989) Comparative aspects of propionate metabolism. Comp Biochem Physiol 92B: 227-231

Hall B (1989) The origin of mutants. Environ Mol Mutagenesis 14 (Suppl 1): 80

Halpern BN (1975) Closing remarks. In: Halpern BN (ed) *Corynebacterium parvum*. Applications in Experimental and Clinical Oncology, pp 419-423. Plenum Press, New York

Halpern BN, Prévot AR, Biozzi G, Stiffel C, Mouton D, Morard JC, Bouthillier Y and Decreusefond C (1964) Stimulation of phagocytic activity of the reticuloendothelial system provoked by *Corynebacterium parvum*. J Reticuloendoth Soc 1: 77-87

Halpern BN, Biozzi G, Stiffel C and Mouton D (1966) Inhibition of tumour growth by administration of *Corynebacterium parvum*. Nature 212: 853-854

Hartl FU (1995) Pathways of chaperone-mediated protein folding. Abstr 7th Eur Congr Biotechnol, Nice, France, v 3, MEC33

Hatanaka H, Wang E, Taniguchi M, Iijima S and Kobayashi T (1988) Production of vitamin B_{12} by a fermentation with a hollow-fiber module. Appl Microbiol Biotechnol 27: 470-473

Hellgren L and Vincent J (1983) New group of prostaglandin-like compounds in *Propionibacterium acnes*. Gen Pharmacol 14: 207-208

Hervé C, Gautier M, Coste A and Rouault A (1998) Lysogeny of propionibacteria. Abstr 2nd Int Symp Propionibacteria, Cork, Ireland

Hettinga DH and Reinbold GW (1972a) The propionic acid bacteria – a review. I. Growth. J Milk Food Technol 35: 295-301

Hettinga DH and Reinbold GW (1972b) The propionic acid bacteria – a review. II. Metabolism. J Milk Food Technol 35: 358-372

Hietaranta M and Antila M (1953) The influence of biological nutrients on the growth of propionic acid bacteria in milk. Proc 13th Int Dairy Congr, v 3, pp 1428-1431

Hietaranta M and Antila M (1954) Some aspects of citric acid breakdown in Emmental cheese. Mejerit Finl Svensk 16: 91-94

Hocman G (1988) Prevention of cancer. Restriction of nutritional energy intake (joules). Comp Biochem Physiol 91A: 209-220

Hocman G (1989) Prevention of cancer: vegetables and plants. Comp Biochem Physiol 93B: 201-212

Höffler U (1977) Enzymatic and hemolytic properties of *Propionibacterium acnes* and related bacteria. J Clin Microbiol 6: 555-558

Höffler U (1979) Production of hyaluronidase by propionibacteria from different origins. Zbl Bakt Hyg, I Abt Orig A 245: 123-129

Hogenkamp HPC (1968) Enzymatic reactions involving corrinoids. Annu Rev Biochem 37: 225-245

Holdeman LV, Cato EP and Moore WEC (1977) *Arachnia* and *Propionibacterium*. In: Anaerobe Laboratory Manual, 4th ed, pp 56-65. The Virginia Polytechnic Institute and State University Anaerobe Laboratory, Blacksburg, Virginia

Holland KT, Cunliffe WJ and Roberts CD (1978) The role of bacteria in acne – a new approach. Clin Exp Dermatol 3: 253-257

Holland KT, Greenman J and Cunliffe WJ (1979) Growth of cutaneous propionibacteria on synthetic medium: growth yields and exoenzyme production. J Appl Bacteriol 47: 383-394

Holland KT, Ingham E and Cunliffe WJ (1981) The microbiology of acne. J Appl Bacteriol 51: 195-215

Holliday R (1956) A new method for the identification of biochemical mutants of microorganisms. Nature 178: 987

Holmberg K and Forsum V (1973) Identification of *Actinomyces, Arachnia, Bacterionema, Rothia* and *Propionibacterium* species by defined immunofluorescence. Appl Microbiol 25: 834-843

Holt JG, Krieg NR, Sneath PH, Staley JT and Williams ST (1994) Genus *Propionibacterium*. In: Holt JG, Krieg NR, Sneath PH, Staley JT and Williams ST (eds) Bergey's Manual of Determinative Bacteriology, 9th ed, pp 580-581. Williams & Wilkins, Baltimore

Holzapfel WH, Haberer P, Snel J, Schillinger U, Huis in't Veld JH (1998) Overview of gut flora and probiotics. Int J Food Microbiol 41: 85-101

Horner SM, Sturridge MF and Swanton RH (1992) *Propionibacterium acnes* causing an aortic root abscess. Br Heart J 68: 218-220

Hsieh H-Y and Glatz BA (1996) Long-term storage stability of the bacteriocin propionicin PLG-1 produced by *Propionibacterium thoenii* and potential as a food preservative. J Food Prot 59: 481-486

Ibragimova SI and Saccharova SV (1972) Influence of excess substrate on some physiological properties of *Propionibacterium shermanii*. Mikrobiologiya 41: 834-837

Ibragimova SI and Saccharova SV (1974) Inhibitory action of sodium propionate on *Propionibacterium shermanii*. Mikrobiologiya 43: 18-23

Ibragimova SI and Shulgovskaya EM (1979) Growth of *Propionibacterium shermanii* at different conditions of aeration. Mikrobiologiya 48: 668-671

Ibragimova SI, Neronova NM and Rabotnova IL (1971) Influence of hydrogen ions on some properties of *Propionibacterium shermanii*. Mikrobiologiya 40: 833-837

Ichikawa J (1955) Microbiological studies on propionic acid bacteria. Formation of propionic acid from succinate. J Agric Chem Soc Japan 29: 357-361

Ikonnikov NP (1985) Production of organic acids, nucleotides and their derivatives by immobilized cells of *Propionibacterium shermanii*. PhD thesis, Moscow State University, Moscow

Ikonnikov NP, Iordan EP and Vorobjeva LI (1982) Production of nucleotides and their derivatives by immobilized cells of *Propionibacterium shermanii*. Prikl Biokhim Mikrobiol 18: 34-40

Iljina KA and Besedina SF (1966) Influence of *Propionibacterium shermanii* on the composition of organic acids in silage. Proc Inst Microbiol Virusol Acad Sci Kasakhstan 9: 29-35

Ingham E, Holland KT, Gowland G and Cunliffe WJ (1979) Purification and partial characterization of hyaluronate lyase (EC 4.2.2.1) from *Propionibacterium acnes*. J Gen Microbiol 115: 411-418

Ingham E, Holland KT, Gowland G and Cunliffe WJ (1980) Purification and partial characterization of an acid phosphatase (EC 3.1.3.2) produced by *Propionibacterium acnes*. J Gen Microbiol 118: 59-65

Ingham E, Holland KT, Gowland G and Cunliffe WJ (1981) Partial purification and characterization of lipase (EC 3.1.1.3) from *Propionibacterium acnes*. J Gen Microbiol 124: 393-401

Ingham E, Holland KT, Gowland G and Cunliffe WJ (1983) Studies of the extracellular proteolytic activity produced by *Propionibacterium acnes*. J Appl Bacteriol 54: 263-271

Iordan EP (1992) Modulation of DNA formation in *Propionibacterium freudenreichii* subsp. *shermanii* under conditions of corrinoid limitation. Mikrobiologiya 61: 341-346

Iordan EP and Petukhova NI (1989) Reorganization of the ribonucleotide reductase system upon inhibition of vitamin B_{12} synthesis in propionic acid bacteria. Mikrobiologiya 58: 533-538

Iordan EP and Petukhova NI (1995) Presence of oxygen-consuming ribonucleotide reductase in corrinoid-deficient *Propionibacterium freudenreichii*. Arch Microbiol 164: 377-381

Iordan EP and Pryanishnikova NI (1994) Intensification of DNA synthesis by 5,6-dimethylbenzimidazole in *Propionibacterium freudenreichii*. Prikl Biokhim Mikrobiol 30: 137-142

Iordan EP, Vorobjeva LI and Gaitan VI (1974) Effect of vitamin B_{12} on the level of sulfhydryl groups and the activity of some dehydrogenases in the cells of *Propionibacterium shermanii*. Mikrobiologiya 43: 596-599

Iordan EP, Antoshkina NV and Vorobjeva LI (1979a) The role of coenzyme B_{12} in the synthesis of DNA by *Propionibacterium shermanii*. In: Zagalak B and Friedrich W (eds) Proc 3rd Eur Symp "Vitamin B_{12} and intrinsic factors", Univ of Zurich, March 5-8, 1979, Zurich, Switzerland, pp 1095-1099. Walter de Gruyter, Berlin

Iordan EP, Ikonnikov NP, Kovrizhnykh VA and Vorobjeva LI (1979b) Synthesis of organic acids by immobilized propionic acid bacteria in continuous system and stabilization of the process. Prikl Biokhim Mikrobiol 15: 515-521

Iordan EP, Novozhilova TY and Vorobjeva LI (1983) Correlation between the DNA synthesis and vitamin B_{12} content in *Propionibacterium shermanii*. Mikrobiologiya 52: 591-596

Iordan EP, Novozhilova TY and Vorobjeva LI (1984) Effects of cellular vitamin B_{12} production on growth and some aspects of the constructive metabolism of *Propionibacterium freudenreichii* subsp. *shermanii*. Prikl Biokhim Mikrobiol 20: 765-772

Iordan EP, Petukhova NI and Vorobjeva LI (1986) Regulatory action of vitamin B_{12} on ribonucleotide reductase system of propionic acid bacteria. Mikrobiologiya 55: 533-538

Israel L (1975) Report on 414 cases of human tumors treated with corynebacteria. In: Halpern B (ed) *Corynebacterium parvum*. Applications in Experimental and Clinical Oncology, pp 389-401. Plenum Press, New York

Iwasaki H, Fujiyama T and Yamashita E (1968) Studies on the red tide dinoflagellates. I. On *Entomosigma* sp. appeared in coastal area of Fukuyama. J Fac Fish Anim Husb Hiroshima Univ 7: 259-267

Jaszewski B, Trojanowska K and Schneider Z (1995) Utilization of whey flavins in biosynthesis of cobalamins in *Propionibacterium shermanii* cultures. Abstr 1st Int Symp Dairy Propionibacteria, Rennes, France, C7

Javanainen PM, Linko Y-Y and Linko P (1987) Propionic acid formation by mixed cultures of *Lactobacillus* sp. and *Propionibacterium* sp. on wheat flour-based substrate. In: Neijssel

OM, van der Meer and Luyben KCAM (eds) Proc 4th European Congr Biotechnol 1987, v 3, pp 313-316. Elsevier Science, Amsterdam

Jeter R, Escalante-Semerena JC, Roof D, Olivera B and Roth J (1987) Synthesis and use of vitamin B_{12}. Cell Mol Biol 1: 551-556

Johns AT (1951) Mechanisms of propionic acid formation by propionibacteria. J Gen Microbiol 5: 337-345

Johnson JL and Cummins CS (1972) Cell wall composition and deoxyribonucleic acid similarities among the anaerobic coryneforms, classical propionibacteria, and strains of *Arachnia propionica*. J Bacteriol 109: 1047-1066

Jong EC, Ko HL and Pulverer G (1975) Studies on bacteriophages of *Propionionibacterium acnes*. Med Microbiol Immunol 161: 263-271

Joseph AA and Wixom RL (1972) Amino acid metabolism in the genus *Propionibacterium*. Proc Soc Exp Biol Med 139: 526-534

Jwanny EM, Chenouda MS and Osman HG (1974) Utilization of hydrocarbons by microorganisms. Z Allg Mikrobiol 14: 205-212

Kabongo ML, Nutini LG and Estes SA (1981) Intradermal infections of bacteria and their relation on acne pathogenesis. J Invest Dermatol 76: 314

Kabongo ML, Sowar MC and Nutini LG (1982) A simplified medium for detecting the effect of lecithin on the growth of *Propionibacterium acnes*. Microbiologica 5: 11-23

Kada T, Inoue T, Ohta T and Shirasu Y (1986) Antimutagens and their modes of action. In: Shankel DM, Hartman PE, Kada T and Hollaender A (eds) Antimutagenesis and Anticarcinogenesis Mechanisms, pp 181-196. Plenum Press, New York

Kalda AH (1984) A study of the effect of immobilization on aspartase activity of propionic acid bacteria and *Escherichia coli*. PhD thesis, University of Leningrad

Kalda AH and Vorobjeva LI (1980) Immobilization of the cells of microorganisms with aspartase activity. Report I. Trudy Tallinsk Polytekh Inst 499: 63-74

Kalda AH and Vorobjeva LI (1981) Immobilization of the cells of microorganisms with aspartase activity. Report II. Trudy Tallinsk Polytekh Inst 510: 37-47

Kamikubo T, Matsuno R, Bieganowski R, Senkpiel B and Friedrich W (1982) Biological activities of new corrinoids. Agric Biol Chem 46: 1673-1674

Kaneko T, Mori H, Iwata M and Meguro S (1994) Growth stimulator for bifidobacteria produced by *Propionibacterium freudenreichii* and several intestinal bacteria. J Dairy Sci 77: 393-404

Kanopkaite SI and Gibavichyute AS (1965) Effects of certain factors on the biosynthesis of vitamin B_{12}, thiamine, riboflavin and folic acid. Trudy Akad Nauk Lit SSSR B(1.36): 185-194

Karlin P (1966) The content of B-group vitamins in kefir and its enrichment by adding *Propionibacterium shermanii*. Prikl Biokhim Mikrobiol 2: 386-391

Kaspar HF (1982) Nitrite reduction to nitrous oxide by propionibacteria: detoxification mechanism. Arch Microbiol 133: 126-130

Kaziro Y and Ochoa S (1964) The metabolism of propionic acid. Adv Enzymol 26: 283

Keenan TW and Bills DD (1968) Volatile compounds produced by *Propionibacterium shermanii*. J Dairy Sci 51: 797-799

Kellermeyer RW, Allen SHG, Stjernholm R and Wood HG (1964) Methylmalonyl isomerase. IV. Purification and properties of the enzyme from propionibacteria. J Biol Chem 239: 2562-2569

Ketterer B (1988) Protective role of glutathione and glutathione transferases in mutagenesis and carcinogenesis. Mutat Res 202: 343-361

Kimball RF, Setlow JK and Liu M (1971) The mutagenic and lethal effects of monofunctional methylating agents in strains of *Haemophilus influenzae* defective in repair processes. Mutat Res 12: 21-28

Kishimoto Y, Williams M, Moser H, Hignite C and Biemann K (1973) Branched-chain and odd-numbered acids and aldehydes in the nervous system of a patient with deranged vitamin B_{12} metabolism. J Lipid Res 14: 69

Klaenhammer TR (1988) Bacteriocins of lactic acid bacteria. Biochimie 70: 337-349

Konoplev EG and Scherbakov LA (1970) Application of mixed starter cultures of propionibacteria and yeasts in maize ensilage. Izvestia Akad Nauk SSSR Ser Biol 1: 142-144

Konovalova LV (1970) Influence of polymyxin M on certain metabolic sites of propionic acid bacteria. PhD Thesis, Moscow State University, Moscow

Konovalova LV and Vorobjeva LI (1969) Biosynthesis of B-group vitamins by propionic acid bacteria. Nauch Dokl Vys Shkoly Biol Nauki 1: 91-93

Konovalova LV and Vorobjeva LI (1970) Influence of antibiotics on the growth and vitamin B_{12} biosynthesis by *Propionibacterium shermanii*. Nauch Dokl Vys Shkoly Biol Nauki 7: 104-108

Konovalova LV and Vorobjeva LI (1972) Influence of polymyxin M on accumulation of lipids and polyphosphates by propionic acid bacteria. Nauch Dokl Vys Shkoly Biol Nauki 7: 101-104

Kornyeva VV (1981) Establishment of propionic acid bacteria in the intestine of infants given propioniacidophilus milk. Dairy Sci Abstr 44: 420

Kosikovsky FV (1977) Cheese and fermented milk foods, 2nd ed. Edwards Brothers, Ann Arbor, Michigan

Krainova OA and Bonarceva GA (1973) Enzymes of tricarboxilic and glyoxylate cycle in propionibacteria. Vestn Mosk Univ Biol 4: 67-68

Kraeva NI and Vorobjeva LI (1981a) Intracellular localization, isolation and characterization of superoxide dismutase from *Propionibacterium globosum*. Prikl Biokhim Mikrobiol 17: 837-843

Kraeva NI and Vorobjeva LI (1981b) Superoxide dismutase, catalase, and peroxidase of propionic acid bacteria. Mikrobiologiya 50: 813-817

Krebs HA and Eggleston LV (1941) Biological synthesis of oxaloacetic acid from pyruvic acid and carbon dioxide. II. The mechanism of carbon dioxide fixation in propionic acid bacteria. Biochem J 35: 676-687

Krecek RC, Els HJ, de Wet SC and Henton MM (1992) Studies on ultrastructure and cultivation of microorganisms associated with zebra nematodes. Microbial Ecol 23: 87-95

Krueger JH and Walker GC (1984) *groEL* and *dnaK* genes of *Escherichia coli* are induced by UV irradiation and nalidixic acid in an $htpR^+$-dependent fashion. Proc Natl Acad Sci USA 81: 1499-1503

Kubatiev AA (1973) Porphyrins, Vitamin B_{12} and Cancer. Priokskoye Knizhnoye Izdatelstvo, Tula, Russia.

Kucheras PV and Gebhardt AG (1972) Influence of amino acids on the cobamide synthesis activity of *Propionibacterium shermanii*. Prikl Biokhim Mikrobiol 8: 341-345

Kulaev IS (1979) The Biochemistry of Inorganic Polyphosphates. John Wiley & Sons, New York

Kulaev IS, Vorobjeva LI, Konovalova LV, Bobik MA and Urison SO (1973) Enzymes of polyphosphate metabolism in *Propionibacterium shermanii* growing under normal conditions and in the presence of polymyxin M. Biokhimiya 38: 595-599

Kupenov LG (1974) Comparative studies on the variability of vitamin B$_{12}$ producers *Actinomyces olivaceus* and *Propionibacterium shermanii* induced by chemical mutagens. PhD thesis, Institute of Genetics of Industrial microorganisms, Moscow

Kurilov NV and Sevastjanova NA (1978) Digestion in Ruminants. Itogi Nauki i Tekhniki. Husbandry and Veterinary (Moscow) 11: 5-78

Kuroda Y and Inoue T (1988) Antimutagenesis by factors affecting DNA repair in bacteria. Mutat Res 202: 387-391

Kurtz FE, Hupfer JA, Corbin EA, Hargrove RE and Walter HE (1958) Interrelationships between pH, populations of *Propionibacterium shermanii*, levels of free fatty acids, and the flavor ratings of Swiss cheese. J Dairy Sci 41: 719, M77

Kusano K, Yamada H, Niwa M and Yamasato K (1997) *Propionibacterium cyclohexanicum* sp. nov., a new acid-tolerant ω-cyclohexyl fatty acid-containing *Propionibacterium* isolated from spoiled orange juice. Int J Syst Bacteriol 47: 825-831

Kusel JP, Fa YH and Demain AL (1984) Betaine stimulation of vitamin B$_{12}$ biosynthesis in *Pseudomonas denitrificans* may be mediated by an increase in activity of δ-aminolevulinic acid synthase (EC 2.3.1.37). J Gen Microbiol 130: 835-842

Labory A (1970) Regulation of Metabolic Processes. Medicina, Moscow

Ladoga JV and Bannikova LA (1970) Some factors influencing the quality and stability of dried starter cultures. Moloch Promyshl 2: 11-15

Lagukas R (1987) The Pasteur effect today. Hoppe-Seyler's Z Physiol Chem 368: 540-541

Laipanov IS (1989) Biotechnological properties of micrococci of the species *Micrococcus caseolyticus*. Proc Conf "Intensification of manufacture and quality improvement of cheese", pp 14-15, Barnaul

Lancelle MA and Asselineau MJ (1968) Sur les lipids de *Propionibacterium freudenreichii*. C R Acad Sci 266: 1901-1903

Langsrud T, Sorhaug T and Vegarud GE (1995) Protein degradation and amino acid metabolism by propionibacteria. Lait 75: 325-330

Lankaputhra WEV and Shah NP (1998) Antimutagenic properties of probiotic bacteria and of organic acids. Mutat Res 397: 169-182

Lara FJS (1959) The succinic dehydrogenase of *Propionibacterium pentosaceum*. Biochim Biophys Acta 33: 565-567

Lascelles J and Hatch TP (1969) Bacteriochlorophyll and heme synthesis in *Rhodopseudomonas spheroides*: possible role of heme in regulation of the branched biosynthetic pathway. J Bacteriol 98: 712-720

Laubinger W and Dimroth P (1987) Characterization of the Na$^+$-stimulated ATPase of *Propionigenium modestum* as an enzyme of the F$_1$F$_0$ type. Eur J Biochem 168: 475-480

Leaver FW, Wood HG and Stjernholm R (1955) The fermentation of three carbon substrates by *Clostridium propionicum* and *Propionibacterium*. J Bacteriol 70: 521-530

Lee PC, Bochner BR and Ames BN (1983a) AppppA, heat shock stress, and cell oxidation. Proc Natl Acad Sci USA 80: 7496-7500

Lee PC, Bochner BR and Ames BN (1983b) Diadenosine-5',5'''-P^{1}P^4-tetraphosphate and related adenylated nucleotides in *Salmonella typhimurium*. J Biol Chem 258: 6827-6834

Lee SY, Vedamuthu ER, Washam CJ and Reinbold GW (1969) Diacetyl production by propionibacteria. J Dairy Sci 52: 893

Lee SY, Vedamuthu ER, Washam CJ and Reinbold GW (1970) Diacetyl production by *Propionibacterium shermanii* in milk cultures. Can J Microbiol 16: 1231-1242

Lee WL, Schalita AR, Pon-Fitzpatrick MB (1978) Comparative studies of porphyrin production in *Propionibacterium acnes* and *Propionibacterium granulosum*. J Bacteriol 133: 811-815

Lemée R, Lortal S and van Heijenoort J (1995) Autolysis of dairy propionibacteria: isolation and renaturing gel electrophoresis of the autolysins of *Propionibacterium freudenreichii* CNRZ 725. Lait 75: 345-365

Leps WT and Ensign JC (1979) Adenosine triphosphate pool levels and endogenous metabolism in *Arthrobacter crystallopoietes* during growth and starvation. Arch Microbiol 122: 61-67

Lewis VP and Yang S-T (1992) A novel extractive fermentation process for propionic acid production from whey lactose. Biotechnol Progr 8: 104-110

Lichstein HC (1955) The presence of bound biotin in purified preparation of oxaloacetic carboxylase. J Biol Chem 212: 217-222

Lichstein HC (1958) On the specificity of biotin in the metabolism of propionibacteria. Arch Biochem Biophys 77: 378-386

Licht S and Stubbe J (1996) Catalysis of carbon-cobalt bond cleavage by the ribonucleotide reductase of *Lactobacillus leichmannii*. Abstr Int Symp Vitamin B_{12}, Zurich, Switzerland, pp 35-36

Lindquist S and Craig EA (1988) The heat-shock proteins. Annu Rev Genet 22: 631-677

Lochhead AG (1958) Soil bacteria and growth-promoting substances. Bacteriol Rev 22: 145

Lochmüller H, Wood HG and Davis JJ (1966) Phosphoenolpyruvate carboxytransphosphorylase. II. Crystallization and properties. J Biol Chem 241: 5678-5691

Lusta KA, Krasnova LA, Kocheenko KA, Fichte BA and Scryabin GK (1976) Ultrastructural changes of *Bacillus megaterium* cells immobilized in polyacrylamide gel. Dokl Akad Nauk SSSR 227: 469-475

Lyon WJ and Glatz BA (1991) Partial purification and characterization of bacteriocin produced by *Propionibacterium thoenii*. Appl Environ Microbiol 57: 701-706

Lyon WJ and Glatz BA (1993) Isolation and purification of propionicin PLG-1, a bacteriocin produced by a strain of *Propionibacterium thoenii*. Appl Environ Microbiol 59: 83-88

Mair-Waldburg H (1974) On the history of cheesemaking in ancient times. In: Mair-Waldburg H (ed) Handbook of Cheese. Volkswirtschaftlicher Verlag, Kempten (Allgäu), Germany

Mantere-Alhonen S (1982) Die Propionibakterien der Molkereiindustrie als Darmkanalmikroben. Meijerit Aikakausk 40: 95S

Mantere-Alhonen S (1995) Propionibacteria used as probiotics – a review. Lait 75: 447-452

Mantere-Alhonen S and Myllymäki H (1985) Meijeri- teollisuuden propionibakteerien massantuottaminen ja ruokinnalliset sovellukset. Elintarvikeylioppilas 1: 14-16

Marcoullis G, Merivuori H and Gräsbeck R (1978) Comparative studies on intrinsic factor and cobalophilin in different parts of the gastrointestinal tract of the pig. Biochem J 173: 705-712

Maron DH and Ames BN (1984) Revised methods for *Salmonella* mutagenicity test. In: Kilbey BJ, Legator M, Nickols W and Ramel C (eds) Handbook of Mutagenicity Test Procedures, 2nd ed, pp 95-130. Elsevier Science, Amsterdam

Maruhashi K, Yada T, Watanabe K and Amari T (1996) Cloning of gene for uroporphyrinogeen III methyltransferase of *Propionibacterium shermanii* and use for recombinant preparation of vitamin B_{12}. Nippon Oil Co Ltd, Japan, p 9

Marwaha SS, Sethi RP, Kennedy JF and Kumar R (1983) Stimulation of fermentation conditions for vitamin B_{12} biosynthesis from whey. Enzyme Microb Technol 5: 449-453

Mashur VA, Vorobjeva LI and Iordan EP (1971) Fermentation caused by propionibacteria not forming coenzyme B_{12}. Prikl Biokhim Mikrobiol 7: 552-555

Matthews RG, Banerjee RV and Ragsdale SW (1990) Cobamide-dependent methyl transferases. Biofactors 2: 147-152

Mäyrä-Mäkinen A and Suomalainen T (1995). *Lactobacillus casei* ssp. *rhamnosus*. Bacterial preparation comprising Said strain, and use of Said strain and preparations for the controlling of yeast and molds. United States Patent US 5 378 458

McCalla DR (1968) Reaction of *N*-methyl-*N'*-nitro-*N*-nitrosoguanidine and *N*-methyl-*N'*-nitroso-*p*-toluenesulfonamide with DNA *in vitro*. Biochim Biophys Acta 155: 114-120

McGinley KJ, Webster GF and Leyden JJ (1978) Regional variation of cutaneous propionibacteria. Appl Environ Microbiol 35: 62-66

Meganathan R and Ensign JC (1976) Stability of enzymes in starving *Arthrobacter crystallopoietes*. J Gen Microbiol 94: 90-96

Melnikova LV (1987) Phospholipase activity of propionic acid bacteria and its influence on the quality of "Soviet" cheese. PhD thesis, Laboratory of Cheesemaking Plant, Uglich

Mengel K and Kirkby EA (1980) Principles of plant nutrition. Int Potash Institute, Worblaufen-Bern, Switzerland

Menon IA and Shemin D (1967) Concurrent decrease of enzymic activities concerned with the synthesis of coenzyme B_{12} and propionic acid in propionibacteria. Arch Biochem Biophys 121: 304-310

Meriläinen V and Antila M (1976) The propionic acid bacteria in Finnish Emmenthal cheese. Meijerit Aikakausk 34: 107-116

Mervin L and Smith EL (1964) The biochemistry of vitamin B_{12} formation. Progr Industr Microbiol 5: 153-201

Mezger-Lallemand V, Loones M-T, Michel E, Rallu M and Morange M (1995) Heat-shock proteins and chaperones in physiology and development. Abstr 7th Eur Congr Biotechnol, Nice, France, v 3, MEC36

Mickelson AM (1977) US Patent No 402981

Miescher S, Stierli M, Dasen G, Smutny J and Meile L (1998a) Molecular analysis of *Propionibacterium* plasmids. Abstr 2nd Int Symp Propionibacteria, Cork, Ireland

Miescher S, Teuber M and Meile L (1998b) Propionicin SM1, a new bacteriocin produced by *Propionibacterium jensenii* DF1. Abstr 2nd Int Symp Propionibacteria, Cork, Ireland

Milner Y and Wood HG (1972) Isolation of a pyrophosphoryl form of pyruvate-phosphate dikinase from propionibacteria. Proc Natl Acad Sci USA 69: 2463-2468

Moat AG and Delwiche EA (1950) Utilization of coenzyme A by *Propionibacterium freudenreichii*. J Bacteriol 60: 757-762

Molinari R and Lara FJS (1960) The lactic dehydrogenase of *Propionibacterium pentosaceum*. Biochem J 75: 57-65

Moore WEC and Cato EP (1963) Validity of *Propionibacterium acnes* (Gilchrist) Douglas and Gunter comb. nov. J Bacteriol 85: 870-874

Moore WEC and Holdeman LV (1974) Genus 1. *Propionibacterium*. In: Buchanan RE and Gibbons NE (eds) Bergey's Manual of Determinative Bacteriology, 8th ed, pp 633-641. Williams & Wilkins, Baltimore

Moss CW, Dowell VR, Farshti D, Raines LJ and Cherry WB (1967) Cultural characteristics and fatty acid composition of *Corynebacterium acnes*. J Bacteriol 94: 1300-1305

Moss CW, Dowell VR, Lewis VJ and Schekter MA (1969) Cultural characteristics and fatty acid composition of propionibacteria. J Bacteriol 97: 561-570

Murakami K, Hashimoto Y and Murooka Y (1993) Cloning and characterization of the gene encoding glutamate 1-semialdehyde 2,1-aminomutase, which is involved in δ-aminolevulinic acid synthesis in *Propionibacterium freudenreichii*. Appl Environ Microbiol 59: 347-350

Murooka Y, Hashimoto Y and Yamashita M (1995) Cloning, characterization and expression of genes involved in the biosynthesis of vitamin B_{12} in *Propionibacterium freudenreichii (shermanii)*. Abstr 1st Int Symp Dairy Propionibacteria, Rennes, France, A7

Müller G, Dieterle W and Siebke G (1970) Production of porphyrins by *Propionibacterium shermanii*. Z Naturforsch B 25: 307-309

Müller G, Hlineny K, Savvidis E, Zipfel F, Schmiedl J and Schneider Z (1990) On the methylation process and cobalt insertion in cobyrinic acid biosynthesis. In: Baldwin TO, Raushel FM and Scott AI (eds) Chemical Aspects of Enzyme Biotechnology, pp 281-298. Plenum Press, New York

Nabukhotnyi TK, Cherevko SA, Sanigullina FI and Grushko AI (1983) Use of adapted propioni-acidophilic "Malyutka" and "Malyush" formulas on complex treatment of acute gastrointestinal diseases of infants. Dairy Sci Abstr 47: 2122

Naud AI, Legault-Démare J and Ryter A (1988) Induction of a stable morphological change in *Propionibacterium freudenreichii*. J Gen Microbiol 134: 283-293

Neronova NM and Ierusalimsky ND (1959) Continuous cultivation of propionic acid bacteria synthesizing vitamin B_{12}. Mikrobiologiya 28: 647-654

Neronova NM, Ibragimova SI and Ierusalimsky ND (1967) Effect of propionate concentration on the specific growth rate of *Propionibacterium shermanii*. Mikrobiologiya 36: 404-409

Neujahr HY and Fries L (1966) On the occurence of light-sensitive corrinoids in axenic cultures of unicellular algae. Acta Chem Scand 20: 347-360

Niederau W, Pape W, Schaal KP, Hoffler V and Pulverer G (1982) Zur Antibiotikehandlung der menschlichen Aktinomykosen. Deutsch Med Wochenschr 107: 1279-1283

Niethammer A and Hitzler M (1960) Synthetic culture of *Propionibacterium* (Orla-Jensen) van Niel. Zbl Bakt Hyg, II Abt 113: 478-479

Nishioka H and Ninoshiba T (1986) Role of enzymes in antimutagenesis of human saliva and serum. In Shankel DM, Hartman PE, Kada T and Hollaender A (eds) Antimutagenesis and Anticarcinogenesis Mechanisms, pp 143-151. Plenum Press, New York

Northrop DB and Wood HG (1969) Transcarboxylase. VII. Exchange reactions and kinetics of oxalate inhibition. J Biol Chem 244: 5820-5827

Novick A and Szilard L (1952) Anti-mutagens. Nature 170: 926-927

O'Brien WE and Wood HG (1974) Carboxytransphosphorylase. VIII. Ligand-mediated interaction of subunits as a possible control mechanism in propionibacteria. J Biol Chem 249: 4917-4925

Odame-Darkwah JK and Marshall DL (1993) Interactive behavior of *Saccharomyces cerevisiae, Bacillus pumilus,* and *Propionibacterium freudenreichii* ssp. *shermanii*. Int J Food Microbiol 19: 259-269

Oh-hama T, Stolowich NJ and Scott A (1993) δ-Aminolevulinic acid biosynthesis in *Propionibacterium shermanii* and *Halobacterium salinarum*: distribution of the two pathways of δ-aminolevulinic acid biosynthesis in prokaryotes. J Gen Microbiol 39: 527-533

Okada J, Murata K and Kimura A (1982) Assimilation of elemental sulfur by a mutant of *Escherichia coli B*. Agric Biol Chem 46: 1915-1919

Orla-Jensen S (1909) Die hauptlinien des natürlichen Bacteriensystems. Zbl Bakt Hyg, II Abt 22: 305-346

Osawa T, Ishibashi H, Namiki M and Kada T (1980) Desmutagenic actions of ascorbic acid and cysteine on a new pyrrole mutagen formed by the reaction between food additives, sorbic acid and sodium nitrite. Biochem Biophys Res Comm 95: 835-841

Osawa T, Namiki M, Udaka S and Kada T (1986) Chemical studies of antimutagens of microbial origin. In: Delbert M and Shankel I (eds) Poster Abstr Int Conf "Mechanisms of antimutagenesis and anticarcinogenesis", October 6-10, 1985, Lawrence, Kansas, p 573. Plenum Press, New York

Oterholm A, Ordal ZJ and Witter LD (1970) Purification and properties of a glycerol ester hydrolase (lipase) from *Propionibacterium shermanii*. Appl Microbiol 20: 16-22

Overath P, Stadtman ER, Kellerman GM and Lynen F (1962) Zum Mechanismus der Umlagerung von Methylmalonyl-CoA in Succinyl-CoA. III. Reinigung und Eigenschaften der Methylmalonyl-CoA Isomerase. Biochem J 336: 77-98

Owais WM and Kleinhofs A (1988) Metabolic activation of the mutagen azide in biological systems. Mutat Res 197: 313-323

Pai SL and Glatz BA (1987) Production, regeneration, and transformation of protoplasts of propionibacterium strains. J Dairy Sci 70 (Suppl 1): 80

Paik H-D and Glatz BA (1995) Purification and partial amino acid sequence of propionicin PLG-1, a bacteriocin produced by *Propionibacterium thoenii* P127. Lait 75: 367-377

Paik H-D and Glatz BA (1997) Enhanced bacteriocin production by *Propionibacterium thoenii* in fed-batch fermentation. J Food Prot 60: 1529-1533

Pankova SV and Abilev SK (1993) Molecular cloning of genes of threonine biosynthesis in the cells of *Escherichia coli*. Genetika 29: 539-542

Pankova SV, Abilev SK and Tarasov VA (1993a) Molecular cloning of genes of *Propionibacterium shermanii* on a plasmid of wide spectrum of action. Genetika 29: 914-921

Pankova SV, Abilev SK, Tarasov VA and Ogarkov OA (1993b) Cloning of *recA* gene of propionic acid bacteria *P. shermanii* in *Escherichia coli*. Genetika 29: 777-784

Park HS, Reinbold GW, Hammond EG and Clark WS (1967) Growth of propionibacteria at low temperatures. J Dairy Sci 50: 589-591

Parmentier Y, Marcoullis G and Nicolas JP (1979) The intraluminal transport of vitamin B_{12} and exocrine pancreatic insufficiency. Proc Soc Exp Biol Med 160: 396-400

Pawelkiewicz J and Legocki AB (1963) Propionate (acetate) kinase of *Propionibacterium shermanii*. Bull Pol Acad Sci 11: 569-572

Pędziwilk F (1962) Biosynthesis of cobalamin by propionic acid bacteria. The Poznan Society of Friends of Science, Department of Agricultural and Sylvicultural Sciences Publications 11: 141-191

Pędziwilk F, Skupin J and Trojanowska K (1984) Biosynthesis of vitamin B_{12} by the mutants of *Propionibacterium shermanii*. I. Induced with ethylmethanesulfonate (EMS) and dimethyl sulfate (DMS). Acta Aliment Pol 9: 113-119

Peltola E (1940) Effect of salt on bacteria of importance in Emmental cheese-making. Meijerit Aikakausk 2: 11-21

Peltola E and Antila M (1953) The effect of oxidizing salts on the oxidation-reduction potential and ripening of Emmental cheese. Proc 13th Int Dairy Congr, v 2, pp 729-731

Perez-Chaia A, De Ruiz Holgado AP and Oliver G (1988) Effect of pH and temperature on the proteolytic activity of propionibacteria. Microbiol Alim Nutr 6: 91-94

Perez-Chaia A, Nader de Macias ME and Oliver G (1995) Propionibacteria in the gut: effect on some metabolic activities of the host. Lait 75: 435-445

Perez-Chaia A, Zárate G and Oliver G (1998) Probiotic properties of propionibacteria. Abstr 2nd Int Symp Propionibacteria, Cork, Ireland

Perlman D (1978) Vitamins. In: Rose AH (ed) Economic Microbiology, v 2: Primary Products of Metabolism, pp 303-326. Academic Press, London

Petkau A (1987) Role of superoxide dismutase in modification of radiation injury. Br J Cancer Suppl 8: 87-95

Pett LB and Wynne AM (1933) The metabolism of propionic acid bacteria. I. The degradation of phosphoric acid esters by *Propionibacterium jensenii* (van Niel). Trans Roy Soc Can V Sec 27: 119-125.

Petukhova NI (1988) Role of vitamin B_{12} in the synthesis and functioning of ribonucleotide reductase system of propionibacteria. PhD thesis, Mosow State University, Moscow

Pfohl-Leszkowicz A, Keith G and Dirheimer G (1991) Effect of cobalamin derivatives on *in vitro* enzymatic DNA methylation: methylcobalamin can act as methyl donor. Biochemistry 30: 8045-8051

Pine L and Georg LK (1969) Reclassification of *Actinomyces propionicus*. Int J Syst Bacteriol 19: 267-272

Piper PW (1995) The heat shock and ethanol stress responses of yeast exhibit extensive similarity and functional overlap. FEMS Microbiol Lett 134: 121-129

Pitcher DG and Collins MD (1991) Phylogenetic analysis of some LL-diaminopimelic acid-containing coryneform bacteria from human skin: description of *Propionibacterium innocuum* sp. nov. FEMS Microbiol Lett 84: 295-300

Piveteau PG, Condon S and Cogan TM (1995a) Co-metabolism of lactate and sugars by propionic acid bacteria. Abstr 1st Int Symp Dairy Propionibacteria, Rennes, France, B9

Piveteau PG, Condon S and Cogan TM (1995b) Interactions between lactic and propionic acid bacteria. Lait 75: 331-343

Plastourgos S and Vaughn RH (1957) Species of *Propionibacterium* associated with Zapateria spoilage of olives. Appl Microbiol 5: 262-271

Pochi PE and Stauss JS (1961) Antibiotic sensitivity of *Corynebacterium acnes* (*Propionibacterium acnes*). J Invest Dermatol 36: 423-429

Polulach OV (1987) Biosynthesis of tetrapyrrole compounds (porphyrins and vitamin B_{12}) by some representatives of genus *Propionibacterium*. PhD thesis, Moscow State University, Moscow

Polulach OV, Zaitseva NI, Rumyantseva VD and Bykhovsky VY (1991) Production of porphyrins from 5-aminolevulinic acid by cell suspensions of propionic acid bacteria. Prikl Biokhim Mikrobiol 27: 91-97

Poroshenko GG and Abilev SK (1988) Anthropogenic Mutagens and Natural Antimutagens. In: Itogi Nauki i Techniki, v 12. Moscow

Postgate JR (1982) Biological nitrogen fixation: fundamentals. Phil Trans R Soc Lond B 296: 375-385

Poston JM and Stadtman TC (1975) Cobamides as cofactors. Methylcobamides and the synthesis of methionine, methane and acetate. In: Babior BM (ed) Cobalamin. Biochemistry and Pathophysiology, pp 111-140. Wiley-Interscience, New York

Poznanskaya AA and Yeliseev SA (1984) Cobalt-porphyrin-protein complexes of bacterial origin. Abstr 4th All-Union Conf "Chemistry and Application of Porphyrins", Yerevan, p 231

Prévot A-R (1960) Les corynébactérioses anaérobics (Anaerobic corynebacterioses). Ergeb Microbiol Immunit Exp Ther 33: 1

Prévot A-R (1975) Bacteriological aspects of anaerobic corynebacteria in relation to RES stimulation. In: Halpern B (ed) *Corynebacterium parvum*. Applications in Experimental and Clinical Oncology, pp 3-10. Plenum Press, New York

Prévot A-R (1976) New concept of the taxonomic position of anaerobic corynebacteria. C R Acad Sci Hebd Seances Acad Sci D 282: 1079-1081

Prévot A-R and Fredette V (1966) Manual for the classification and determination of the anaerobic bacteria, pp 345-355. Lea & Febiger, Philadelphia

Prévot A-R and Thouvenot H (1961) Essai de lysotypie des *Corynebacterium anaerobies*. Ann Inst Pasteur 101: 966-970

Prévot A-R, Tam N-D and Thouvenot H (1968) Influence of the cell wall of *Corynebacterium parvum* (strain 936B) on the reticuloendothelial system of mice. C R Acad Sci Hebd Seances Acad Sci D 267: 1061-1062

Pritchard GG and Asmundson RV (1980) Aerobic electron transport in *Propionibacterium shermanii*. Effects of cyanide. Arch Microbiol 126: 167-173

Pritchard GG, Wimpenny JWT, Morris HA, Lewis MWA and Hughes DE (1977) Effects of oxygen on *Propionibacterium shermanii* grown in continuous culture. J Gen Microbiol 102: 223-233

Probst H, Schiffer H, Gekeler V and Scheffler K (1989) Oxygen-dependent regulation of mammalian ribonucleotide reductase *in vivo* and possible significance for replication initiation. Biochem Biophys Res Comm 163: 334-340

Prottey C and Ballou CE (1968) Diacylmyoinositol monomannoside from *Propionibacterium shermanii*. J Biol Chem 243: 6196-6201

Pryor WA (1986) Oxy-radicals and related species. Their formation, lifetimes and reactions. Annu Rev Physiol 48: 657-667

Puhvel SM and Reisner RM (1972) The production of hyaluronidase (hyaluronate lyase) by *Corynebacterium acnes*. J Invest Dermatol 58: 66-70

Pulverer G and Ko HL (1973) Fermentative and serological studies on *Propionibacterium acnes*. Appl Microbiol 25: 222-229

Pulverer G, Sorgo W and Ko HL (1973) Bacteriophagen von *Propionibakterium acnes*. Zbl Bakt Hyg, I Abt Orig A 225: 353-363

Puvion F, Fray A and Halpern B (1975) A comparative, scanning electron microscope study of the interaction between stimulated or unstimulated mouse peritoneal macrophages and tumor cells. In: Halpern B (ed) *Corynebacterium parvum*. Applications in Experimental and Clinical Oncology, pp 137-144. Plenum Press, New York

Quastel JH and Webley DM (1942) Vitamin B$_1$ and bacterial oxidations. II. The effects of magnesium, potassium and hexosediphosphate ions. Biochem J 36: 8-33

Racine M, Dumont J, Champagne CP and Morin A (1991) Production and characterization of the polysaccharide from *Propionibacterium acidipropionici* on whey-based medium. J Appl Bacteriol 71: 233-238

Ramniece VE, Marauska MK and Beker ME (1981) Propionic acid bacteria and perspectives of their use in fermentation of plant sap. Latv Psr Zin Akad Vestis: 96-104

Ramos JM, Esteban J and Soriano F (1995) Isolation of *Propionibacterium acnes* from central nervous system infections. Anaerobe 1: 17-20

Rapp P (1968) Wachstum, Corrinoidbildung und Biogenese des Cobinamides aus Cobyrsäure bei *Propionibacterium shermanii*. PhD thesis, University of Stuttgart

Ratnam P, Barefoot SF, Prince LD, Bodine AB and McCaskill LH (1998) Partial purification and characterization of the *Propionibacterium* bacteriocin jensenin P. Abstr 2nd Int Symp Propionibacteria, Cork, Ireland

Reddy GV, Shahani KM and Benerjee MR (1973) Inhibitory effect of yogurt on Erlich ascites tumor cell proliferation. J Natl Cancer Inst 50: 815-817

Reddy MS, Reinbold GW and Williams FD (1973) Inhibition of propionibacteria by antibiotic and antimicrobial agents. J Milk Food Technol 36: 564-569

Reeves RE, Menzies RA and Hsu DS (1968) The pyruvate-phosphate dikinase reaction. J Biol Chem 243: 5486-5491

Rehberger TG and Glatz BA (1987a) Characterization of plasmid DNA in *Propionibacterium freudenreichii* subsp. *globosum* P93: evidence for plasmid-linked lactose utilization. J Dairy Sci (Suppl 1) 70, D73

Rehberger TG and Glatz BA (1987b) Restriction endonuclease analysis of *Propionibacterium* plasmids. J Dairy Sci (Suppl 1) 70, P11

Rehberger TG and Glatz BA (1987c) Southern hybridization analysis of *Propionibacterium* plasmids. J Dairy Sci (Suppl 1) 70, P48

Rehberger TG and Glatz BA (1990) Characterization of *Propionibacterium* plasmids. Appl Environ Microbiol 56: 864-871

Rehberger TG and Glatz BA (1995) Response of propionibacteria to acid: pH tolerance and maintenance of a proton motive force. Abstr 1st Int Symp Dairy Propionibacteria, Rennes, France, C6

Reichard P (1962) Enzymatic synthesis of deoxyribonucleotides. I. Formation of deoxycytidine diphosphate from cytidine diphosphate with enzymes from *Escherichia coli*. J Biol Chem 237: 3513-3519

Reichard P (1985) Ribonucleotide reductase and deoxyribonucleotide pools. In: Genetic consequences of nucleotide pool imbalance. Proceedings of Conference, Research Triangle Park, NC, May 9-11 1983, New York, London, pp 33-45

Reichard P (1993) From RNA to DNA: why so many ribonucleotide reductases. Science 260: 1773-1777

Rétey J (1998) Coenzyme B_{12}-dependent enzymes and their models. In: Kräutler B, Arigoni D and Golding BT (eds) Vitamin B_{12} and B_{12}-proteins, pp 273-288. Wiley-VCH, Weinheim

Rickard TR, Bigger GW and Elliot JM (1975) Effect of 5,6-dimethylbenzimidazole, adenine and riboflavin on ruminal vitamin B_{12} synthesis. J Anim Sci 40:1199-1204

Rickert DA, Glatz CE and Glatz BA (1998) Improved organic acid production by calcium alginate-immobilized propionibacteria. Enzyme Microbiol Technol 22: 409-414

Riedel K-HJ and Britz TJ (1992) Differentiation of "classical" *Propionibacterium* species by numerical analysis of electrophoretic protein profiles. System Appl Microbiol 15: 567-572

Riedel K-HJ and Britz TJ (1993) *Propionibacterium* species diversity in anaerobic digesters. Biodiv Conserv 2: 400-411

Riedel K-HJ and Britz TJ (1996) Justification of the "classical" *Propionibacterium* species concept by ribotyping. System Appl Microbiol 19: 370-380

Riedel K-HJ, Wingfield BD and Britz TJ (1994) Justification of the "classical" *Propionibacterium* species consept by restriction analysis of 16S ribosomal RNA genes. System Appl Microbiol 17: 536-542

Riedel K-HJ, Wingfield BD and Britz TJ (1998) Identification of classical *Propionibacterium* species using 16S rDNA-restriction fragment length polymorphisms. System Appl Microbiol 21: 419-428

Ritter P, Schwab H and Holzer H (1967) Testing the stimulatory or inhibitory effect of micrococci on propionic bacteria. Schweiz Milchztg 93 (34) Wiss Beil 113: 929-930

Robinson NA and Wood HG (1986) Polyphosphate kinase from *Propionibacterium shermanii*. Demonstration that the synthesis and utilization of polyphosphate is by a processive mechanism. J Biol Chem 261: 4481-4485

Roland N, Bouglé D, Lebeurrier F, Arhan P and Maubois JL (1997) *Propionibacterium freudenreichii* stimulates *in vitro* the growth of *Bifidobacterium bifidum* and increases faecal bifidobacteria in healthy human volunteers. Abstr 1st Int Symp "Functual Foods: Design Foods for Future", Cork, Ireland

Roland N, Avice JC, Ourry A and Maubois JL (1998) Production of nitric oxide by propionibacteria: a new criterion for their utilization as probiotics? Abstr 2nd Int Symp Propionibacteria, Cork, Ireland

Rollman N and Sjöstrom G (1946) The behavior of some propionic acid bacteria strains against NaCl, $NaNO_3$, and heating. Svenska Mejerit 38(19): 199-201; (20): 209-212

Romanskaya NN, Diment GS and Vorobjeva LI (1985) Method of production of dairy beverages. Avt svidet No 1184506; Bull No 38

Rosenberg LE (1983) Disorders of propionate, methylmalonate and cobalamin metabolism. In: Stanbury JB, Wyngaarden JB and Fredrickson DS (eds) The Metabolic Basis of Inherited Disease, 4th ed, pp 411-429. McGraw-Hill, New York

Roslyakova NV (1974) Natural and induced variability of *Propionibacterium shermanii*. PhD thesis, Institut Mikrobiol Kasakh Akad Nauk, Alma-Ata

Roszkowski K, Roszkowski W, Ko HL, Szmigielski S, Pulverer G and Jeljaszewicz J (1982) Clinical experience in treatment of cancer by propionibacteria. In: Jeljaszewicz J, Pulverer G and Roszkowski W (eds) Bacteria and Cancer, pp 331-357. Academic Press, London

Roszkowski W, Roszkowski K, Ko HL, Beuth J and Jeljaszewicz J (1990) Immunomodulation by propionibacteria. Zbl Bakt Hyg 274: 289-298

Saino Y, Eda J, Nagoya T, Yoshimura Y, Yamaguchi M and Kobayashi F (1976) Anaerobic coryneforms isolated from human bone marrow and skin. Jap J Microbiol 20: 17-25

Salton MRJ and Owen P (1976) Bacterial membrane structure. Annu Rev Microbiol 30: 451-482

Samain E, Albagnac G, Dubourguier HC and Touzel JP (1982) Characterization of a new propionic acid bacterium that ferments ethanol and displays a growth factor-dependent association with Gram-negative homoacetogen. FEMS Microbiol Lett 15: 69-74

Samoilov PM, Baranova NA, Vorobjeva LI and Fedulova IE (1968) Influence of carbon source on the secretion of amino acids and protein synthesis by *Mycobacterium luteum*. Mikrobiologiya 37: 264-268

Samoilova KA (1967) Action of ultraviolet radiation on the cell. Nauka, Leningrad

Sarada R and Joseph R (1994) Characterization and enumeration of microorganisms associated with anaerobic digestion of tomato-processing waste. Bioresource Technol 49: 261-265

Sasaki T, Uchida NA, Uchida H, Takasuka N, Kamiya H, Endo Y, Tanaka M, Hayashi T and Shimizu Y (1985) Antitumor activity of aqueous extracts of marine animals. J Pharmacobiodyn 8: 969-974

Sattler I, Roessner CA, Stolowich NJ, Hardin SH, Harris-Haller LW, Yokubaitis NT, Murooka Y, Hashimoto Y and Scott AI (1995) Cloning, sequencing, and expression of the uroporphyrinogen III methyltransferase *cobA* gene of *Propionibacterium freudenreuchii* (*shermanii*). J Bacteriol 177: 1564-1569

Schaal KP (1986a) Genus *Actinomyces* Harz 1877, 133[AL]. In: Sneath PHA, Mair NS, Sharpe ME and Holt JG (eds) Bergey's Manual of Systematic Bacteriology, 1st ed, v 2, pp 1383-1418. Williams & Wilkins, Baltimore

Schaal KP (1986b) Genus *Arachnia* Pine and Georg 1969, 269[AL]. In: Sneath PHA, Mair NS, Sharpe ME and Holt JG (eds) Bergey's Manual of Systematic Bacteriology, 1st ed, v 2, pp 1332-1342. Williams & Wilkins, Baltimore

Schaal KP and Pape W (1980) Special methodological problems in antibiotic testing of fermentative actinomyces. Infection 8 (Suppl 2): 176-182

Schiff J and Fankhauser H (1981) Assimilatory sulfate reduction. In: Bothe H and Trebst AB (eds) Biology of Inorganic Nitrogen and Sulfur, pp 153-168. Springer-Verlag, Berlin

Schleifer KH and Kandler O (1972) Peptidoglycan types of bacterial cell walls and their taxonomic implications. Bacteriol Rev 36: 407-477

Schleifer KH, Plapp R and Kandler O (1968) Glycine as cross-linking bridge in the LL-diaminopimelic acid containing murein of *Propionibacterium petersonii*. FEMS Microbiol Lett 1: 287-290

Schneider Z (1987a) Biosynthesis of vitamin B_{12}. In: Schneider Z and Stroinski A (eds) Comprehensive B_{12}. Chemistry. Nutrition. Ecology. Medicine, pp 93-110. Walter de Gruyter, Berlin, New York

Schneider Z (1987b) Cobamide dependent enzymes. In: Schneider Z and Stroinsky A (eds) Comprehensive B_{12}. Chemistry. Biochemistry. Nutrition. Ecology. Medicine, pp 225-250. Walter de Gruyter, Berlin, New York

Schneider Z, Trojanowska K, Jaszewski B and Nowak T (1995) Cobalt binding by *Propionibacterium arabinosum*. Lait 75: 379-389

Schofield GM and Schaal KP (1981) A numerical taxonomic study of members of the *Actinomycetaceae* and related taxa. J Gen Microbiol 127: 237-259

Scholl K (1976) Zur immobilisierung von *Propionibacterium shermanii*. PhD thesis, University of Stuttgart

Schrauzer GN (1968) Organocobalt chemistry of vitamin B_{12} model compounds (cobaloximes). Acc Chem Res 1: 97-103

Schwartz AC (1973) Anaerobiosis and oxygen consumption of some strains of *Propionibacterium* and a modified method for comparing the oxygen sensitivity of various anaerobes. Z Allg Mikrobiol 13: 681-691

Schwartz AC and Sporkenbach J (1975) The electron transport system of the anaerobic *Propionibacterium shermanii*. Cytochrome and inhibitor studies. Arch Microbiol 102: 261-273

Schwartz AC, Mertens B, Voss KW and Hahn H (1976) Inhibition of acetate and propionate formation upon aeration of resting cells of the anaerobic *Propionibacterium shermanii*: evidence of the Pasteur reaction. Z Allg Mikrobiol 16: 123-131

Scott AI (1994) Recent studies of enzymically controlled steps in vitamin B_{12} biosynthesis. In: Chadwick DJ and Ackrill K (eds) The Biosynthesis of the Tetrapyrrole Pigments, pp 285-308. John Wiley & Sons, New York

Scott AI (1998) How nature synthesizes B_{12} without oxygen: discoveries along the ancient, anaerobic pathway. In: Kräutler B, Arigoni D and Golding BT (eds) Vitamin B_{12} and B_{12}-proteins, pp 81-100. Wiley-VCH, Weinheim

Scott AI, Stolowich NJ, Wang J, Gawatz O, Fridrich E and Müller G (1996) Biosynthesis of vitamin B_{12}: Factor IV, a new intermediate in the anaerobic pathway. Proc Natl Acad Sci USA 93: 14316-14319

Seiler JP (1973) The mutagenicity of benzimidazole and benzimidazole derivatives. II. Incorporation of benzimidazole into the nucleic acids of *Escherichia coli*. Mutat Res 17: 21-25

Sergeeva GY, Pushkareva VI, Malachova TI and Melnikov EH (1959) Application of biological preparations of vitamin B_{12} in husbandry. Veterinariya 5: 49-54

Serzedello A, Molinary R and Lara FJS (1969) Further studies on the lactic dehydrogenase from *Propionibacterium pentosaceum*. Ann Acad Brasil Cienc 41: 137-140

Shaw N and Baddiley J (1968) Structure and distribution of glycosyl diglycerides in bacteria. Nature 217: 142-144

Shinohara K, Kuroki S, Miwa M, Kong Z-L and Hosoda H (1988) Antimutagenicity of dialyzates of vegetables and fruits. Agric Biol Chem 52: 1369-1375

Shpokauskas A, Kierszenbaum F and Daculite YL (1965) Role of microelements in plant nutrition and enhancing the effectiveness of fertilizers. Nauka, Moscow

Shuster S (1976) Biological purpose of acne. Lancet 1: 1328-1329

Sibley JA and Leninger AL (1949) Determination of aldolase in animal tissues. J Biol Chem 177: 859-872

Sidermene AA, Germane SK and Meirena DV (1985) Combination of antitumor preparations with metabolites of nucleic acid metabolism. Zinatne, Riga

Siewert R (1989) Zum auftreten von blinden Emmentaler käsen. Teil 2. Milchforsch Milchprax 131: 118-119

Silverman M and Werkman CH (1939) Adaptation of the propionic acid bacteria to vitamin B_1 synthesis including method of assay. J Bacteriol 38: 25-32

Siu PML and Wood HG (1962) Phosphoenolpyruvic carboxytransphosphorylase, a CO_2 fixation enzyme from propionic acid bacteria. J Biol Chem 237: 3044-3051

Sizova AV and Arkadjeva ZA (1968) Propionic acid bacteria of rumen and their capacity for vitamin B_{12} biosynthesis. Mikrobiol Sintez 10: 8-13

Sizova AV and Volkova HT (1974) Production of dry bacterial vitaminised preparation and its biological activity. In: Complex use of biological substances in feeding of husbandry animals. Proc All-Union Conf, pp 229-232. Gorki, USSR

Skerman VBD (1967) A guide to the identification of the genera of bacteria, 2nd ed. Williams & Wilkins, Baltimore

Skogen LO, Reinbold GW and Vedamuthu ER (1974) Capsulation of *Propionibacterium*. J Milk Food Technol 37: 314-321

Skupin J, Pędziwilk F and Jaszewski B (1970) Identification of CH_3-B_{12} in light-sensitive corrinoid-polypeptide complex from propionic acid bacteria. Bull Acad Polon Ser Sci Biol 18: 511-515

Slack CR and Hatch MD (1967) Comparative studies on the activity of carboxylases and other enzymes in relation to the new pathway of photosynthetic carbon dioxide fixation in tropical grasses. Biochem J 103: 660-665

Slack JM and Gerencser MA (1975) *Actinomyces*, Filamentous Bacteria. Biology and Pathogenicity. Burgess Publishing Co, Minneapolis

Smith EL (1960) Vitamin B_{12}, pp 81-96. John Wiley & Sons, Methuen, New York

Smith RM and Marston HR (1970) Production, absorption, distribution and excretion of vitamin B_{12} in sheep. Br J Nutr 24: 857-867

Sobczak E and Komorowska Z (1984) Propionic fermentation of some wastes of food industry. Ann Warsaw Agr Univ SGGW-AR Food Technol Nutr 16: 45-51

Sone N (1972) The redox reactions in propionic acid fermentation. I. Occurrence and nature of an electron tranfer system in *Propionibacterium arabinosum*. J Biochem 71: 931-940

Spicher G (1983) Baked goods. In: Rehm H-J and Reed G (eds) Biotechnology, v 5: Food and Feed Production with Microorganisms, pp 1-115. Verlag-Chemie, Weinheim

Stackebrandt E and Woese CR (1981) The evolution of prokaryotes. In: Carlile MJ, Collins JF and Moseley BEB (eds) Molecular and Cellular Aspects of Microbial Evolution, pp 1-31. Cambridge University Press, Cambridge

Stackebrandt E, Rainey FA and Ward-Rainey NL (1997) Proposal for a new hierarchic classification system *Actinobacteria* classis nov. Int J Syst Bacteriol 47: 479-491

Stadtman TC (1971) Vitamin B_{12}. Science 171: 859-867

Stasinska B (1977) Estimation of the nutritive value of propionic bacterial biomass proteins on application of chemical indicators. Bull Acad Polon Sci 25: 711-717

Steffen C, Eberhard P, Bosset JO and Ruegg M (1993) Swiss-type varieties. In: Fox PF (ed) Cheese: Chemistry, Physics and Microbiology, v 2. Chapman & Hall, London

Stevenson IL (1968) The fine structure of *Arthrobacter pascens* and the development of mesosomes during the growth cycle. Can J Microbiol 14: 1029-1034

Stjernholm R and Flanders F (1962) Metabolism of D-ribose-1-C^{14} and C^{14}-labeled D-gluconate in an enzyme system of the genus *Propionibacterium*. J Bacteriol 84: 563-568

Stjernholm R and Wood HC (1963) The symmetrical C_3 in the propionic acid fermentation and the effect of avidin on propionate formation. Iowa State J Sci 38: 123-140

Stoltz DR, Stavric B, Starley R, Klassen R, Bendall R and Krewski D (1984) Mutagenicity screening of food. II. Results with fruits and vegetables. Environ Mutagen 6: 343-354

Stone RW, Wood HG and Werkman CH (1937) Phosphorylation and first stages in glucose breakdown by propionic acid bacteria. J Bacteriol 33: 101

Stoyanova LG, Vorobjeva LI and Lobzov KI (1979) Physiological and biochemical peculiarities of growth and development of *Propionibacterium shermanii* in egg-white. Mikrobiologiya 48: 1011-1015

Strehler BJ and McElroy WD (1957) Assay of adenosine triphosphate. In: Colowick SP and Kaplan NO (eds) Methods in Enzymology, v 3, p 871. Academic Press, New York

Stroinski A (1987) Medical aspects of vitamin B_{12}. In: Schneider Z and Stroinski A (eds) Comprehensive B_{12}. Chemistry. Biochemistry. Nutrition. Ecology. Medicine, pp 335-370. Walter de Gruyter, Berlin, New York

Stubbe J, Licht S, Gerfen G, Silva D and Booker S (1998) Adenosylcobalamin-dependent ribonucleotide reductases: still amazing but no longer confusing. In: Kräutler B, Arigoni D and Golding BT (eds) Vitamin B_{12} and B_{12}-proteins, pp 321-332. Wiley-VCH, Weinheim

Stupperich E and Eisinger HJ (1990) Characterization of two novel corrinoids from the acetogenic bacterium *Sporomusa ovata*. Forum Microbiol 13: 65

Sutcliffe IC and Shaw N (1993) The phosholipids of *Propionibacterium freudenreichii*: absence of phosphatidylinositol mannosides. System Appl Microbiol 16: 9-12

Swart R, Britz TJ and Riedel K-HJ (1995) Influence of environmental factors on the reduction of nitrate by the type strains of *Propionibacterium*. Abstr 1st Int Symp Dairy Propionibacteria, Rennes, France, C2

Swick RW and Wood HG (1960) The role of transcarboxylation in propionic acid fermentation. Proc Natl Acad Sci USA 46: 28-41

Szmigielski S, Roszkowski W, Roszkowski K, Ko HL, Jeljaszewicz J and Pulverer G (1982) Experimental immunostimulation by propionibacteria. In: Jeljaszewicz J, Pulverer G and Roszkowski W (eds) Bacteria and Cancer, pp 129-147. Academic Press, London

Tagg JR, Dajani AS and Wannamaker LW (1976) Bacteriocins of Gram-positive bacteria. Bacteriol Rev 40: 722-756

Tamura T, Takeuchi M and Yokota A (1994) *Luteococcus japonicus* gen. nov., sp. nov., a new gram-positive coccus with LL-diaminopimelic acid in the cell wall. Int J Syst Bacteriol 44: 348-356

Tatum EL, Fred EB, Wood HG and Peterson WH (1936) Essential growth factors for propionic acid bacteria. II. Nature of the Neuberg precipitate fraction of potato: replacement by ammonium sulfate or by certain amino acids. J Bacteriol 32: 157-174

Taylor MJ and Richardson T (1979) Application of microbial enzymes in food systems and in biotechnology. Adv Appl Microbiol 25: 7-35

Thelander L, Gräslund A and Thelander M (1983) Continual presence of oxygen and iron required for mammalian ribonucleotide reduction: possible regulation mechanism. Biochem Biophys Res Comm 110: 859-865

Tomka G (1949) Acetoin and diacetyl production of the rod shaped propionic acid bacteria. Proc 12th Int Dairy Congr, v 2, pp 619-622

Toohey JI, Perlman D and Barker HA (1961) Purification and properties of cobamide coenzymes obtained from *Propionibacterium arabinosum*. J Biol Chem 236: 2119-2127

Torray JC (1916) Bacteria associated with certain types of abnormal lymph glands. J Med Res 34: 65-80

Toujas L, Dazord L and Guelfi J (1975) Kinetics of proliferation of bone-marrow cell lines after injections of immunostimulant bacteria. In: Halpern B (ed) *Corynebacterium parvum*. Applications in Experimental and Clinical Oncology, pp 117-125. Plenum Press, New York

Travers AA and Mace HAF (1982) The heat-shock phenomenon in bacteria – a protection against DNA relaxation? In: Schlesinger MJ, Ashburner M and Tissieres A (eds) Heat Shock: from Bacteria to Man, pp 127-130. Cold Spring Harbor, New York

Trojanowska K, Jaszewski B, Schneider Z and Czaczyk K (1995) Biosynthesis of vitamin B_{12} in mixed cultures of propionic and lactic acid bacteria. Abstr 1st Int Symp Dairy Propionibacteria, Rennes, France, C11

Tyree RW, Clausen EC and Gaddy JL (1991) The production of propionic acid from sugars by fermentation through lactic acid as an intermediate. J Chem Technol Biotechnol 50: 157-166

Umansky MS and Melnikova LV (1986) Influence of milk phospholipid hydrolysis products on the quality of "Soviet" cheese. Information Bulletin no. 195-86, Barnaul

Van Bogelen RA, Kelley PM and Neidhardt FC (1987) Differential induction of heat shock, SOS, and oxidation stress regulons and accumulation of nucleotides in *Escherichia coli*. J Bacteriol 169: 26-32

Van Demark PJ and Fukui GM (1956) An enzymatic study of the utilization of gluconic acid by *Propionibacterium pentosaceum*. J Bacteriol 72: 610-614

van Gent-Ruijters MLW, de Vries W and Stouthamer AH (1975) Influence of nitrate on fermentation pattern, molar growth yields and synthesis of cytochrome *b* in *Propionibacterium pentosaceum*. J Gen Microbiol 88: 36-48

van Gent-Ruijters MLW, Meijere FA, de Vries W and Stouthamer AH (1976) Lactate metabolism in *P. pentosaceum* growing with nitrate or oxygen as hydrogen acceptor. J Microbiol Serol 42: 217-228

Van Niel CB (1928) The propionic acid bacteria. Boissevain JW & Co, Haarlem, The Netherlands (thesis: Delft)

Van Niel CB (1957) The genus *Propionibacterium*. In: Breed RS, Murray EGD and Smith NR (eds) Bergey's Manual of Determinative Bacteriology, 7th ed, pp 569-575. Williams & Wilkins, Baltimore

Vogt JRA and Renz P (1988) Biosynthesis of vitamin B_{12} in anaerobic bacteria. Experiments with *Eubacterium limosum* on the origin of the amide groups of the corrin ring and of *N-3* of the 5,6-dimethylbenzimidazole part. Eur J Biochem 171: 655-659

Volk WA (1954) The effect of fluoride on the permeability and phosphatase activity of *Propionibacterium pentosaceum*. J Biol Chem 208: 777-784

Volkova NT (1980) A study of physiological and biochemical properties of *Propionibacterium acnes* for the production of a dried preparation for husbandry. PhD thesis, Moscow State University, Moscow

von Drews G (1960) Elektronenmikroskopische Untersuchungen an *Mycobacterium phlei*. Archiv Mikrobiol 35: 53-62

von Nicolai H, Höffler U and Zilliken F (1980) Isolation, purification and properties of neuraminidase from *Propionibacterium acnes*. Zbl Bakt Hyg, I Abt Orig A 247: 84-94

Vorobjeva LI (1958a) Fermentation of different carbon sources by propionic acid bacteria. PhD thesis, Moscow State University, Moscow

Vorobjeva LI (1958b) Balance of carbon in the fermentation of lactic and pyruvic acids by propionic acid bacteria. Mikrobiologiya 27: 287-294

Vorobjeva LI (1959) Influence of aeration on propionic acid fermentation. Mikrobiologiya 28: 224-229

Vorobjeva LI (1972) Propionic acid fermentation and formation of vitamin B_{12}. Uspekhi Mikrobiologii 8: 182-207

Vorobjeva LI (1976) Propionic Acid Bacteria and the Formation of Vitamin B_{12}. Moscow State University, Moscow

Vorobjeva LI (1978). Fermentation by immobilized cells of propionic acid bacteria. In: Koscheenko KA (ed) Immobilized cells of microorganisms, pp 127-134. Akad Nauk SSSR, Puschino

Vorobjeva LI and Baranova NA (1966) Stimulatory action of *Mycobacterium luteum* on vitamin B_{12} production by propionic acid bacteria. Mikrobiologiya 35: 250-252

Vorobjeva LI and Baranova NA (1969) On the variability of propionic acid bacteria. Mikrobiologiya 38: 114-117

Vorobjeva LI and Charakhchyan IA (1983) Utilization of different sulfur sources by propionic acid bacteria. Mikrobiologiya 52: 875-879

Vorobjeva LI and Golozubova GA (1968) Catalase of propionic acid bacteria. Prikl Biokhim Mikrobiol 4: 654-658

Vorobjeva LI and Iordan EP (1976) Functions of cobamide coenzymes in metabolism of propionic acid bacteria. In: Chagovets RV (ed) Vitaminy, pp 16-20. Naukova Dumka, Kiev

Vorobjeva LI and Kozyreva LF (1967) Influence of temperature on vitamin B_{12} formation by *Propionibacterium shermanii*. Vestn Mosk Univ Biol: 52-54

Vorobjeva LI and Kraeva NI (1982) Superoxide radicals and antiradical defence of propionic acid bacteria. Arch Microbiol 133: 110-113

Vorobjeva LI, Baranova NA and Chan TT (1973) Mutagenic effects of *N*-methyl-*N'*-nitro-*N'*-nitrosoguanidine on propionic acid bacteria. Mikrobiologiya 42: 301-306

Vorobjeva LI, Alekseeva MA, Surkova IG and Gaitan VI (1977) Synthesis of volatile acids by immobilized cells of propionic acid bacteria. Prikl Biokhim Mikrobiol 13: 531-537

Vorobjeva LI, Kraeva NI, Ebringer L and Olsinskaya NL (1979a) Oxidation of *n*-alkanes by propionic acid bacteria. Mikrobiologiya 48: 33-38

Vorobjeva LI, Stoyanova LG and Alekseeva MA (1979b) Various applications of propionibacteria. In: Zvjagintsev DG (ed) Microbial Metabolites, pp 88-101. Moscow State University, Moscow

Vorobjeva LI, Kraeva NI and Charakhchyan IA (1982) New data on the metabolism of propionic acid bacteria. Abstr All-Union Conf Anaerobes, Puschino, pp 4-5

Vorobjeva LI, Turova TP, Kraeva NI and Alekseeva MA (1983) Propionic cocci and their taxonomic position. Mikrobiologiya 52: 465-471

Vorobjeva LI, Egorov NS, Kraeva NI and Al-Sudani S (1984) Production of physiologically active compounds by propionic acid bacteria. Proc 3rd Eur Congr Biotechnol, v 3, pp 690-695, Verlag-Chemie, Weinheim

Vorobjeva LI, Al-Sudani S and Kraeva NI (1986) Peroxidase of *Propionibacterium shermanii*. Mikrobiologiya 55: 750-753

Vorobjeva LI, Alekseeva MA and Vorobjeva NV (1990) Characteristics of newly isolated strains of propionic acid bacteria. Abstr 4th Eur Actinomycetes Group Meet, Udine, Italy, p 20

Vorobjeva LI, Cherdinceva TA, Abilev SK and Vorobjeva NV (1991) Antimutagenicity of propionic acid bacteria. Mutat Res 251: 233-239

Vorobjeva LI, Ganicheva TV, Rusinov SF, Shishchenko LD, Vorobjeva NV, Cherdinceva TA and Shironosova LA (1992) Stabilization of microbial fermentations by using cell extracts of propionic acid bacteria. Priklad Biokhim Mikrobiol 28: 416-422

Vorobjeva LI, Cherdinceva TA, Averjanov AA and Abilev SK (1993a) Antimutagenic action of superoxide dismutase on sodium azide-induced mutagenesis in *Salmonella typhimurium* TA 1535. Genetika 29: 760-766

Vorobjeva LI, Altukhova EA, Naumova ES and Abilev SK (1993b) Antimutagenic effect of culture liquid obtained in propionic acid fermentation. Microbiology 62: 634-638

Vorobjeva LI, Nikitenko GV, Khodzhaev EY and Ponomareva GM (1993c) Cell extracts of propionic acid bacteria reactivate cells of *Escherichia coli* inactivated by ultraviolet radiation. Microbiology 62: 657-661

Vorobjeva LI, Cherdinceva TA and Abilev SK (1995a) Antimutagenic effect of bacteria on mutagenesis induced by 4-nitroquinoline-1-oxide in *Salmonella typhimurium*. Microbiology 64: 187-192

Vorobjeva LI, Khodzhaev EY and Cherdinceva TA (1995b) Antimutagenic and reactivative activities of dairy propionibacteria. Lait 5: 473-487

Vorobjeva LI, Khodzhaev EY and Ponomareva GM (1995c) Mechanism of reactivation of the UV-inactivated cells of *Escherichia coli* by cell extracts of propionic acid bacteria. Microbiology 64: 555-559

Vorobjeva LI, Khodzhaev EY and Cherdinceva TA (1996a) The study of induced antimutagenesis of propionic acid bacteria. J Microbiol Methods 24: 249-258

Vorobjeva LI, Khodzhaev EY, Ponomareva GM, Chernyshov DV and Cherdinceva TA (1996b) Protective and reactivative action of bacterial peptides in organisms inactivated by different stress factors. Res Conserv Recycl 18: 151-159

Vorobjeva LI, Khodzhaev EY and Ponomareva GM (1997) Antistress effects of a protein fraction from *Propionibacterium freudenreichii* subsp. *shermanii*. Microbiology 66: 375-379

Vuorinen A and Mantere-Alhonen S (1982) On trace elements in *Propionibacterium freudenreichii* mass. Meijerit Aikakausk 40: 283-290

Vygovskaya MV, Daciuk NM and Yeliseev CA (1990) On the ability of some microorganisms to synthesize porphyrins on water-insoluble substrates. Mikrobiol Zh 52: 28-30

Wagner F and Bernhauer K (1964) New aspects of the structure of corrinoid coenzymes. Ann N Y Acad Sci 112: 580-584

Wagner F, Pfeifner H and Rapp P (1968) Statische und kontinuirliche kultur von *Propionibacterium*. Zbl Bakt Hyg, I Abt (Suppl 2): 85-89

Wakayama EJ, Dillwith JW, Howard RW and Blomquist GJ (1984) Vitamin B$_{12}$ levels in selected insects. Insect Biochem 14: 175-180

Walker GC (1984) Mutagenesis and inducible responses to deoxyribonucleic acid damage in *Escherichia coli*. Microbiol Rev 48: 60-93

Warnecke D, Schwartz AS and Höfer M (1982) Properties of the transport system for glucose in *Propionibacterium freudenreichii* ssp. *shermanii*. Zbl Bakt Hyg, Abt IC 3: 547-548

Wawszkiewicz EL and Barker HA (1968) Erythritol metabolism by *Propionibacterium pentosaceum*. The overall reaction sequence. J Biol Chem 243: 1948-1956

Webster GF and Cummins CS (1978) Use of bacteriophage typing to distinguish *Propionibacterium acnes* types I and II. J Clin Microbiol 7: 84-90

Wegner WS, Reeves HC, Rabin R and Ajl SJ (1968) Alternative pathways of metabolism of short chain fatty acids. Bacteriol Rev 32: 1-26

Weinberg ZG, Ashbell G, Bolsen KK, Pahlow G, Hen Y and Azrieli A (1995a) The effect of a propionic acid bacterial inoculant applied at ensiling, with or without lactic acid bacteria, on the aerobic stability of pearl-millet and maize silages. J Appl Bacteriol 78: 430-436

Weinberg ZG, Ashbell G, Hen Y and Azrieli A (1995b) The effect of propionic acid bacterial inoculant applied at ensiling on the aerobic stability of wheat and sorghum silages. J Ind Microbiol 15: 493-497

Werkman CH and Wood HG (1942) Heterotrophic assimilation of carbon dioxide. Adv Enzymol 2: 135-182

Werner H (1967) Untersuchungen über die lipase und lecithinase-aktivität von aeroben und anaeroben Corynebacterium- und Propionibacterium-arten. Zbl Bakt Hyg 204: 127-138

White RG (1975) The macrophage-stimulating properties of a variety of anaerobic coryneforms. In: Halpern B (ed) *Corynebacterium parvum*. Applications in Experimental and Clinical Oncology, pp 148-161. Plenum Press, New York

Wilkinson PC (1975) Macrophage-stimulating effects of anaerobic coryneform bacteria *in vitro*. In: Halpern B (ed) *Corynebacterium parvum*. Applications in Experimental and Clinical Oncology, pp 162-170. Plenum Press, New York

Winder FG, Brennan PJ and McDonnell I (1967) Effects of isoniazid on the composition of mycobacteria, with particular reference to soluble carbohydrates and related substances. Biochem J 104: 385-393

Wixom RL, Joseph HA and Hwang SU (1971) Studies in valine biosynthesis. VIII. Dihydroxyacid dehydratase activity in microorganisms with diverse fermentation patterns. Proc Soc Exp Biol Med 137: 292-298

Wolberg G, Duncan GS, Adlam C and Wishnant JK (1977) Antibody to *Corynebacterium parvum* in normal human and animal sera. Infect Immunol 15: 1004-1007

Wood WA (1961) Fermentation of carbohydrate and related compounds. In: Gunsalus IC and Stanier RY (eds) The Bacteria, v 2, pp 59-150. Academic Press, New York

Wood HG and Goss NH (1985) Phosphorylation enzymes of the propionic acid bacteria and the roles of ATP, inorganic pyrophosphate, and polyphosphates. Proc Natl Acad Sci USA 82: 312-315

Wood HG and Leaver FW (1953) CO_2 turnover in the fermentation of 3, 4, 5 and 6 carbon compounds by the propionic acid bacteria. Biochim Biophys Acta 12: 207-222

Wood HG and Werkman CH (1934) The propionic acid bacteria: on the mechanism of glucose dissimilation. J Biol Chem 105: 63-72

Wood HG and Werkman CH (1936) The utilization of CO_2 in the dissimilation of glycerol by the propionic acid bacteria. Biochem J 30: 48-53

Wood HG and Werkman CH (1938) The utilization of CO_2 by the propionic acid bacteria. Biochem J 32: 1262-1271

Wood HG and Werkman CH (1940) The relationship of bacterial utilization of CO_2 to succinic acid formation. Biochem J 34: 129-137

Wood HG, Stone RW and Werkman CH (1937) The intermediate metabolism of the propionic acid bacteria. Biochem J 31: 349-359

Wood HG, Allen SHG, Stjernholm R and Jacobson B (1963) Transcarboxylase. III. Purification and properties of methylmalonyl-oxaloacetic transcarboxylase containing tritiated biotin. J Biol Chem 238: 547-556

Wood HG, Davis JJ and Willard JM (1969) Phosphoenolpyruvate carboxytransphosphorylase. V. Mechanism of the reaction and the role of metal ions. Biochemistry 8: 3145-3155

Wood HG, Robinson NA, Pepin CA and Clarc JE (1987) Polyphosphate kinase and polyphosphate glucokinase of *Propionibacterium shermanii*. In: Torriani-Gorini A, Rothman FG, Silver S, Wright A and Yagil E (eds) Phosphate Metabolism and Cellular Regulation in Microorganisms, pp 225-232. American Society for Microbiology, Washington, DC

Woolford MK (1973) *In vitro* techniques in microbiological studies of the ensiling process. PhD thesis, University of Edinburg

Woolford MK (1975) The significance of *Propionibacterium* species and *Micrococcus lactilyticus* to the ensiling process. J Appl Bacteriol 39: 301-306

Yamamoto I, Takahashi M, Tamura E and Maruyama H (1982) Antitumor activity of crude extracts from edible marine algae against L-1210 leukemia. Bot Marine 25: 455-457

Yamamoto I, Maruyama H, Takahashi M and Komiyama K (1986) The effect of dietary or intraperitoneally injected seaweed preparations on the growth of sarcoma-180 cells subcutaneously implanted into mice. Cancer Lett 30: 125-131

Yeliseev AA and Bykhovsky VY (1990) Influence of light illumination on the tetrapyrrole biosynthesis by propionic acid bacteria. Forum Microbiol 13: 58

Yeliseev SA, Dacyuk NM, Girna OV, Nakonechnaya LI and Khlyan AM (1985) Localization of vitamin B_{12} in *Propionibacterium shermanii*. Mikrobiol Zh 47: 52-56

Yeliseev AA, Pushkin AV, Belozerova EV and Bykhovsky VY (1986a) Role of glutamine in the biosynthesis of cobinamide by *Propionibacterium shermanii*. Biokhimiya 51: 1072-1076

Yeliseev AA, Pushkin AV, Zaitseva NI, Evstigneeva ZG, Bykhovsky VY and Kretovich WL (1986b) Glutamine as the source of amide groups in vitamin B_{12} biosynthesis. Biochim Biophys Acta 880: 131-138

Yeliseev AA, Pushkin AV, Zaitseva NI and Bykhovsky VY (1988a) Interrelations in biosynthesis of corrinoids and glutamine by *Propionibacteria*. Prikl Biokhim Mikrobiol 24: 765-773

Yeliseev AA, Zaitseva NI and Bykhovsky VY (1988b) Amidation as a stage of regulation in vitamin B_{12} biosynthesis in *Propionibacterium shermanii*. Biokhimiya 53: 1086-1092

Yokota A, Tamura T, Takeuchi M, Weiss N and Stackebrandt E (1994) Transfer of *Propionibacterium innocuum* Pitcher and Collins 1991 to *Propioniferax* gen. nov. as *Propioniferax innocua* comb. nov. Int J Syst Bacteriol 4: 579-582

Yongsmith B and Cole JA (1986) Cloning and expression of propionibacterial genes in *Escherichia coli*. Abstr XIV Int Congr Microbiol, Manchester, p 244

Yongsmith B, Sonomoto K, Tanaka A and Fukui S (1982) Production of vitamin B_{12} by immobilized cells of propionic acid bacterium. Eur J Appl Microbiol Biotechnol 16: 70-74

Yoshida Y and Yuki S (1968) Action of N-methyl-N'-nitro-N-nitrosoguanidine in *Bacillus subtilis*. Jap J Genet 43: 173-179

Zagalak B, Rétey J and Sund H (1974) Studies on methylmalonyl-CoA mutase from *Propionibacterium shermanii*. Eur J Biochem 44: 529-535

Zierdt CH (1974) Properties of *Corynebacterium acnes* bacteriophage and description of an interference phenomenon. J Virol 14: 1268-1273

Zinchenko AA, Vorobjeva LI, Gordeeva EA, Khodzhaev EY and Ponomareva GM (1998) Isolation and purification of a protective protein from *Propionibacterium freudenreichii* subsp. *shermanii* cells. Microbiology 67: 433-437

Zodrow K and Stefaniak O (1963) Effect of temperature on the growth and production of corrinoids by *Propionibacterium shermanii*. Acta Microbiol Pol 12: 271-280

Zodrow K, Chelkowski J, Stefaniak O and Czarnecka D (1963a) The effect of different casein hydrolysates on the growth and biosynthesis of corrinoids by propionibacteria. Acta Microbiol Pol 12: 259-262

Zodrow K, Stefaniak O, Chelkowski J and Szesepska K (1963b) Influence of Ca-pantothenate and biotin on the growth and biosynthesis of corrinoids by propionibacteria. Acta Microbiol Pol 12: 263-266

Zodrow K, Stefaniak O and Kaczmarek W (1967) Influence of incubation temperature on the content of different corrinoids in the cells of *Propionibacterium shermanii*. Acta Microbiol Pol 16: 223-226

Index

Acetaldehyde, 88, 106, 131, 213
Acetate, 14, 91-93, 98-100, 103, 105,
 107, 108, 112-114, 123, 138, 149, 150,
 176, 228, 229
 synthase, 176
Acetic acid, 30, 39, 52, 64, 65, 80, 93,
 103-105, 149, 169, 179, 226-229, 231,
 233, 234
Acetobacter
 lovaniense, 232, 233
 woodii, 176
Acetoin, 88, 106, 213
Acetyl-CoA, 93, 96-98
Acetyl kinase, 99, 100
Acetylphosphate, 93, 99, 100
Acne vulgaris, 32, 33, 36
Actinobacteria, 4
Actinomyces, 2, 8, 41, 42, 74, 245
Actinomycetes, 4, 23, 41, 42, 44, 45, 74,
 130, 244-246
Active factor (AF), 56, 83-86, 247
Adenine, 63, 64, 164, 189, 204, 205
Adenosine triphosphate, *see* ATP
Adenosylcobalamin (AdoCbl), 71, 88,
 159, 164, 166, 172, 174, 176, 178,
 179, 181, 184-186, 188, 189, 191-194,
 215-218, 220, 246
 biological functions, 98, 174, 176,
 186, 193, 194
AdoCbl, *see* Adenosylcobalamin

AdoMet (S-adenosylmethionine), 161,
 164, 189, 194
Adonitol, 29, 43, 53, 105
Aerobe, 4
Aerobic cells, 2, 4, 5, 9, 12, 19, 25, 27,
 36, 38, 39, 41, 51, 53, 77, 97, 106-111,
 113, 114, 116, 117, 119, 121, 124,
 126, 129, 148-150, 161, 163-166, 169,
 172, 179, 196, 225, 240, 246
Alanine, 5, 20, 21, 29, 30, 129, 151, 234
Alkanes, 111, 138, 140-142, 221, 246
Ames test, 70, 73, 75
Amino acids, 34, 49, 69, 76, 88, 105, 128,
 129, 132, 137, 138, 147, 151, 163,
 170, 171, 176, 183, 209, 213, 214,
 216, 217, 221, 234, 236
p-Aminobenzoic acid, 23, 131, 132
Aminoglycosides, 43, 54
Aminohexuronic acid, 21
Aminolevulinic acid (ALA), 50, 64, 65,
 107, 158, 160, 161, 164, 172, 173,
 207, 221
 dehydratase, 65, 107, 161, 164, 172,
 173, 207
 synthetase, 65, 107, 164, 172
Ammonia, 26, 148, 163, 225
Ammonium, 72, 83, 86, 99, 128, 138,
 151, 156, 157, 199, 206, 215, 224,
 234, 238
Anabolic reactions, 182
Anaerobe, 4, 8, 40, 160, 206, 223, 237